T0325863

The Material Point Method

The Material Point Method

A Continuum-Based Particle Method for Extreme Loading Cases

Xiong Zhang
Zhen Chen
Yan Liu

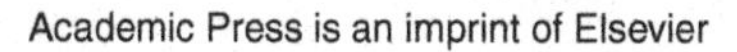
AMSTERDAM • BOSTON • HEIDELBERG • LONDON
NEW YORK • OXFORD • PARIS • SAN DIEGO
SAN FRANCISCO • SINGAPORE • SYDNEY • TOKYO
Academic Press is an imprint of Elsevier

Academic Press is an imprint of Elsevier
125 London Wall, London EC2Y 5AS, United Kingdom
525 B Street, Suite 1800, San Diego, CA 92101-4495, United States
50 Hampshire Street, 5th Floor, Cambridge, MA 02139, United States
The Boulevard, Langford Lane, Kidlington, Oxford OX5 1GB, United Kingdom

Library of Congress Cataloging-in-Publication Data
A catalog record for this book is available from the Library of Congress

British Library Cataloguing-in-Publication Data
A catalogue record for this book is available from the British Library

ISBN: 978-0-12-407716-4

For information on all Academic Press publications
visit our website at https://www.elsevier.com

Publisher: Glyn Jones
Acquisition Editor: Glyn Jones
Editorial Project Manager: Naomi Robertson
Production Project Manager: Kiruthika Govindaraju
Designer: Christian Bilbow

Typeset by VTeX

Dedication

To Yun and Ximing for their company and support!

Xiong Zhang

To Jessie and Ethan for their understanding and support!

Zhen Chen

To my parents!

Yan Liu

To my parents.

Contents

About the Authors

Xiong Zhang received his Ph.D. in Computational Mechanics from the Dalian University of Technology in 1992. He is now a professor in the School of Aerospace Engineering at Tsinghua University, and the Associate Editor of the *International Journal of Mechanics and Materials in Design*. His selected honors include the New Century Excellent Talents in University (Ministry of Education of China, 2004), the First prize (2008) and Second prize (2009) for Natural Science from the Ministry of Education of China, the ICACM Fellows Award (2011), the Beijing Municipal Famous Teacher Award for Higher Education (2015) and the Qian Lingxi Computational Mechanics Award (Achievement Award, 2016). His current research interests focus on numerical modeling of extreme events, such as hypervelocity impact, blast, bird impact, penetration, perforation and fluid–structure interaction. He has published 3 monographs and 3 textbooks in Chinese. He was included in the list of Elsevier's "Most Cited Chinese Researchers" in 2015.

Zhen Chen is C.W. LaPierre Professor of Engineering at the University of Missouri (MU). His research area is in Computational Mechanics with a recent focus on multiscale modeling and simulation of the multiphysical phenomena involved in structural failure subjected to extreme loading conditions. Before joining MU in 1995, he was a professional staff member at New Mexico Engineering Research Institute as well as in the Department for the Waste Isolation Pilot Project/Performance Analysis Code Development at Sandia National Laboratories. Among his honors and awards are the Fellow of ASME, the Fellow of the ICACM, the Yangtze visiting professor and Qianren-Plan visiting professor appointed by the Ministry of Education in China, the Faculty Research Award in the College of Engineering at MU, the Outstanding Youth Award (Oversea) from the National Natural Science Foundation of China, and the NSF-CAREER Award. He received his Ph.D. in solid and computational mechanics from the University of New Mexico in 1989.

Yan Liu received his Ph.D. in Mechanics from Tsinghua University in 2007. He did his postdoctoral research at Northwestern University from 2008 to 2010. He has been a faculty member at Tsinghua University since 2010, and is now an associate professor in the School of Aerospace Engineering. His research interests include multiscale simulation, meshfree particle methods and impact dynamics. He received the Natural Science Award from Chinese Ministry of Education in 2009, Du Qing-Hua Medal & Young Researcher Award of Computational Methods in Engineering in 2012, the ICACM Young Investigator Award in 2013.

Preface

Simulation-based Engineering Science (SBES) has become the third pillar of modern science and technology, a peer alongside theory and physical experiment [1]. Computer modeling and simulation are now an indispensable tool for resolving a multitude of scientific and technological problems we are facing [2]. To model and simulate those extreme loading events such as hypervelocity impact, penetration, blast, machining, transient crack propagation and multiphase (solid–liquid–gas) interactions involving failure evolution, however, how to effectively describe localized large deformations, the transition from continuous to discontinuous failure modes, and fragmentation remains a challenging task.

Both Lagrangian and Eulerian approaches have been used in SBES to tackle different kinds of extreme events. Lagrangian methods have a computational grid embedded and deformed with the material [3,4]. As a result, material interfaces can be easily tracked, and history-dependent constitutive models can be readily implemented. However, Lagrangian methods suffer from the difficulties associated with grid distortion and element entanglement, which make Lagrangian methods unsuitable for solving problems involving localized large deformation, fragmentation, melting and vaporization. By contrast, in Eulerian methods, the computational grid is fixed in space, and mass flows through the grid. There is no difficulty associated with grid distortion and element entanglement in Eulerian methods so that they can easily solve the problems involving extreme deformation, fragmentation, melting and vaporization. However, special procedures are required to identify the material interfaces and history-dependency, which are very computationally intensive as compared with Lagrangian methods.

To take advantage of both Eulerian and Lagrangian methods while avoiding the shortcomings of each, the Material Point Method (MPM) has evolved over more than twenty years since its first journal paper was published in 1994 [5]. The MPM is an extension of the particle-in-cell (PIC) method in computational fluid dynamics to computational solid dynamics, formulated using the weak formulation and including the history-dependency of constitutive models.

It discretizes a continuum body into a set of material points (particles) moving through an Eulerian background grid. Hence, the MPM is a continuum-based particle method. The particles carry all material properties such as mass, velocity, stress, strain and state variables so that it is easy to track material interfaces and to implement history-dependent constitutive models. As the equations of motion are solved on the Eulerian background grid, there is no grid distortion or element entanglement, which makes the MPM robust in dealing with various types of extreme loading events.

After providing the necessary background information, this book describes the fundamental theory, implementation and application of the MPM as well as its recent extensions. It contains eight chapters. Chapter 1 briefly introduces the basic ideas and features of the Lagrangian methods, Eulerian methods, hybrid methods and meshfree methods, respectively. Chapter 2 reviews the Lagrangian and Eulerian descriptions of deformation and motion, as well as the strain and stress measures in large deformation theory. The governing equations of motion in an updated Lagrangian framework are given. Based on the updated Lagrangian description, Chapter 3 establishes the MPM formulation by discretizing a continuum body into a set of particles. Both explicit and implicit formulations are presented. The Generalized Interpolation Material Point (GIMP) method, contact algorithm, adaptive MPM, incompressible MPM and non-reflection boundary are discussed in detail. The computer implementation of the MPM and corresponding source codes are described in Chapter 4 based on our open source MPM code, MPM3D-F90. A user's guide and several numerical examples of the MPM3D-F90 code are also presented, for which the input data files can be downloaded from our web site: http://mpm3d.comdyn.cn. Chapter 5 first reviews the explicit finite element method, and then presents the material point finite element method, coupled material point finite element method, adaptive material point finite element method and hybrid material point finite element method as developed in the Computational Dynamics Lab of the School of Aerospace Engineering at Tsinghua University. Chapter 6 discusses the constitutive models which describe different types of material behaviors, with a focus on the extreme events. The computer implementation of these constitutive models is specified in detail, and corresponding source codes are provided. Chapter 7 introduces a multiscale MPM that could couple discrete forcing functions as used in molecular dynamics with constitutive models as used in the continuous approaches in a single computational domain. The mapping and remapping process in the MPM could effectively coarse-grain fine details. Chapter 8 describes the applications of the MPM and its extensions in those extreme events such as transient crack propagation, impact/penetration, blast, fluid–structure interaction, and biomechanical responses to extreme loading.

The most materials of this book were based on our MPM book in Chinese [6] with significant extensions and revisions. Zhen Chen added Sect. 3.2.3 and Chapter 7 while Yan Liu drafted Chapter 8. The remaining chapters were drafted by Xiong Zhang. Xiong Zhang and Zhen Chen have revised the whole book.

Finally, the first author wishes to acknowledge his students, S. Ma, P. Huang, Z.T. Ma, Y.P. Lian, H.K. Wang, W.W. Gong, S.Z. Zhou, P.F. Yang, X.X. Cui, P. Liu, Y.T. Zhang, X.J. Wang, Z.X. Hu, J.G. Li, Z.P. Chen and F. Zhang, for their contributions to the algorithm development and programming related to the book. Especially, Tamás Benedek who implemented a subroutine in MPM3D-F90 to output simulation results to ParaView [7] for postprocessing when he worked on his master thesis at Tsinghua University.

Chapter 1

Introduction

Contents

Simulation-based Engineering Science (SBES) [2] is the third pillar of the modern science and engineering, a peer alongside theory and physical experiment [1]. Compared with physical experiment, SBES has the advantages of low cost, safety, and efficiency in solving various kinds of challenging problems. To better simulate those extreme events such as hypervelocity impact, penetration, blast, crack propagation, and multi-phase (solid–liquid–gas) interactions involving failure evolution, yet effectively discretize localized large deformation, the transition among different types of failure modes and fragmentation remains a very difficult task. Based on the way how deformation and motion are described, existing spatial discretization methods can be classified into Lagrangian, Eulerian, and hybrid ones, respectively.

1.1 LAGRANGIAN METHODS

In Lagrangian methods the computational grid is embedded and deformed with the material. Since there is no advection between the grid and material, no advection term appears in the governing equations, which significantly simplifies the solution process. The mass of each material element keeps constant during the solution process, but the element volume varies due to element deformation. Lagrangian methods have the following advantages:

1. They are conceptually more simple and efficient than Eulerian methods. Because there is no advection term that describes the mass flow across element boundaries, the conservation equations for mass, momentum, and energy are simple in form, and can be efficiently solved.

The Material Point Method. http://dx.doi.org/10.1016/B978-0-12-407716-4.00001-6

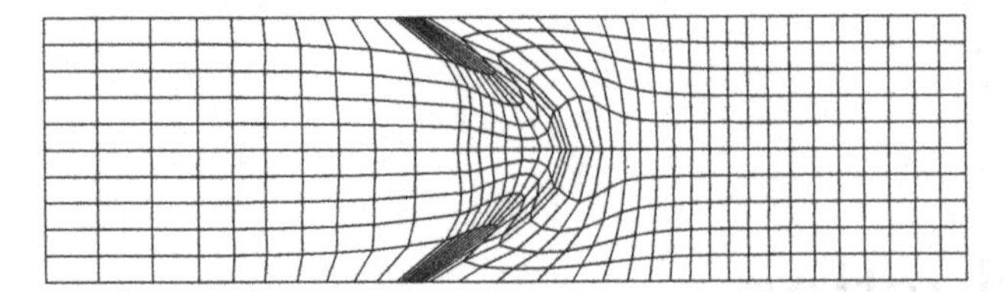

FIGURE 1.1 Lagrangian grid.

2. Element boundaries coincide with the material interfaces during the solution process so that it is easy to impose boundary conditions and to track material interfaces.
3. Since Lagrangian methods track the flow of individual masses, it is easy to implement history-dependent constitutive models.

Fig. 1.1 shows a typical Lagrangian grid which is embedded and deformed with the material. Severe element distortion results in significant errors in numerical solution, and even leads to a negative element volume or area which would cause abnormal termination of the computation. To obtain a stable solution with an explicit time integration scheme, the time step must be smaller than a critical time step which is controlled by the minimum characteristic length of all elements in the grid. Because severe element distortion would significantly decrease the characteristic element length, the time step in a Lagrangian calculation could become smaller and smaller, and finally approach zero, which makes the computation impossible to be completed. To complete a Lagrangian computation for an extreme loading case, a distorted grid must be remeshed and its result must be interpolated to the remeshed grid. The remesh or rezone technique has been successfully used in solving many 1D and 2D problems, but rezoning a complicated 3D material domain is still a challenging task. For a history-dependent material, the history variables are also required to be interpolated from the old grid to the new grid, which may further cause numerical error in stress calculation.

Another way to eliminate the element distortion is to use the erosion technique, which simply deletes the heavily distorted elements. An element is considered to be heavily distorted if its equivalent plastic strain exceeds a user-defined erosion strain value, or the critical time step size is less than a prescribed value. Introducing element erosion can resolve some of the issues related to the severe element distortion, but also introduce new issues. The global system will lose both mass and energy, which can severely affect the simulation outcome. Furthermore, the erosion technique cannot model the formation process of debris cloud and its interaction with other panels in hypervelocity impact simulation.

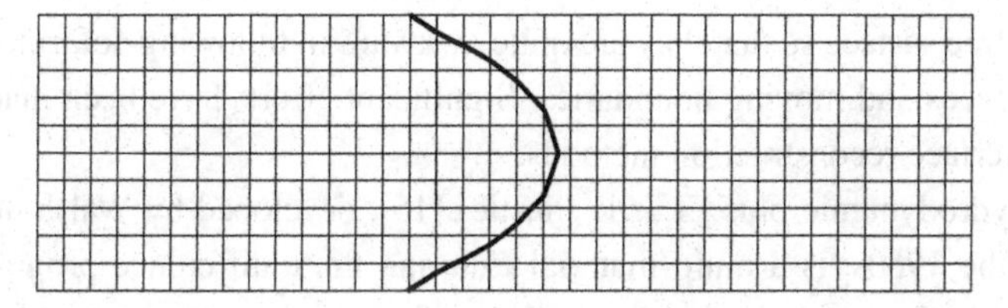

FIGURE 1.2 Eulerian grid.

Many Lagrangian codes have been developed, as shown in the open literature. The HEMP [8] was developed in the early 1960s by Wilkins at the Lawrence Livermore National Laboratory. The HEMP was an explicit Lagrangian finite-difference code that could handle large strains, elastic–plastic flow, wave propagation, and sliding interfaces. The EPIC code [9] was an explicit Lagrangian finite element code developed in the 1970s by Johnson. Both the rezoning and erosion techniques were employed in the EPIC to simulate high velocity impact and blast problems. The PRONTO3D code [10] was a 3D transient solid dynamics code developed at the Sandia National Laboratory for analyzing large deformations of highly nonlinear materials subjected to extremely high strain rates. This code was based on an explicit finite element formulation, and had been coupled with the smoothed particle hydrodynamics (SPH) method through a contact-like algorithm [11]. The DYNA2D and DYNA3D codes were developed in the 1970s at the Lawrence Livermore National Laboratory as explicit Lagrangian finite element codes and were successfully commercialized [12–14].

1.2 EULERIAN METHODS

For problems in which a material domain could become heavily distorted or different materials are mixed, an Eulerian method is more appropriate. In Eulerian methods, the computational grid is fixed in space and does not move with the material such that the material flows through the grid, as shown in Fig. 1.2.

There is no element distortion in Eulerian methods, but the physical variables, such as mass, momentum, and energy, advect between adjacent elements across their interface. The volume of each element keeps constant during the simulation, but its density varies due to the advection of mass. Eulerian methods are suited for modeling large deformations of materials so that most of computational fluid dynamics codes and early hydrocodes for impact and blast simulation employ Eulerian methods.

Eulerian methods only calculate the material quantities advected between elements without explicitly and accurately determining the position of material interface and free surface so that they are quite awkward in following deforming material interfaces and moving boundaries. Significant efforts have been made to develop interface reconstruction methods.

HELP (Hydrodynamic plus ELastic Plastic) [15], developed by Walsh and Hageman in the 1960s, is a multi-material Eulerian finite difference program for compressible fluid and elastic–plastic flows. To treat the material interface or free surface, massless tracer particles are used, which define the surface position and move across the Eulerian grid. CTH [16] is an Eulerian finite volume code developed at Sandia National Laboratories to model multi-dimensional, multi-material, large deformation, and strong shock wave physics. The CTH code employs a two-step Eulerian solution scheme, a Lagrangian step in which the cells distort to follow the material motion, and a remesh step where the distorted cells are mapped back to the Eulerian mesh. Material interfaces are reconstructed using the Sandia Modified Young's Reconstruction Algorithm. The CTH has adaptive mesh refinement and uses second-order accurate numerical methods to reduce numerical dispersion and dissipation. It is still under development at Sandia National Laboratories [17].

The Zapotec developed at Sandia National Laboratories is a framework that tightly couples the CTH and PRONTO codes [18,19]. In a Zapotec analysis, both CTH and PRONTO are run concurrently. For a given time step, the Zapotec maps the current configuration of a Lagrangian body onto the fixed Eulerian mesh. Any overlapping Lagrangian material is inserted into the Eulerian mesh with the updated mesh data passed back to the CTH. After that the external loading on the Lagrangian material surfaces is determined from the stress state in the Eulerian mesh. These loads are passed back to PRONTO as a set of external nodal forces. After the coupled treatment is completed, both CTH and PRONTO are run independently over the next time step.

1.3 HYBRID METHODS

Both purely Lagrangian and purely Eulerian methods possess different shortcomings and advantages so that it is desirable to find new approaches to take advantage of both methods to better tackle challenging problems. The arbitrary Lagrangian–Eulerian (ALE) method [20] and the particle-in-cell (PIC) method [21,22] are two representatives.

1.3.1 Arbitrary Eulerian–Lagrangian Method and Its Variations

The ALE method was first proposed in the finite difference and finite volume context [23,24], and was subsequently adopted in the finite element context

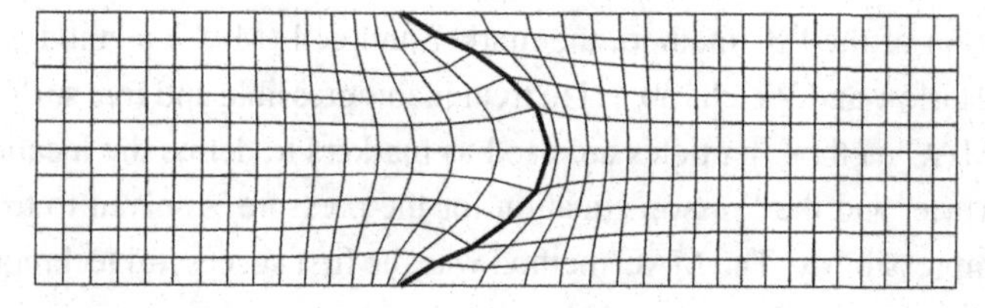

FIGURE 1.3 ALE grid.

[25–27]. The mixed Eulerian–Lagrangian method [24] involves the Eulerian set-up with respect to one dimension and the Lagrangian one to the other dimension which corresponds to the direction of fluid flow. The coupled Eulerian–Lagrangian [23] code employs an Eulerian mesh for the entire region and Lagrangian meshes for the subregions of fluids with nonstationary boundaries approximated by Lagrangian lines.

In the above methods, the computational mesh may be moved with the material in Lagrangian manner, or be held fixed in Eulerian manner, or be moved independently of material deformation to optimize element shapes and to describe the boundaries accurately [20], as shown in Fig. 1.3. Because these methods offer great flexibility in moving the computational mesh, they can handle a much greater distortion of the material than a Lagrangian method, with a higher resolution than that afforded by an Eulerian method. However, the convective terms still pose some problems. Furthermore, designing an efficient and effective mesh-moving algorithm for complicated 3D problems remains a challenging task.

1.3.2 Particle-In-Cell Method and Its Variations

The PIC method was proposed and developed at Los Alamos National Laboratory by Harlow in the late 1950s [21,22,28]. PIC makes use of both Lagrangian and Eulerian descriptions, namely, the fluid is discretized as a set of Lagrangian particles that carry material position, mass, and species information, but the computational mesh is a uniform Eulerian one. A computational cycle is divided into two phases, a Lagrangian phase and an Eulerian (remap or rezone) phase. In the Lagrangian phase, all the variables, including the mesh coordinates and the particle positions, are advanced. In the Eulerian phase, the mesh is mapped back into its original configuration, leaving the particles at their new locations. This process can also be viewed in a time splitting way, namely, the Lagrangian phase updates the quantities by all the processes except for advection, while the Eulerian phase moves the particles and accomplishes all of the advective fluxing [29].

As a variation of the PIC method, the marker-and-cell (MAC) method was developed by Harlow and Welch [30,31] to treat incompressible and free surface flows. In the MAC method, particles are used as markers to define the location of the free-surface, and the Poisson equation for the pressure is solved to treat the fluid incompressibility. The MAC method was the first successful technique for simulating incompressible flows [32].

The original version of PIC is not a fully Lagrangian particle method because only the material position, mass, and species information is carried by the particles, while the remaining quantities are still stored in the computational grid. The transfer of information between the particles and the underlying grid leads to significant numerical diffusion. There are two strategies to reduce the numerical diffusion, namely, second-order accuracy advection scheme [33] and fully Lagrangian particle method. Brackbill et al. developed a fully Lagrangian particle method, FLuid-Implicit-Particle (FLIP) method [34,35], in which each particle carries all of the properties of the fluid, including momentum and energy. FLIP preserves the ability of the original PIC to resolve contact discontinuities, but eliminates the major source of numerical diffusion.

1.3.3 Material Point Method

When working on the penetration problems in the early 1990, Zhen Chen and his former PhD advisor, Buck Schreyer, faced a challenging task to improve the computational fidelity and efficiency of the finite element method (FEM), due to its limitation in the required use of a pin-hole in the mesh design. In a seminar at University of New Mexico, Deborah Sulsky presented the advances of the PIC method, based on her collaborative research on computational fluid dynamics with the scientists at Los Alamos National Laboratory. Since the particle motion in fluid is similar to the penetrator's motion in solid from the viewpoint of hard–soft body interaction, Sulsky's seminar opened Chen's eyes to a new direction of research so that he initiated an interdisciplinary discussion. In collaboration with Sandia National Laboratories, the team of three folks with diversified tastes then started to combine computational fluid dynamics with computational solid dynamics to develop a continuum-based particle method with its first journal paper published in 1994 [5], which was later named as the Material Point Method (MPM). Over the last two decades, many research teams in the world have further developed the MPM and combined the MPM with other numerical methods for multiphase, multiphysics, and multiscale simulations to advance SBES.

The MPM is an extension of the FLIP method from computational fluid dynamics to computational solid dynamics with two key differences. First, the constitutive equations are solved at the particles (material points) rather than the grid cell centers such that the MPM can readily model history-dependent materials. Second, the MPM is formulated in the weak form consistent with the FEM so that the FEM and MPM could be effectively combined together [36–41] for large-scale simulations.

The MPM is a fully Lagrangian particle method which utilizes the advantages of both Eulerian and Lagrangian methods. As compared with Eulerian methods, the numerical dissipation normally associated with a Eulerian method is eliminated, while the complete deformation history of material points are tracked. Compared with Lagrangian methods, mesh distortion and element entanglement are avoided. Therefore, the MPM has demonstrated obvious advantages in tackling those extreme events such as impact, blast, penetration, perforation, machining, fragmentation, and multi-phase interaction involving failure evolution, as demonstrated in Chapter 8.

1.4 MESHFREE METHODS

In addition to the evolution of the MPM, different types of meshfree and particle methods for improved spatial discretization in different problems have also been proposed and developed in the SBES community [42]. Since all these meshfree and particle methods do not use a rigid mesh connectivity compared with the conventional mesh-based methods such as the FEM, they have been applied to many challenging problems of current interests such as impact/contact, localization, crack propagation, penetration, perforation, and fragmentation. Nevertheless, many of the meshfree methods suffer from higher computational costs, and the accuracy of some meshfree methods is still dependent on the node regularities to some extent.

Smoothed particle hydrodynamics (SPH) [43–46] is one of the earliest meshfree Lagrangian particle methods. The SPH was first proposed by Lucy [43] and Gingold and Monaghan [44] in 1977 to solve astrophysical problems in the 3D open space, and has been extensively studied and extended to solid and fluid dynamics problems with large deformations. The SPH and its improved versions have been successfully applied to the hypervelocity impact simulations, and become some of the most popular meshfree methods in this area. Because of their good performance, several commercial softwares, such as AUTODYN, PAM-CRASH, and LS-DYNA, have incorporated the SPH into their solvers. However, the SPH is limited in simulating multiphase interactions involving failure evolution.

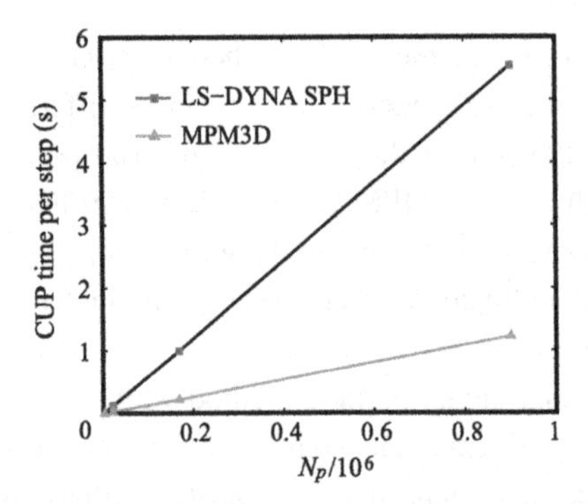

FIGURE 1.4 Efficiency comparison between the MPM and SPH [47].

Ma et al. [47] compared the basic formulation and features of the MPM with SPH from the following aspects: neighbor searching, approximation functions, consistency of shape functions, tensile instability, time integration, boundary conditions, and contact algorithm. A comparative study showed that the MPM possesses many prominent features. The formulation of the MPM is simple and similar to the traditional FEM. The time consuming neighbor searching, which is compulsory in most meshfree methods, is not required in the MPM. The MPM shape functions exactly satisfy the constant and linear consistency. The MPM avoids tensile instability that is annoying in the SPH. The boundary conditions can be applied in the MPM as easily as in the FEM, and the contact algorithm can be efficiently implemented whose cost is linear in the number of material points involved. Because the same regular computational grid can be used in all time steps, the time step keeps constant in the MPM simulations. Numerical studies have showed that the computational efficiency and stability of our MPM3D code are much higher than those of LS-DYNA SPH module.

Fig. 1.4 compares the CPU time per step as used by LS-DYNA SPH module and our MPM3D code in the simulation of the translation motion of a cubic block [48]. It demonstrates that the CPU time per step used by both methods increases linearly with the increase of number of particles, but the rate of increase of the SPH is much higher than that of the MPM.

Ma et al. [47] also investigated the accuracy and efficiency of the SPH and MPM by simulating the impact of a copper cylinder to a rigid wall with an impact velocity of 190 m/s. In the SPH simulations, the constant associated with the smoothing length was set to 1.2 (SPH1) and 1.4 (SPH2), respectively. The value of 1.2 is the default value used in LS-DYNA, and a larger value will increase the computational time but may improve the result with more neighbors for each particle. Fig. 1.5 compares the final configurations of the bar obtained

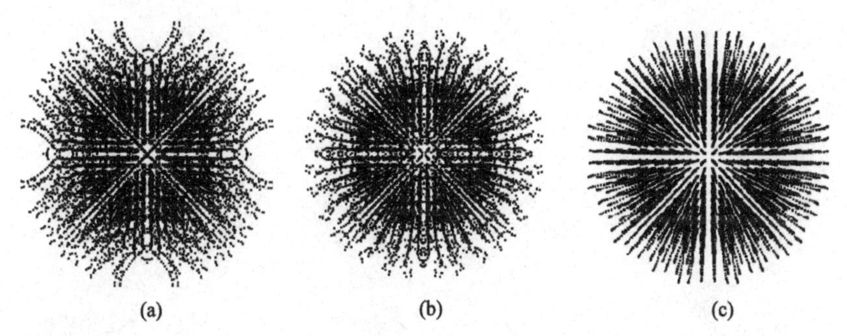

FIGURE 1.5 Final configurations of the Taylor bar impact (top view): (a) SPH1, (b) SPH2, and (c) MPM3D [47].

by the SPH and MPM, which shows that the SPH algorithm suffers from numerical fracture due to tensile instability. Enlarging the smoothing length can alleviate the numerical fracture, but particle clumps may still exist. Furthermore, enlarging the smoothing length increases the time step size, which raises the one-step computational cost significantly.

Chapter 2

Governing Equations

Contents

For large deformation problems, such as the evolution of localized failure under extreme loading conditions, the finite strain theory should be used to establish the governing equations. This chapter briefly introduces the descriptions of motion in both Lagrangian and Eulerian frameworks, the deformation gradient, the rate of deformation, the Cauchy stress, and the Jaumann rate of Cauchy stress. The equations governing the motion and deformation of materials are formulated in the updated Lagrangian frame that is used in the MPM and FEM formulations in the book. For continuum-based variables or equations, bold-faced letters denote tensors of 1st or higher orders while indicial notation is adopted to represent their components. Matrix notation is employed to describe discrete variables or equations for numerical implementation.

2.1 DESCRIPTION OF MOTION

Considering the motion of a continuum body within which a material point (particle) moves from its original position P at time $t = 0$ to its current position p at time t, as shown in Fig. 2.1. The region of three-dimensional (3D) Euclidean space occupied by the body at time $t = 0$ is called the **initial or undeformed configuration** (Ω_0), while the region of 3D Euclidean space occupied by the body at time t is called the **current or deformed configuration** (Ω). To measure the motion of the body, one particular configuration should be selected as the **reference configuration** to which the motion of the body will be referred.

The Material Point Method. http://dx.doi.org/10.1016/B978-0-12-407716-4.00002-8

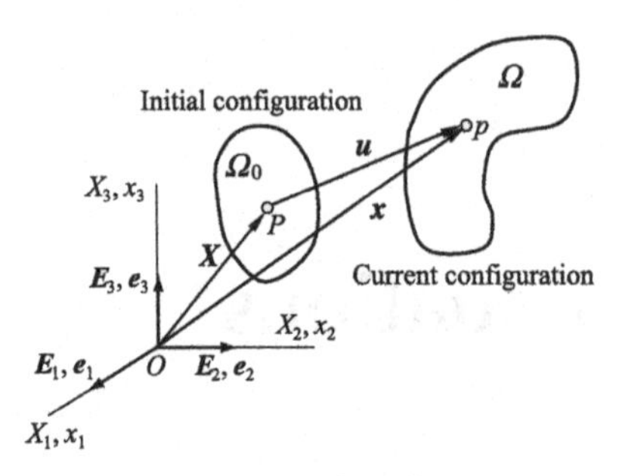

FIGURE 2.1 Initial configuration and current configuration.

Any possible configuration can be chosen as the reference configuration, but a common practice is to choose the initial configuration as the reference configuration.

The original position vector $\boldsymbol{X}$ of a particle in the reference configuration can be expressed as

$$\boldsymbol{X} = X_i \boldsymbol{E}_i, \quad i = 1, 2, 3 \tag{2.1}$$

where $\boldsymbol{E}_i$ is the unit vector that defines the basis of the material (body-frame) coordinate system, and the components X_i are the coordinates of the particle in the material coordinate system. In this book, the subscripts with lowercase Latin alphabet, such as i, j, and k, indicate the spatial components following Einstein summation convention, namely, repeated indices imply the summation over all the values of the index. The original position vector $\boldsymbol{X}$ of the particle P serves as a label to the particle. The coordinates X_i are called the **material coordinates** or **Lagrangian coordinates**.

The position vector $\boldsymbol{x}$ of the particle $\boldsymbol{X}$ in the current configuration, as shown in Fig. 2.1, can be written as

$$\boldsymbol{x} = x_i \boldsymbol{e}_i, \quad i = 1, 2, 3 \tag{2.2}$$

where e_i is the unit vector that defines the basis of the spatial coordinate system, and the components x_i are the coordinates of the particle in the spatial coordinate system. The coordinates x_i define the spatial position of the particle such that they are called the **spatial coordinates** or **Eulerian coordinates**.

The motion and deformation of a continuum body can be described by the evolution of configuration with time. There are two types of descriptions for the motion, namely, **material** or **Lagrangian description** and **spatial** or **Eulerian description**. The Lagrangian description is in terms of the material coordinates so that the position and physical properties of the particle are described in terms of the material coordinates $\boldsymbol{X}$ and time t. The reference configuration is the initial configuration at $t = 0$. In the Lagrangian description, an observer standing in the body-frame observes the changes in the position and physical properties as the continuum body moves in space with time, which is commonly used in solid mechanics.

The Eulerian description is in terms of the spatial coordinates $\boldsymbol{x}$ and time t, in which the current configuration is selected as the reference configuration. The Eulerian description is focused on what is occurring at a fixed point in space as time advances, rather than individual particles as they move through space with time. No history-dependency is considered in the Eulerian description. Therefore, this approach is commonly used in fluid mechanics.

In the Lagrangian description, the motion of a continuum body can be expressed as

$$\boldsymbol{x} = \boldsymbol{\chi}(\boldsymbol{X}, t) = \boldsymbol{x}(\boldsymbol{X}, t) \tag{2.3}$$

which is a mapping from the initial configuration Ω_0 onto the current configuration Ω. It states that a particle with a position vector $\boldsymbol{X}$ in the initial configuration Ω_0 will occupy the position $\boldsymbol{x}$ in the current configuration Ω at time t. In continuum mechanics, the mapping operation $\boldsymbol{\chi}(\cdot,\cdot)$ is assumed to be invertible at each time t and differentiable as many times as necessary with respect to both $\boldsymbol{X}$ and t.

In the Eulerian description, the motion of a continuum body can be expressed as

$$\boldsymbol{X} = \boldsymbol{\chi}^{-1}(\boldsymbol{x}, t) = \boldsymbol{X}(\boldsymbol{x}, t) \tag{2.4}$$

which states that the particle, which occupies the position $\boldsymbol{x}$ in the current configuration, is the one with the position vector $\boldsymbol{X}$ in the initial configuration.

In the Lagrangian description, the displacement field is expressed in terms of the material coordinates as

$$\boldsymbol{u} = \boldsymbol{x}(\boldsymbol{X}, t) - \boldsymbol{X}. \tag{2.5}$$

In the Eulerian description, the displacement field is expressed in terms of the spatial coordinates as

$$\boldsymbol{u} = \boldsymbol{x} - \boldsymbol{X}(\boldsymbol{x}, t). \tag{2.6}$$

The velocity of the particle $\boldsymbol{X}$ is the time rate of change of its instantaneous position vector $\boldsymbol{x}$, namely, the partial derivative of $\boldsymbol{x}$ with respect to time t, holding $\boldsymbol{X}$ constant. The partial derivative of any quantity for a moving continuum body with respect to time t, holding $\boldsymbol{X}$ constant, is called the **material derivative** of that quantity, which is also known as the **total derivative** or **Lagrangian**

derivative. It can be considered as the time rate of change of the quantity following a particle. Therefore, the velocity field of the body can be obtained from the material derivative of Eq. (2.5) as

$$\boldsymbol{v} = \dot{\boldsymbol{x}} = \frac{\partial \boldsymbol{x}(\boldsymbol{X}, t)}{\partial t} = \frac{\partial \boldsymbol{u}(\boldsymbol{X}, t)}{\partial t}, \tag{2.7}$$

and the acceleration field can be obtained from the material derivative of the velocity filed (2.7) as

$$\boldsymbol{a} = \dot{\boldsymbol{v}} = \frac{\partial \boldsymbol{v}(\boldsymbol{X}, t)}{\partial t} = \frac{\partial^2 \boldsymbol{u}(\boldsymbol{X}, t)}{\partial t^2}. \tag{2.8}$$

In the Eulerian description, the material derivative of a quantity $F(\boldsymbol{x}, t)$ can be obtained by using the chain rule and Eq. (2.3) as

$$\begin{aligned} \frac{DF(\boldsymbol{x}, t)}{Dt} &= \frac{\partial F(\boldsymbol{x}, t)}{\partial t} + \frac{\partial F(\boldsymbol{x}, t)}{\partial \boldsymbol{x}} \cdot \frac{\partial \boldsymbol{x}(\boldsymbol{X}, t)}{\partial t} \\ &= \frac{\partial F(\boldsymbol{x}, t)}{\partial t} + \boldsymbol{v} \cdot \nabla F(\boldsymbol{x}, t). \end{aligned} \tag{2.9}$$

The first term on the right-hand side of Eq. (2.9) is the **local derivative** or **Eulerian derivative**, which yields the local rate of change of the quantity occurring at position $\boldsymbol{x}$. The second term of the right-hand side is the **convective derivative** that expresses the rate of change of the quantity contributed by the particle motion due to the nonuniformity of the quantity in space. The material derivative (2.9) establishes a link between the Eulerian description and Lagrangian one of continuum deformation.

By following the motion of material points, the Lagrangian description can readily track material interfaces and implement history-dependent material models so that it is commonly used in solid mechanics. Only the strain and stress measures that use the Lagrangian description are considered in this chapter.

2.2 DEFORMATION GRADIENT

The partial derivative of $\boldsymbol{x}$ with respect to $\boldsymbol{X}$,

$$\boldsymbol{F} = \frac{\partial \boldsymbol{x}(\boldsymbol{X}, t)}{\partial \boldsymbol{X}} = \frac{\partial x_i}{\partial X_j} \boldsymbol{e}_i \otimes \boldsymbol{E}_j,$$

is called the **deformation gradient** that is a second-order, asymmetric, and two-point tensor as it is related to both the reference and current configurations. Consider an infinitesimal linear element $\mathrm{d}\boldsymbol{X}$ joining a particle P with the position vector $\boldsymbol{X}$ and its neighbor Q with the position vector $\boldsymbol{X} + \mathrm{d}\boldsymbol{X}$ in the reference configuration Ω_0. In the current configuration, the linear element becomes $\mathrm{d}\boldsymbol{x}$, which can be expressed as

$$\mathrm{d}\boldsymbol{x} = \boldsymbol{x}(\boldsymbol{X} + \mathrm{d}\boldsymbol{X}, t) - \boldsymbol{x}(\boldsymbol{X}, t). \tag{2.10}$$

Expanding $\boldsymbol{x}(\boldsymbol{X} + \mathrm{d}\boldsymbol{X}, t)$ in Eq. (2.10) into Taylor series about $\boldsymbol{X}$ and neglecting higher order terms results in

$$\mathrm{d}\boldsymbol{x} = \frac{\partial \boldsymbol{x}}{\partial \boldsymbol{X}} \cdot \mathrm{d}\boldsymbol{X}. \tag{2.11}$$

Eq. (2.11) implies that the material deformation gradient tensor $\boldsymbol{F}$ is a linear mapping operator, which maps each infinitesimal linear element $\mathrm{d}\boldsymbol{X}$ in the reference configuration into an infinitesimal linear element $\mathrm{d}\boldsymbol{x}$ in the current configuration. Therefore, the deformation gradient tensor $\boldsymbol{F}$ describes the mapping from the infinitesimal neighborhood of $\boldsymbol{X}$ to the infinitesimal neighborhood of $\boldsymbol{x}$ such that $\boldsymbol{F}(\boldsymbol{X}, t)$ is a measure of both the stretch and rotation in the infinitesimal neighborhood of $\boldsymbol{X}$ as it deforms to $\boldsymbol{x}$ at time t.

A necessary and sufficient condition for the motion (2.3) to be invertible for all $\boldsymbol{X}$ and times t is that the determinant of the deformation gradient, often referred to as the Jacobian, be nonzero, namely

$$J = \left| \frac{\partial \boldsymbol{x}}{\partial \boldsymbol{X}} \right| = \begin{vmatrix} \frac{\partial x_1}{\partial X_1} & \frac{\partial x_1}{\partial X_2} & \frac{\partial x_1}{\partial X_3} \\ \frac{\partial x_2}{\partial X_1} & \frac{\partial x_2}{\partial X_2} & \frac{\partial x_2}{\partial X_3} \\ \frac{\partial x_3}{\partial X_1} & \frac{\partial x_3}{\partial X_2} & \frac{\partial x_3}{\partial X_3} \end{vmatrix} = e_{ijk} \frac{\partial x_i}{\partial X_1} \frac{\partial x_j}{\partial X_2} \frac{\partial x_k}{\partial X_3} \neq 0 \tag{2.12}$$

where e_{ijk} is the Levi-Civita symbol (also called permutation symbol or alternating symbol), which is equal to 1 if (i, j, k) is an even permutation of $(1, 2, 3)$, -1 if it is an odd permutation, and 0 if any index is repeated.

Consider a parallelepiped formed by three linear elements $\mathrm{d}\boldsymbol{X}$, $\delta\boldsymbol{X}$, and $\Delta\boldsymbol{X}$ at point $\boldsymbol{X}$ in the reference configuration, which deforms to the parallelepiped formed by the linear element $\mathrm{d}\boldsymbol{x}$, $\delta\boldsymbol{x}$, and $\Delta\boldsymbol{x}$ at point $\boldsymbol{x}$ in the current configuration, as shown in Fig. 2.2. According to Eq. (2.11), the linear elements $\mathrm{d}\boldsymbol{x}$, $\delta\boldsymbol{x}$, and $\Delta\boldsymbol{x}$ are related to the linear elements $\mathrm{d}\boldsymbol{X}$, $\delta\boldsymbol{X}$, and $\Delta\boldsymbol{X}$ by

$$\mathrm{d}\boldsymbol{x} = \frac{\partial \boldsymbol{x}}{\partial \boldsymbol{X}} \cdot \mathrm{d}\boldsymbol{X}, \quad \delta\boldsymbol{x} = \frac{\partial \boldsymbol{x}}{\partial \boldsymbol{X}} \cdot \delta\boldsymbol{X}, \quad \Delta\boldsymbol{x} = \frac{\partial \boldsymbol{x}}{\partial \boldsymbol{X}} \cdot \Delta\boldsymbol{X}.$$

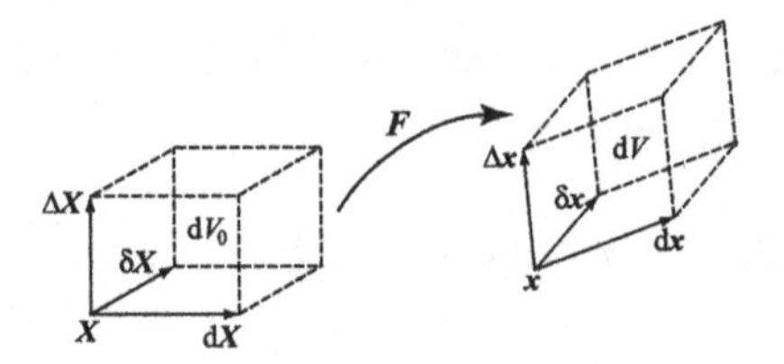

FIGURE 2.2 The volume elements $\mathrm{d}V_0$ and $\mathrm{d}V$.

In the current configuration, the volume of the parallelepiped formed by the linear element $\mathrm{d}\boldsymbol{x}$, $\delta\boldsymbol{x}$, and $\Delta\boldsymbol{x}$ at point $\boldsymbol{x}$ is

$$\mathrm{d}V = \begin{vmatrix} \mathrm{d}x_1 & \mathrm{d}x_2 & \mathrm{d}x_3 \\ \delta x_1 & \delta x_2 & \delta x_3 \\ \Delta x_1 & \Delta x_2 & \Delta x_3 \end{vmatrix} = J\mathrm{d}V_0. \tag{2.13}$$

Therefore, the Jacobian of the deformation gradient is related to the local volume change experienced by an infinitesimal parallelepiped during the deformation of the continuum body. The dilatation or normalized volume change is defined as

$$\mu = \frac{\mathrm{d}V - \mathrm{d}V_0}{\mathrm{d}V_0} = J - 1. \tag{2.14}$$

2.3 RATE OF DEFORMATION

Consider a particle P and its neighboring particle P'. Their position vectors at time t are $\boldsymbol{x}$ and $\boldsymbol{x} + \mathrm{d}\boldsymbol{x}$, respectively, while their velocity vectors at time t are $\boldsymbol{v}(\boldsymbol{x}, t)$ and $\boldsymbol{v}(\boldsymbol{x} + \mathrm{d}\boldsymbol{x}, t)$, respectively. The difference between the velocity of particle P' and that of particle P then becomes

$$\mathrm{d}\boldsymbol{v} = \boldsymbol{v}(\boldsymbol{x} + \mathrm{d}\boldsymbol{x}, t) - \boldsymbol{v}(\boldsymbol{x}, t) = \boldsymbol{L} \cdot \mathrm{d}\boldsymbol{x} \tag{2.15}$$

where

$$\boldsymbol{L} = \frac{\partial \boldsymbol{v}}{\partial \boldsymbol{x}} \tag{2.16}$$

is the **velocity gradient** that can be decomposed into a symmetric tensor $\boldsymbol{D}$ and a skew-symmetric tensor $\boldsymbol{\Omega}$, i.e.,

$$\boldsymbol{L} = \boldsymbol{D} + \boldsymbol{\Omega} \tag{2.17}$$

where

$$\boldsymbol{\Omega} = \frac{1}{2}\left(\boldsymbol{L} - \boldsymbol{L}^{\mathrm{T}}\right), \tag{2.18}$$

$$\boldsymbol{D} = \frac{1}{2}\left(\boldsymbol{L} + \boldsymbol{L}^{\mathrm{T}}\right) \tag{2.19}$$

are referred to as the **spin tensor** (or **vorticity tensor**) and the **rate of deformation** tensor, respectively. The rate of deformation tensor $\boldsymbol{D}$ represents the rate of stretching of a line element, while the spin tensor $\boldsymbol{\Omega}$ indicates the rate of rotation or vorticity of the line element.

Similar to the decomposition of the velocity gradient tensor, the relative velocity $\mathrm{d}\boldsymbol{v}$ in Eq. (2.15) can also be decomposed as

$$\mathrm{d}\boldsymbol{v} = \mathrm{d}\boldsymbol{v}^{*} + \mathrm{d}\boldsymbol{v}^{**} \tag{2.20}$$

where

$$\mathrm{d}\boldsymbol{v}^{*} = \boldsymbol{\Omega} \cdot \mathrm{d}\boldsymbol{x}, \quad \mathrm{d}\boldsymbol{v}^{**} = \boldsymbol{D} \cdot \mathrm{d}\boldsymbol{x}. \tag{2.21}$$

The spin tensor $\boldsymbol{\Omega}$ is a skew-symmetric tensor so that there exists a unique vector $\boldsymbol{\omega}$ such that

$$\boldsymbol{\Omega} \cdot \boldsymbol{s} = \boldsymbol{\omega} \times \boldsymbol{s} \tag{2.22}$$

for any vector $\boldsymbol{s}$. The vector $\boldsymbol{\omega}$ is the axial vector of the spin tensor $\boldsymbol{\Omega}$, called the **angular velocity** vector.

Substituting Eq. (2.22) into the first equation in Eq. (2.21) leads to

$$\mathrm{d}\boldsymbol{v}^{*} = \boldsymbol{\omega} \times \mathrm{d}\boldsymbol{x}. \tag{2.23}$$

Eq. (2.23) states that the relative velocity $\mathrm{d}\boldsymbol{v}^{*}$ equals the vector cross-product of $\boldsymbol{\omega}$ and $\mathrm{d}\boldsymbol{x}$. That is to say, within the neighbor of particle P, the relative velocity $\mathrm{d}\boldsymbol{v}^{*}$ corresponds to the rigid body rotation of the neighbor of the particle about an axis with the angular velocity $\boldsymbol{\omega}$.

Note that the rate of deformation $\boldsymbol{D}$ is the rate of true strain $\dot{\boldsymbol{\varepsilon}}$, i.e.,

$$\boldsymbol{D} = \dot{\boldsymbol{\varepsilon}}, \tag{2.24}$$

which is the true strain measure that corresponds to the current configuration of the continuum body.

2.4 CAUCHY STRESS

Consider an infinitesimal area element $\boldsymbol{n}\mathrm{d}A$ oriented in an arbitrary direction specified by a unit normal vector $\boldsymbol{n}$, as shown in Fig. 2.3. The contact force

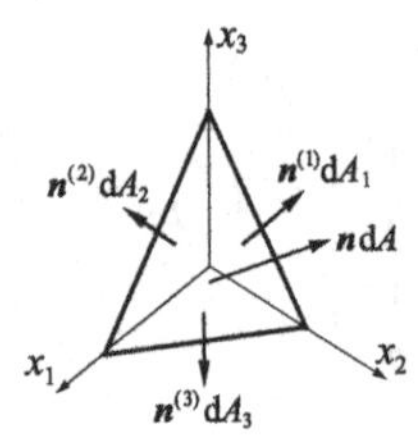

FIGURE 2.3 Infinitesimal tetrahedron.

acting on the area element due to the interaction between the two portions of a continuum body separated by the area element is denoted as $\mathrm{d}\boldsymbol{T}$. Thus, the **stress vector $\boldsymbol{t}^{(n)}$**, also called **traction vector**, acting on the area is defined as the ratio of the contact force $\mathrm{d}\boldsymbol{T}$ applied on the area element to its area $\mathrm{d}A$, i.e.,

$$\boldsymbol{t}^{(n)} = \frac{\mathrm{d}\boldsymbol{T}}{\mathrm{d}A}. \tag{2.25}$$

The infinitesimal area element $\boldsymbol{n}\mathrm{d}A$ forms a tetrahedron with three faces $\boldsymbol{n}^{(k)}\mathrm{d}A_k$ $(k = 1, 2, 3)$ perpendicular to the coordinate axis x_k, as shown in Fig. 2.3. The area of each face, $\mathrm{d}A_k$ $(k = 1, 2, 3)$, is related to the area $\mathrm{d}A$ by $\mathrm{d}A_k = \mathrm{d}An_k$, where n_k is the component of the unit normal vector $\boldsymbol{n}$ along the axis $\boldsymbol{e}_k$. The stress vectors acting on the faces of the tetrahedron are denoted as $\boldsymbol{t}^{(k)}$ $(k = 1, 2, 3)$. The equilibrium of forces results in

$$\mathrm{d}\boldsymbol{T} = \boldsymbol{t}^{(n)}\mathrm{d}A = \boldsymbol{t}^{(k)}\mathrm{d}A_k = \boldsymbol{t}^{(k)}\mathrm{d}An_k = \boldsymbol{n}\cdot\boldsymbol{\sigma}\mathrm{d}A \tag{2.26}$$

where $\sigma_{ki} = \boldsymbol{t}^{(k)}\boldsymbol{e}_i$ is the component of the stress vector $\boldsymbol{t}^{(k)}$ acting on the area element $\boldsymbol{n}^{(k)}dA_k$ in the direction along the ith coordinate. The nine components of σ_{ki} define a 2nd-order tensor, called the **Cauchy stress tensor** or **true stress tensor**. The Cauchy stress tensor completely defines the state of stress at a point inside a continuum body in the deformed configuration. It can be further shown from the equilibrium of moments that the Cauchy stress tensor is symmetric if there is no body couple, i.e.,

$$\boldsymbol{\sigma} = \boldsymbol{\sigma}^{\mathrm{T}}. \tag{2.27}$$

Substituting Eq. (2.26) into Eq. (2.25) leads to

$$\boldsymbol{t}^{(n)} = \boldsymbol{n}\cdot\boldsymbol{\sigma}. \tag{2.28}$$

Eq. (2.28) shows that the Cauchy stress tensor relates a unit normal vector $\boldsymbol{n}$ to the stress vector $\boldsymbol{t}^{(n)}$ across an imaginary surface perpendicular to $\boldsymbol{n}$.

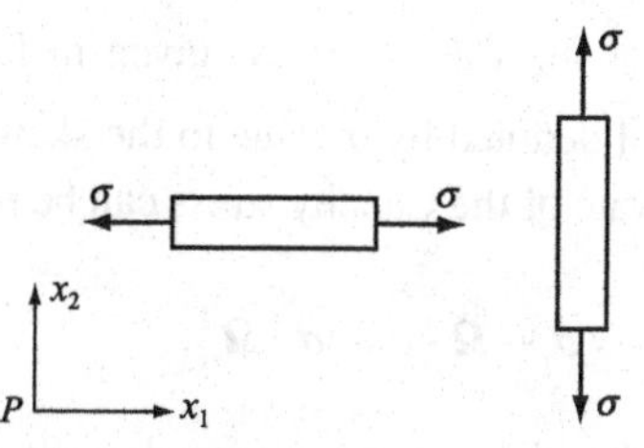

FIGURE 2.4 Rotation of a bar under initial uniaxial stress.

2.5 JAUMANN STRESS RATE

The material response should not depend on the frame of reference. Thus, any constitutive model should be objective, namely, not depend on the frame of reference. Many constitutive equations relate a stress-rate to a strain-rate or the rate of deformation, but the time derivative of the Cauchy stress tensor is not objective.

To demonstrate the above point, consider the rotation of a bar under uniaxial stress in the initial configuration, as shown in Fig. 2.4. When the bar is parallel to the axial axis x_1, $\sigma_{11} = \sigma$ and $\sigma_{22} = 0$. However, when the bar rotates to the position parallel to the axis x_2, $\sigma_{11} = 0$ and $\sigma_{22} = \sigma$. Although no deformation occurs during the rigid body rotation, the Cauchy stress expressed in the fixed coordinate system has changed so that the time derivative of the Cauchy stress is not objective, and should not be used in any constitutive model.

Fig. 2.4 illustrates that although the rate of the Cauchy stress in the fixed coordinate system Px_1x_2 is not objective, the rate of the Cauchy stress in the co-rotational coordinate system rotating with the bar is objective. Therefore, we can obtain an objective stress rate by taking the co-rotational stress as the stress measure. The co-rotational Cauchy stress $\hat{\boldsymbol{\sigma}}$ can be obtained as

$$\hat{\boldsymbol{\sigma}} = \boldsymbol{R}^{\mathrm{T}} \cdot \boldsymbol{\sigma} \cdot \boldsymbol{R} \tag{2.29}$$

where $\boldsymbol{R}$ represents the rotation of the co-rotational coordinate system. Considering the co-rotational coordinate system coincident with the reference coordinate system at time t but rotating with the spin tensor $\boldsymbol{\Omega}$, we have

$$\boldsymbol{R} = \boldsymbol{I}, \quad \dot{\boldsymbol{R}} = \boldsymbol{\Omega}. \tag{2.30}$$

Taking the material derivative in Eq. (2.29) and then invoking Eq. (2.30) results in

$$\begin{aligned}\dot{\hat{\boldsymbol{\sigma}}} &= \dot{\boldsymbol{R}}^{\mathrm{T}} \cdot \boldsymbol{\sigma} \cdot \boldsymbol{R} + \boldsymbol{R}^{\mathrm{T}} \cdot \dot{\boldsymbol{\sigma}} \cdot \boldsymbol{R} + \boldsymbol{R}^{\mathrm{T}} \cdot \boldsymbol{\sigma} \cdot \dot{\boldsymbol{R}} \\ &= \dot{\boldsymbol{\sigma}} + \boldsymbol{\Omega}^{\mathrm{T}} \cdot \boldsymbol{\sigma} + \boldsymbol{\sigma} \cdot \boldsymbol{\Omega}.\end{aligned} \tag{2.31}$$

The **co-rotational rate** of the Cauchy stress given in Eq. (2.31) is often called the **Jaumann rate** and denoted by $\overset{\nabla}{\boldsymbol{\sigma}}$. Due to the skew symmetry of the spin tensor $\boldsymbol{\Omega}$, the Jaumann rate of the Cauchy stress can be rewritten as

$$\overset{\nabla}{\boldsymbol{\sigma}} = \dot{\boldsymbol{\sigma}} - \boldsymbol{\Omega} \cdot \boldsymbol{\sigma} - \boldsymbol{\sigma} \cdot \boldsymbol{\Omega}^{\mathrm{T}}. \tag{2.32}$$

The Jaumann rate is used widely because it is relatively easy to be implemented in numerical procedures, and leads to symmetric tangent moduli. There are many other objective stress rates, such as the Truesdell rate of the Cauchy stress and the Green–Naghdi rate of the Cauchy stress. However, the Jaumann rate will be used in this book.

2.6 UPDATED LAGRANGIAN FORMULATION

The governing equations in continuum mechanics are based on the conservation laws for mass, momentum, and energy, constitutive models, and kinematic conditions. To discretize the governing equations in Lagrangian meshes, the mesh nodes must move with the continuum body so that material boundaries and interfaces always coincide with element edges. Thus material interfaces can be precisely followed and boundary conditions can be easily applied. Quadrature points also move with the material and follow the same material points such that it is straightforward to implement history-dependent constitutive models.

The finite element discretization with Lagrangian meshes can be classified into two categories, the **updated Lagrangian** approach and **total Lagrangian** one. Both approaches use Lagrangian description, namely, the independent variables in both approaches are the Lagrangian (material) coordinate system $\boldsymbol{X}$ and time t. In the total Lagrangian approach, the stress and strain measures are defined with respect to the initial (reference) configuration, namely, the nominal stress and Green strain are used. Derivatives and integrals are evaluated with respect to the Lagrangian coordinate system $\boldsymbol{X}$. In the updated Lagrangian approach, however, the stress and strain measures are defined with respect to the current configuration, namely, the Cauchy stress and rate of deformation are used. Derivatives and integrals are computed with respect to the Eulerian coordinate system $\boldsymbol{x}$. The updated Lagrangian approach has been widely used in the MPM such that the governing equations are only given in the updated Lagrangian formulation in this book.

2.6.1 Reynolds' Transport Theorem

The material derivative of the volume integral of function $\mathbf{f}(\boldsymbol{x}, t)$ over the time-dependent region $\Omega(t)$ that has boundary $\partial\Omega(t)$ can be evaluated by using

Eq. (2.13) as follows:

$$\frac{\mathrm{D}}{\mathrm{D}t}\int_{\Omega} f(\boldsymbol{x},t)\mathrm{d}V = \frac{\mathrm{D}}{\mathrm{D}t}\int_{\Omega_0} f(\boldsymbol{x},t)J\mathrm{d}V_0 = \int_{\Omega_0}\left[\dot{f}(\boldsymbol{x},t)J + f(\boldsymbol{x},t)\frac{\mathrm{D}J}{\mathrm{D}t}\right]\mathrm{d}V_0 \tag{2.33}$$

where $\dot{f}(\boldsymbol{x},t) = \mathrm{D}f(\boldsymbol{x},t)/\mathrm{D}t$ is the material derivative of function $f(\boldsymbol{x},t)$ with respective to time t.

Taking the material derivative of Eq. (2.12) with respective to time t and making use of the property of the determinant that the determinant of a matrix with repeated rows is zero gives

$$\frac{\mathrm{D}J}{\mathrm{D}t} = J\nabla\cdot\boldsymbol{v}. \tag{2.34}$$

Substituting Eq. (2.34) into Eq. (2.33) results in the Reynold's transport theorem

$$\frac{\mathrm{D}}{\mathrm{D}t}\int_{\Omega} f(\boldsymbol{x},t)\mathrm{d}V = \int_{\Omega}\left[\dot{f}(\boldsymbol{x},t) + f(\boldsymbol{x},t)\nabla\cdot\boldsymbol{v}\right]\mathrm{d}V. \tag{2.35}$$

2.6.2 Conservation of Mass

The total mass of a continuum body in the current configuration is

$$m = \int_{\Omega}\rho(\boldsymbol{x},t)\mathrm{d}V \tag{2.36}$$

where $\rho(\boldsymbol{x},t)$ is the local density of the continuum body in the current configuration. The conservation of mass requires that the material derivative of the total mass equal zero, namely

$$\frac{\mathrm{D}}{\mathrm{D}t}\int_{\Omega}\rho(\boldsymbol{x},t)\mathrm{d}V = \int_{\Omega}(\dot{\rho} + \rho\nabla\cdot\boldsymbol{v})\mathrm{d}V = 0. \tag{2.37}$$

Thus

$$\dot{\rho} + \rho\nabla\cdot\boldsymbol{v} = 0. \tag{2.38}$$

Eq. (2.38) is the **continuity equation** expressed with respect to the current configuration. Eq. (2.36) can be rewritten in the initial configuration as

$$\int_{\Omega}\rho\mathrm{d}V = \int_{\Omega_0}\rho_0\mathrm{d}V_0 \tag{2.39}$$

where ρ_0 is the local density of the continuum body in the initial configuration. By using Eq. (2.13), Eq. (2.39) becomes

$$\int_{\Omega_0} (\rho J - \rho_0) \mathrm{d}V_0 = 0. \tag{2.40}$$

Therefore, the conservation of mass can also be written as

$$\rho(\boldsymbol{X}, t) J(\boldsymbol{X}, t) = \rho_0(\boldsymbol{X}). \tag{2.41}$$

Although the continuity equation (2.38) can be used to obtain the current density in the Lagrangian approach, it is simpler and more accurate to use Eq. (2.41).

2.6.3 Conservation of Linear Momentum

The conservation of linear momentum requires that the material derivative of the linear momentum of the continuum body Ω equal the resultant force acting on it, i.e.,

$$\frac{\mathrm{D}}{\mathrm{D}t} \int_{\Omega} \rho \boldsymbol{v}(\boldsymbol{x}, t) \mathrm{d}V = \int_{\Omega} \rho \boldsymbol{b}(\boldsymbol{x}, t) \mathrm{d}V + \int_{\Gamma} \boldsymbol{t}(\boldsymbol{x}, t) \mathrm{d}A \tag{2.42}$$

where $\boldsymbol{b}$ is the body force per unit mass acting on the continuum body, $\boldsymbol{t} = \boldsymbol{n} \cdot \boldsymbol{\sigma}$ is the external traction acting on the surface Γ with the normal $\boldsymbol{n}$ that bounds Ω.

Using the Reynold's transport theorem (2.35), the left side of Eq. (2.42) can be reduced to

$$\begin{aligned} \frac{\mathrm{D}}{\mathrm{D}t} \int_{\Omega} \rho \boldsymbol{v}(\boldsymbol{x}, t) \mathrm{d}V &= \int_{\Omega} \left[\frac{\mathrm{D}(\rho \boldsymbol{v})}{\mathrm{D}t} + \rho \boldsymbol{v} \nabla \cdot \boldsymbol{v} \right] \mathrm{d}V \\ &= \int_{\Omega} [\rho \dot{\boldsymbol{v}} + \boldsymbol{v} (\dot{\rho} + \rho \nabla \cdot \boldsymbol{v})] \mathrm{d}V. \end{aligned} \tag{2.43}$$

Substituting the continuity equation (2.38) into Eq. (2.43) results in

$$\frac{\mathrm{D}}{\mathrm{D}t} \int_{\Omega} \rho \boldsymbol{v}(\boldsymbol{x}, t) \mathrm{d}V = \int_{\Omega} \rho \dot{\boldsymbol{v}} \mathrm{d}V. \tag{2.44}$$

Similarly, for any function $\phi(\boldsymbol{x}, t)$, we have

$$\frac{\mathrm{D}}{\mathrm{D}t} \int_{\Omega} \rho \phi(\boldsymbol{x}, t) \mathrm{d}V = \int_{\Omega} \rho \dot{\phi} \mathrm{d}V. \tag{2.45}$$

Applying the divergence theorem (Gauss' theorem) to the second term on the right side of Eq. (2.42) gives

$$\int_{\Gamma} \boldsymbol{t}(\boldsymbol{x}, t) \mathrm{d}A = \int_{\Gamma} \boldsymbol{n} \cdot \boldsymbol{\sigma} \mathrm{d}A = \int_{\Omega} \boldsymbol{\sigma} \cdot \nabla \mathrm{d}V. \tag{2.46}$$

Substituting Eqs. (2.44) and (2.46) into (2.42) leads to

$$\int_{\Omega} (\rho\dot{\boldsymbol{v}} - \rho\boldsymbol{b} - \boldsymbol{\sigma}\cdot\nabla)\,\mathrm{d}V = \boldsymbol{0}. \tag{2.47}$$

Therefore, the conservation of linear momentum for the updated Lagrangian approach takes the form of

$$\rho\dot{\boldsymbol{v}} - \rho\boldsymbol{b} - \boldsymbol{\sigma}\cdot\nabla = \boldsymbol{0}. \tag{2.48}$$

2.6.4 Conservation of Energy

The first law of thermodynamics requires that the material derivative of the total energy of a continuum body Ω in the current configuration equal the sum of the net rate of change in mechanical work done on the body and the net heat flux into the body, i.e.,

$$\frac{\mathrm{D}}{\mathrm{D}t}\int_{\Omega}\left(\rho e + \frac{1}{2}\rho\boldsymbol{v}\cdot\boldsymbol{v}\right)\mathrm{d}V = \int_{\Omega}\rho s\mathrm{d}V - \int_{\Gamma}\boldsymbol{n}\cdot\boldsymbol{q}\mathrm{d}A + \int_{\Omega}\rho\boldsymbol{v}\cdot\boldsymbol{b}\mathrm{d}V + \int_{\Gamma}\boldsymbol{v}\cdot\boldsymbol{t}\mathrm{d}A \tag{2.49}$$

where e is the specific internal energy (internal energy per unit mass), s is the heat supply, and $\boldsymbol{q}$ is the heat flux vector that represents the thermal energy flow per unit time and per unit area in the deformed body. The Fourier's Law of Heat Conduction states that the local heat flux vector is proportional to the negative local temperature gradient, i.e.,

$$\boldsymbol{q} = -k\frac{\partial T}{\partial \boldsymbol{x}} \tag{2.50}$$

where k denotes the material's conductivity, and T is the temperature.

The two terms on the left side of Eq. (2.49) are the rate of the total internal energy and the rate of the total kinetic energy, respectively. The last two terms on the right side are the rate of work done by the body force and the external traction, respectively. Using Eq. (2.45), the left side of Eq. (2.49) can be reduced to

$$\frac{\mathrm{D}}{\mathrm{D}t}\int_{\Omega}\left(\rho e + \frac{1}{2}\rho\boldsymbol{v}\cdot\boldsymbol{v}\right)\mathrm{d}V = \int_{\Omega}(\rho\dot{e} + \rho\boldsymbol{v}\cdot\dot{\boldsymbol{v}})\,\mathrm{d}V. \tag{2.51}$$

The last term on the right side of Eq. (2.49) can be rewritten with the use of Gauss's theorem and $\boldsymbol{\Omega} : \boldsymbol{\sigma} = 0$ due to the antisymmetry of spin tensor $\boldsymbol{\Omega}$ as follows:

$$\int_{\Gamma} \boldsymbol{v} \cdot \boldsymbol{t} \mathrm{d}A = \int_{\Gamma} \boldsymbol{n} \cdot \boldsymbol{\sigma} \cdot \boldsymbol{v} \mathrm{d}A = \int_{\Omega} \nabla \cdot (\boldsymbol{\sigma} \cdot \boldsymbol{v}) \mathrm{d}V$$
$$= \int_{\Omega} (\nabla \boldsymbol{v} : \boldsymbol{\sigma} + \boldsymbol{v} \cdot (\boldsymbol{\sigma} \cdot \nabla)) \mathrm{d}V$$
$$= \int_{\Omega} (\boldsymbol{D} : \boldsymbol{\sigma} + \boldsymbol{v} \cdot (\boldsymbol{\sigma} \cdot \nabla)) \mathrm{d}V. \tag{2.52}$$

Substituting Eqs. (2.51) and (2.52) into Eq. (2.49) and applying Gauss' theorem in the second term on the right side of Eq. (2.49) yields

$$\int_{\Omega} \left[\rho \dot{e} - \boldsymbol{D} : \boldsymbol{\sigma} - \rho s - \nabla \cdot (k \nabla T) + \boldsymbol{v} \cdot (\rho \dot{\boldsymbol{v}} - \boldsymbol{\sigma} \cdot \nabla - \rho \boldsymbol{b})\right] \mathrm{d}V = 0. \tag{2.53}$$

Substituting the conservation of linear momentum (2.48) into Eq. (2.53) gives the conservation of energy for the updated Lagrangian approach as below

$$\rho \dot{e} = \rho s + \nabla \cdot (k \nabla T) + \boldsymbol{D} : \boldsymbol{\sigma}. \tag{2.54}$$

2.6.5 Governing Equations

The governing equations consist of the conservation equations as derived above for the updated Lagrangian approach, constitutive equation, kinematic condition, and boundary/initial data, summarized as follows:

(Conservation of mass) $\rho J = J_0,$ (2.55)

(Conservation of momentum) $\boldsymbol{\sigma} \cdot \nabla + \rho \boldsymbol{b} = \rho \dot{\boldsymbol{v}},$ (2.56)

(Conservation of energy) $\rho \dot{e} = \boldsymbol{D} : \boldsymbol{\sigma} + \rho s + \nabla \cdot (k \nabla T),$ (2.57)

(Constitutive equation) $\boldsymbol{\sigma}^{\nabla} = \sigma^{\nabla}(\boldsymbol{D}, \sigma, \text{etc.}),$ (2.58)

(Rate of deformation) $\boldsymbol{D} = \frac{1}{2}\left(\boldsymbol{L} + \boldsymbol{L}^{\mathrm{T}}\right),$ (2.59)

(Boundary conditions) $\begin{cases} (\boldsymbol{n} \cdot \boldsymbol{\sigma})|_{\Gamma_t} = \bar{\boldsymbol{t}}, \\ \boldsymbol{v}|_{\Gamma_u} = \bar{\boldsymbol{v}}, \end{cases}$ (2.60)

(Initial conditions) $\boldsymbol{v}(\boldsymbol{X}, 0) = \boldsymbol{v}_0(\boldsymbol{X}), \quad \boldsymbol{u}(\boldsymbol{X}, 0) = \boldsymbol{u}_0(\boldsymbol{X})$ (2.61)

where Γ_t denotes the traction boundary, Γ_u denotes the displacement boundary, $\boldsymbol{\sigma}$ is the Cauchy stress, ρ is the current density, $\boldsymbol{b}$ is the body force per unit mass acting on the continuum, $\dot{\boldsymbol{v}}$ is the acceleration, and $\boldsymbol{n}$ is the unit normal of the boundary Γ_t.

2.7 WEAK FORM OF THE UPDATED LAGRANGIAN FORMULATION

The governing equations presented in Sect. 2.6.5 result in a set of partial differential equations (PDEs) which describe the motion of a continuum. These governing equations could be solved exactly for some simple problems, but have to be solved numerically for most of modern engineering problems with complicated domains and boundary conditions.

There are two kinds of numerical methods for solving partial differential equations. The first kind of methods obtain directly the approximate solution of the PDEs with their initial and boundary conditions. Fox example, the finite difference method (FDM) converts PDEs into a set of linear equations by approximating the derivatives in the PDEs with corresponding differences at grid points. The FDM has been widely used in the computational fluid dynamics to tackle fluid flows in space, but it is cumbersome in solving the problems with arbitrarily shaped domains.

The second kind of methods first establish a week form equivalent to the original PDEs with their initial and boundary conditions, and then solve the weak form numerically. For example, the method of weighted residual (MWR) minimizes the integral error of numerical solutions in a certain way. Similar to the FEM, the MPM is also formulated based on the weak form.

For the isothermal problems to be considered in this book, conservation of mass and momentum implies that of energy. Conservation of mass is inherent in the MPM, as shown later. To discretize the governing equations in space, hence, the momentum equation (2.56) must be satisfied everywhere within the solution domain Ω. In the MWR, the error (residual) due to the spacial discretization is forced to be zero in an average sense over the solution domain. Taking the virtual displacements $\delta u_j \in \Re_0$, $\Re_0 = \{\delta u_j | \delta u_j \in C^0,\ \delta u_j|_{\Gamma_u} = 0\}$ as the test functions, the weak forms equivalent to the momentum equation (2.56) and the traction boundary conditions (2.60) are given by

$$\int_{\Omega} \delta u_i \left(\sigma_{ij,j} + \rho b_i - \rho \ddot{u}_i\right) \mathrm{d}V = 0, \tag{2.62}$$

$$\int_{\Gamma_t} \delta u_i \left(\sigma_{ij} n_j - \bar{t}_i\right) \mathrm{d}V = 0, \tag{2.63}$$

respectively. The first term on the left side of Eq. (2.62) can be rewritten with the use of integration by parts, Eq. (2.63), and $\delta u_i|_{\Gamma_u} = 0$ (due to the condition

$u_i|_{\Gamma_u}$ being prescribed) as follows:

$$\begin{aligned}\int_\Omega \delta u_i \sigma_{ij,j} \mathrm{d}V &= \int_\Omega [(\delta u_i \sigma_{ij})_{,j} - \delta u_{i,j} \sigma_{ij}] \mathrm{d}V \\ &= \int_\Gamma \delta u_i \sigma_{ij} n_j \mathrm{d}A - \int_\Omega \delta u_{i,j} \sigma_{ij} \mathrm{d}V \\ &= \int_{\Gamma_t} \delta u_i \bar{t}_i \mathrm{d}A - \int_\Omega \delta u_{i,j} \sigma_{ij} dV.\end{aligned} \tag{2.64}$$

Substituting Eq. (2.64) into Eq. (2.62) leads to

$$\int_\Omega \rho \ddot{u}_i \delta u_i \mathrm{d}V + \int_\Omega \sigma_{ij} \delta u_{i,j} \mathrm{d}V - \int_\Omega \rho b_i \delta u_i \mathrm{d}V - \int_{\Gamma_t} \bar{t}_i \delta u_i \mathrm{d}A = 0. \tag{2.65}$$

Eq. (2.65) is the weak form equivalent to the momentum equation and the traction boundary condition, or the virtual work equation. The highest order of derivatives of the displacement u_i with respect to the coordinate system in Eq. (2.65) is one, which is one order lower than that appearing in the strong form (2.56). Thus, the trial function u_i could only be C_0. The weak form (2.65) will be used in Sect. 5.1 to establish the FEM formulation.

Eq. (2.65) can be rewritten as

$$\delta w = \delta w^{\text{int}} - \delta w^{\text{ext}} + \delta w^{\text{kin}} = 0 \tag{2.66}$$

where

$$\delta w^{\text{int}} = \int_\Omega \delta u_{i,j} \sigma_{ij} \mathrm{d}V, \tag{2.67}$$

$$\delta w^{\text{ext}} = \int_\Omega \delta u_i \rho b_i \mathrm{d}V + \int_{\Gamma_t} \delta u_i \bar{t}_i \mathrm{d}A, \quad \text{and} \tag{2.68}$$

$$\delta w^{\text{kin}} = \int_\Omega \delta u_i \rho \ddot{u}_i \mathrm{d}V \tag{2.69}$$

are the virtual work of the internal force, external force, and inertial force, respectively.

The weak form (2.65) can also take the form of

$$\int_\Omega \rho \ddot{u}_i \delta u_i \mathrm{d}V + \int_\Omega \rho \sigma^s_{ij} \delta u_{i,j} \mathrm{d}V - \int_\Omega \rho b_i \delta u_i \mathrm{d}V - \int_{\Gamma_t} \rho \bar{t}^s_i \delta u_i \mathrm{d}A = 0 \tag{2.70}$$

in which $\sigma^s_{ij} = \sigma_{ij}/\rho$ is the specific stress, and $\bar{t}^s_i = \bar{t}_i/\rho$ is the specific traction. The weak form (2.70) will be used in Chapter 3 to formulate the MPM.

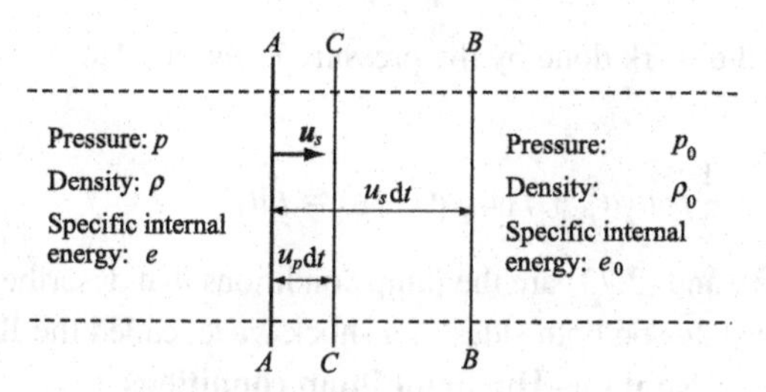

FIGURE 2.5 A steady shock wave propagating with velocity u_s.

2.8 SHOCK WAVE

A shock wave is a type of propagating disturbance. The thickness of the shocked layer is negligibly small as compared with other physically relevant dimensions so that the shock front may be viewed as a moving surface across which the state variables change discontinuously [49].

2.8.1 Rankine–Hugoniot Equations

Consider the case of a uniform pressure suddenly propagating with velocity u_s into a plate of compressible material that is initially at rest with density ρ_0, pressure p_0, and specific internal energy e_0, which imparts a particle velocity u_p, density ρ, pressure p, and specific internal energy e to the plate, as shown in Fig. 2.5. At time t, the shock front is located at *AA*, but propagates to the position *BB* at time $t + \mathrm{d}t$. The particles located at *AA* at time t move to the position *CC* at time $t + \mathrm{d}t$.

Consider a segment of the material with a unit cross-sectional area. At time t, the density ahead of the shock front *AA* is ρ_0 such that the mass of the material segment in the region *AABB* equals $\rho_0 u_s \mathrm{d}t$. After time interval $\mathrm{d}t$, the material in the region *AABB* is compressed into the region *CCBB* with density of ρ and mass of $\rho(u_s - u_p)\mathrm{d}t$. The conservation of mass gives

$$\rho_0 u_s = \rho(u_s - u_p). \tag{2.71}$$

During the time interval $\mathrm{d}t$, the material in the region *AABB* is accelerated to velocity u_p from rest such that the change of momentum in the material segment is $\rho_0 u_s \mathrm{d}t u_p$, with the net pressure impulse being $(p - p_0)\mathrm{d}t$. The conservation of momentum gives

$$p - p_0 = \rho_0 u_s u_p \tag{2.72}$$

where $\rho_0 u_s$ is often called the **shock impedance**.

During the time interval $\mathrm{d}t$, the incremental kinetic energy of the material segment in the region *AABB* is $\frac{1}{2}\rho_0 u_s \mathrm{d}t u_p^2$, the incremental internal energy is

$\rho_0 u_s dt(e - e_0)$, and the work done by the pressure is $p u_p dt$. The conservation of energy yields

$$\frac{1}{2}\rho_0 u_s u_p^2 + \rho_0 u_s (e - e_0) = p u_p. \tag{2.73}$$

Eqs. (2.71), (2.72), and (2.73) are the jump conditions that describe the relationship between the states on both sides of a shock wave, called the **Rankine–Hugoniot equations** or **Rankine–Hugoniot jump conditions.**

Using the Rankine–Hugoniot equations for the conservation of mass and momentum to eliminate the particle velocity u_p results in the relationship among the shock velocity u_s, pressure p, and density ρ. From the conservation of mass, (2.71), we have

$$u_s = \frac{v_0}{v_0 - v} u_p \tag{2.74}$$

where $v_0 = 1/\rho_0$ and $v = 1/\rho$ are the uncompressed and compressed specific volumes. Solving u_s from the conservation of momentum, (2.72), we get

$$u_p = \frac{p - p_0}{\rho_0 u_s}. \tag{2.75}$$

Substituting Eq. (2.75) into Eq. (2.74) leads to

$$u_s^2 = \frac{1}{\rho_0^2} \frac{p - p_0}{v_0 - v}. \tag{2.76}$$

Eq. (2.76) is the equation of a line with slope $-(\rho_0 u_s)^2$ in the p–v plane. This line is called the **Rayleigh line**. Eq. (2.76) represents a series of thermodynamic paths which the continuum follows when being shocked from its initial to final state.

Eliminating the particle velocity u_p and shock velocity u_s by using Eqs. (2.75) and (2.76) from the conservation of energy, (2.73), the conservation laws reduce to a single equation, knowns as **the Rankine–Hugoniot relation,** namely

$$e - e_0 = \frac{1}{2}(p + p_0)(v_0 - v). \tag{2.77}$$

When the initial state (v_0, p_0) ahead of the shock front and p behind the shock front are given, the above relation can be used in the p–v plane if the equation of state (EOS), $p = p(e, v)$, of the material is known. The Hugoniot curve can be plotted using the Rankine–Hugoniot relation (2.77) and the EOS, as shown in Fig. 2.6. The compressed specific volume v can be found because p is given. The shock velocity u_s can be obtained from the slope of the Rayleigh line, and the specific energy e can be obtained from Eq. (2.77).

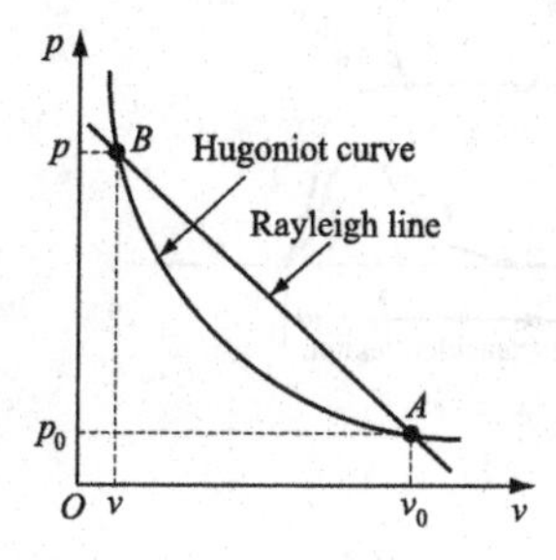

FIGURE 2.6 Hugoniot curve.

The Hugoniot curve describes the locus of all possible thermodynamic states that can be achieved from a given initial state in a material behind a shock. However, these states should follow a thermodynamic path dictated by the Rayleigh line, as shown in Fig. 2.6. Thus, successive states along the Hugoniot curve cannot be achieved from each other as a shock propagates.

For most of solid materials, experimental studies have shown that the shock velocity u_s and the particle velocity u_p can be empirically described in the regions where a substantial phase change in the continuum does not occur, via the following equation:

$$u_s = c_0 + s u_p \tag{2.78}$$

where c_0 is the bulk sound velocity at ambient pressure, and s is a material constant. The monograph of Meyers [50] listed the value of parameters c_0, s, and γ_0 for many different materials.

The shock velocity u_s and particle velocity u_p can be expressed as a function of the specific volume v by solving Eqs. (2.74) and (2.78) for u_s and u_p, namely

$$u_p = \frac{c_0(v_0 - v)}{v_0 - s(v_0 - v)}, \tag{2.79}$$

$$u_s = \frac{c_0 v_0}{v_0 - s(v_0 - v)}. \tag{2.80}$$

Substituting Eqs. (2.79) and (2.80) into Eq. (2.72) gives the pressure-specific volume relationship, called the shock Hugoniot, as follows:

$$p = p_0 + \frac{c_0^2(v_0 - v)}{[v_0 - s(v_0 - v)]^2}. \tag{2.81}$$

2.8.2 Artificial Bulk Viscosity

A shock front mathematically represents a traveling surface of discontinuity which can be described as a moving boundary condition inside the continuum.

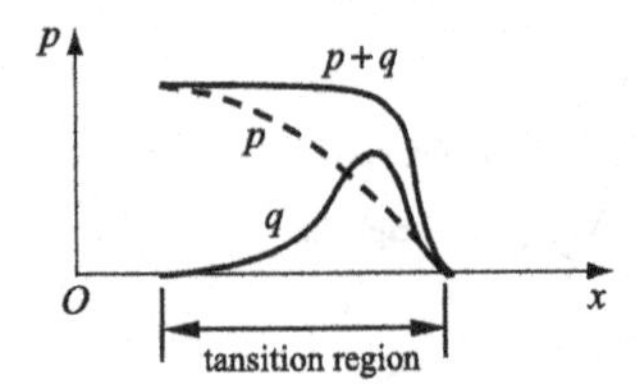

FIGURE 2.7 Artificial bulk viscosity.

Therefore, this problem could be solved with the use of a **shock fitting** technique that explicitly locates and tracks the motion of the shock front. The tracked discontinuities are treated as interior boundaries at which the Rankine–Hugoniot jump conditions are imposed. This technique is cumbersome and inefficient when applied to multidimensional problems so that it has been almost completely abandoned nowadays in favor of the **shock-capturing** approach.

Shock capturing schemes make use of numerical damping to remove the oscillations near the discontinuities, and any shock waves or discontinuities are computed as a part of the solution. **Artificial bulk viscosity** proposed by von Neumann and Richtmyer in 1950 [51] is an attractive approach to capture the shock wave, in which a viscous term q is added to the pressure p within a transition region in the vicinity of the shock wave to smear out the discontinuities into a rapidly varying but continuous transitioning region, as shown in Fig. 2.7. The Rankine–Hugoniot jump conditions are satisfied across the transition region such that no special treatment is required to take care of the shocks themselves.

The viscosity proposed by von Neumann and Richtmyer [51] has the quadratic form as below

$$q = \begin{cases} c_0 \rho l^2 \mathrm{tr}(\dot{\boldsymbol{\varepsilon}})^2 & \text{if } \mathrm{tr}(\dot{\boldsymbol{\varepsilon}}) < 0, \\ 0 & \text{if } \mathrm{tr}(\dot{\boldsymbol{\varepsilon}}) \geqslant 0 \end{cases} \tag{2.82}$$

where c_0 is a dimensionless constant, $l = \sqrt[3]{V}$ is the characteristic length of a grid cell, ρ is the density in the current configuration, $\mathrm{tr}(\dot{\boldsymbol{\varepsilon}}) = \dot{\varepsilon}_{11} + \dot{\varepsilon}_{22} + \dot{\varepsilon}_{33}$ is the trace of the strain rate tensor, i.e., volumetric strain rate.

To damp out numerical oscillations behind the shock front, Landshoof suggested the addition of a linear term [52] as follows:

$$q = \begin{cases} -c_1 \rho l c \mathrm{tr}(\dot{\boldsymbol{\varepsilon}}) & \text{if } \mathrm{tr}(\dot{\boldsymbol{\varepsilon}}) < 0, \\ 0 & \text{if } \mathrm{tr}(\dot{\boldsymbol{\varepsilon}}) \geqslant 0 \end{cases} \tag{2.83}$$

where c_1 is a dimensionless constant, and c is the local sound speed. Combination of the linear and quadratic terms yields

$$q = \begin{cases} c_0 \rho l^2 \mathrm{tr}(\dot{\boldsymbol{\varepsilon}})^2 - c_1 \rho l c \mathrm{tr}(\dot{\boldsymbol{\varepsilon}}) & \text{if } \mathrm{tr}(\dot{\boldsymbol{\varepsilon}}) < 0, \\ 0 & \text{if } \mathrm{tr}(\dot{\boldsymbol{\varepsilon}}) \geqslant 0. \end{cases} \tag{2.84}$$

This artificial viscosity disappears as cell size $h \to 0$ and does not affect the solution in smooth solution regimes. However, the shock waves predicted by the shock-capturing schemes are generally not sharp, and smeared over several grid cells.

After adding the artificial viscosity term q, the stress tensor can be calculated as

$$\boldsymbol{\sigma} = \boldsymbol{s} - (p + q)\mathbf{1} \tag{2.85}$$

where $\boldsymbol{s}$ is the deviatoric stress tensor, $\mathbf{1}$ is the identity tensor, and p is the pressure (positive in compression) such that

$$p = -\frac{1}{3}\mathrm{tr}(\boldsymbol{\sigma}) - q. \tag{2.86}$$

The conservation of energy, (2.57), can be rewritten for isothermal cases as follows:

$$\dot{E} = J\boldsymbol{D} : \boldsymbol{s} - J(p + q)\mathrm{tr}(\dot{\boldsymbol{\varepsilon}}) \tag{2.87}$$

where $\dot{E} = \rho \dot{e}$ is the rate of internal energy per initial volume.

The artificial bulk viscosity introduces into the system a damping term with the damping ratio given by

$$\xi = -\frac{q}{\rho l c \mathrm{tr}(\dot{\boldsymbol{\varepsilon}})} = \frac{Q}{c} \tag{2.88}$$

with

$$Q = \begin{cases} c_1 c - c_0 l \mathrm{tr}(\dot{\boldsymbol{\varepsilon}}) & \text{if } \mathrm{tr}(\dot{\boldsymbol{\varepsilon}}) < 0, \\ 0 & \text{if } \mathrm{tr}(\dot{\boldsymbol{\varepsilon}}) \geqslant 0. \end{cases} \tag{2.89}$$

Thus, after adding the artificial viscosity term into the system of discrete equations, the critical time step size for an explicit time integration becomes

$$\Delta t_e = \frac{l}{Q + (Q^2 + c^2)^{1/2}}. \tag{2.90}$$

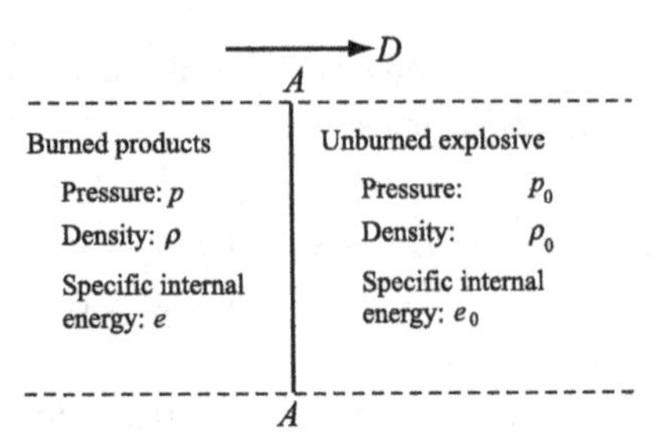

FIGURE 2.8 CJ detonation model.

2.9 DETONATION WAVE

A detonation wave is a combustion wave propagating at a supersonic speed. It is composed of a leading shock followed by a chemical reaction zone. The leading shock compresses the high-explosive material, and converts the high-explosive into gaseous products with very high pressure and temperature in the reaction zone. During this process, an enormous amount of energy is released in just billionths of a second, which sustains the shock wave traveling at the supersonic velocity. Thus, the detonation wave is a shock wave in a reactive medium that is sustained by the energy released in chemical reactions triggered by the shock wave itself [53].

2.9.1 CJ Detonation Model

The **CJ detonation model** was developed by David Chapman [54] and Émile Jouguet [55,56] independently around the turn of the 20th century. The CJ model neglects the thickness of the chemical reaction zone and simplifies the detonation wave as a one-dimensional steady discontinuity. As a result, the detonation shock front raises the pressure from zero to the detonation pressure instantaneously with a complete chemical reaction. The thickness of the reaction zone is usually about 10^{-7} m, which is negligible as compared with the size of a typical explosive charge. Furthermore, the time required for a detonation wave to travel through the reaction zone is about 10^{-7} s, which is also much smaller than the time (about 10^{-5} s) required for the explosion process of the explosive charge to complete. Therefore, the assumption of the CJ model is reasonable in most of engineering cases.

The shock separates the upstream unburned explosive and downstream burned products, as shown in Fig. 2.8. The unburned explosive is at rest with density ρ_0, pressure p_0, and specific energy e_0. The detonation velocity D and downstream state (ρ, p, e) need to be determined.

Across the shock front, the Rankine–Hugoniot conditions must hold. In Eqs. (2.71)–(2.73), replacing the shock velocity u_s with the detonation velocity

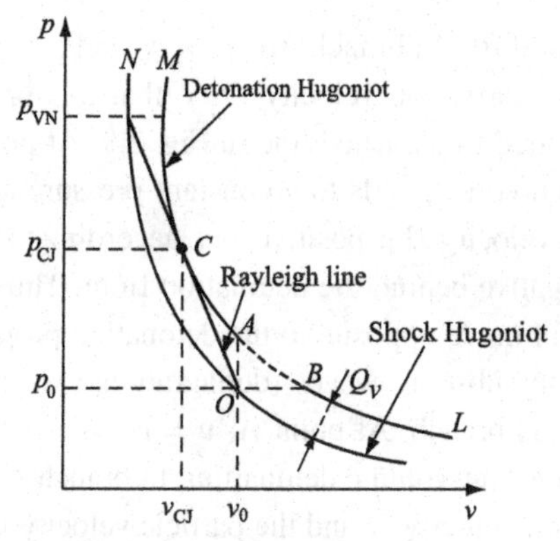

FIGURE 2.9 Detonation Hugoniot and shock Hugoniot.

D and adding the specific chemical energy release Q_v of the detonation reaction to the conservation of energy leads to

$$\text{(Conservation of mass)} \qquad \rho_0 D = \rho(D - u_p), \tag{2.91}$$

$$\text{(Conservation of momentum)} \qquad p - p_0 = \rho_0 D u_p, \tag{2.92}$$

$$\text{(Conservation of energy)} \qquad e - e_0 = \frac{1}{2}(p + p_0)(v_0 - v) + Q_v. \tag{2.93}$$

To fully describe the thermodynamic state of the gaseous product, five unknowns (p, ρ, e, u_p, and D) must be determined. However, we have only four equations, i.e., three equations of conservation (2.91)–(2.93), plus an EOS of the product. To make this problem solvable, Chapman and Jouguet established the fifth equation by studying the condition of a detonation wave propagating steadily within a high-explosive [54–56].

When the EOS $p = p(e, v)$ of the detonation product is known, the Hugoniot curve of the detonation product, called the **detonation Hugoniot**, can be plotted in the p–v plane using Eq. (2.93) and EOS, as shown in Fig. 2.9, in which the shock Hugoniot of the unreacted explosive and the Rayleigh line are also plotted. Due to the energy release, the detonation Hugoniot does not pass through the initial state point $O(p_0, v_0)$ of the unreacted explosive, but is located in the upper right of the shock Hugoniot of the unreacted explosive, as shown in Fig. 2.9.

The Rayleigh line is emanated from the initial state point $O(p_0, v_0)$ of the unreacted explosive in the shock Hugoniot with the slope of $-\rho_0^2 D^2$. The Rayleigh lines with different slopes correspond to different values of detonation velocity D. The horizontal line OB corresponds to $D = 0$, while the vertical line OA corresponds to $D = \infty$. The detonation Hugoniot can be divided into three branches: MA, AB, and BL. In branch AB, $p > p_0$ and $v > v_0$, which corresponds to an imaginary detonation velocity D so that this branch does not exist physically and is plotted as a dashed line in Fig. 2.9. At point B, $v > v_0$, $p = p_0$, and $D = 0$, which corresponds to a constant-pressure combustion. In branch BL, the detonation velocity D is positive, but, according to Eq. (2.75), the particle velocity u_p is negative behind the detonation front. Thus, the particles behind the detonation front move opposite to the detonation propagation direction, which corresponds to subsonic waves (deflagrations) so that the branch BL is called the deflagration branch. At point A, $v = v_0$, $p > p_0$, and $D = \infty$, which corresponds to a constant-volume detonation. In branch AM, $v < v_0$ and $p > p_0$, both the detonation velocity D and the particle velocity u_p are positive such that the particles behind the detonation front move in the same direction as detonation propagation. Hence, the branch AM corresponds to detonation, and it is called the detonation branch.

Fig. 2.9 shows that the leading shock wave traveling at the detonation velocity D compresses the unreacted explosive from the initial state point $O(p_0, v_0)$ along the Rayleigh line to its intersection point N with the shock Hugoniot curve. The pressure at the intersection point N is called the von Neumann spike, and is denoted as p_{VN}. At the von Neumann spike point N, the explosive remains unreacted. The exothermic chemical reaction starts from the spike, and completes at the Chapman–Jouguet state. Because the chemical reaction energy Q_v has been fully released behind the detonation front, the final state of the detonation product behind the detonation front corresponds to the intersection point or tangent point of the Rayleigh line with the detonation Hugoniot. In Fig. 2.9, the detonation Hugoniot represents the locus of all possible thermodynamic states that could be achieved by the detonation product behind the detonation front from the initial state (p_0, v_0), but these states have to be achieved via a thermodynamical path given by the Rayleigh line. Thus, when and only when the denotation product achieves the thermodynamic state corresponding to the tangent point C of the detonation Hugoniot and the Rayleigh line, the detonation wave can travel steadily. The tangent point C is called the CJ point, which corresponds to the CJ state, denoted by the subscript "CJ".

The slope of the detonation Hugoniot is given by

$$\frac{\partial p}{\partial v} = -\rho^2 \frac{\partial p}{\partial \rho} = -\rho^2 c^2 \tag{2.94}$$

where $c = \sqrt{\partial p / \partial \rho}$ is the local sound speed. At the CJ point, the Rayleigh line is tangent to the detonation Hugoniot so that the slope $-\rho^2 D^2$ of the Rayleigh line equals the slope $-\rho^2 c^2$ of the detonation Hugoniot. In other words, the detonation velocity D corresponding to the CJ state is equal to the local sound

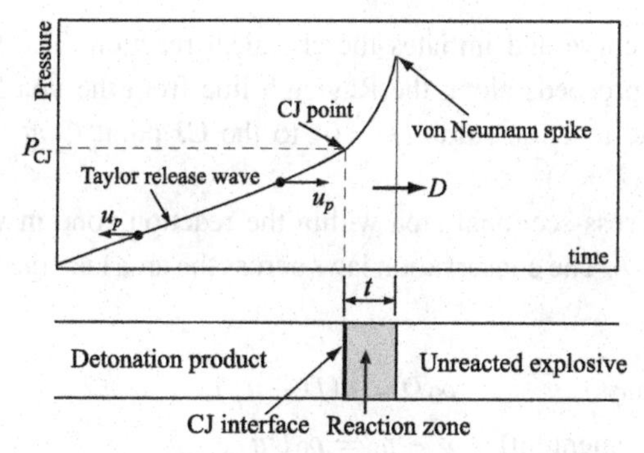

FIGURE 2.10 The structure of detonation wave and pressure profile [50].

speed c of the detonation product behind the detonation front. Thus, it follows that

$$D = c_{CJ} + u_{pCJ} \tag{2.95}$$

where c_{CJ} is the sound speed in the detonation products behind the detonation front, and u_{pCJ} is the particle velocity of the detonation product. Eq. (2.95) is called the CJ condition.

Given the EOS of the detonation product, the final thermodynamic state of the product behind the detonation front can be completely determined from Eqs. (2.91)–(2.93), Eq. (2.95), and the EOS.

2.9.2 ZND Detonation Model

Chapman–Jouguet theory assumes that chemical reactions take place instantaneously inside the zero-thickness, steady-traveling wave at a velocity $D = c + u_p$. Chapman–Jouguet theory is quite successful in explaining detonations, but real chemical detonations show a complex three-dimensional structure, with parts of the wave traveling faster than the average velocity.

The **ZND detonation model**, proposed independently by Y.B. Zel'dovich, John von Neumann, and Werner Döring, describes the detonation wave as a leading shock followed by a finite-thickness zone of chemical reaction proceeding at a finite velocity that depends on the local chemical composition and thermodynamic state, as shown in Fig. 2.10. The leading shock wave adiabatically compresses the explosive to the von Neumann spike and initiates the exothermic chemical reaction. The chemical reaction completes at the CJ state. In the ZND detonation model, hence, the chemical reaction zone is assumed to be traveling at the detonation velocity D.

With the ZND model, the leading shock wave compresses the unreacted explosive from the initial state point $O(p_0, v_0)$ to the von Neumann spike N in the shock Hugoniot curve and initiates the chemical reaction. The following exothermic reaction proceeds along the Rayleigh line from the von Neumann spike N at which the reaction starts ($\lambda = 0$) to the CJ point C at which the reaction is complete ($\lambda = 1$).

Consider a unit cross-sectional area within the reaction zone in which the fraction of reaction is λ. The conservation laws across the area take the following forms:

$$\text{(Conservation of mass)} \qquad \rho_0 D = \rho(D - u_p), \tag{2.96}$$

$$\text{(Conservation of momentum)} \qquad p - p_0 = \rho_0 D u_p, \tag{2.97}$$

$$\text{(Conservation of energy)} \qquad e - e_0 = \frac{1}{2}(p + p_0)(v_0 - v) + \lambda Q_v. \tag{2.98}$$

The EOS is a function of pressure p, specific volume v, and fraction of reaction λ, i.e.,

$$e = e(p, v, \lambda). \tag{2.99}$$

The ZND model assumes that the fraction of reaction λ increases continuously from the leading shock to the CJ interface, which is determined from the reaction rate equation as follows:

$$\frac{\mathrm{d}\lambda}{\mathrm{d}t} = r(p, v, \lambda). \tag{2.100}$$

The final state, p, ρ, e, λ, and u_p, can be obtained by solving the above five equations.

Chapter 3

The Material Point Method

Contents

This chapter establishes the MPM formulation by discretizing a continuum body into a set of material points (particles). Both explicit and implicit formulations are presented. The Generalized Interpolation Material Point (GIMP) method, contact algorithm, adaptive MPM, incompressible MPM, and non-reflecting boundary are discussed in detail.

3.1 MATERIAL POINT DISCRETIZATION

The MPM discretizes the material domain Ω with a set of particles (material points) moving through an Eulerian background grid, as shown in Fig. 3.1. Each particle represents a subdomain Ω_p with all its information such as mass, momentum, energy, strain, stress, and internal state variables for history-dependent

The Material Point Method. http://dx.doi.org/10.1016/B978-0-12-407716-4.00003-X

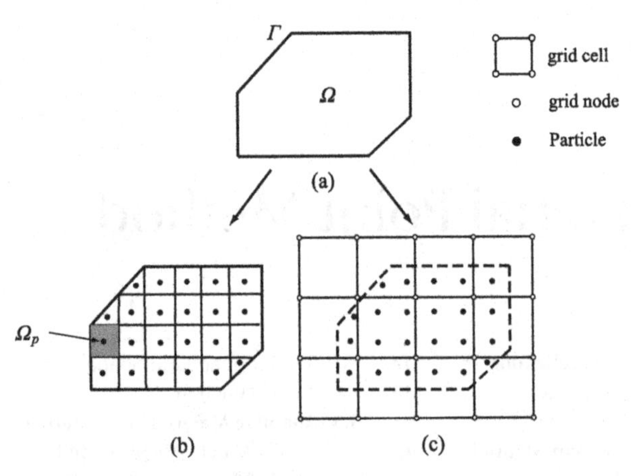

FIGURE 3.1 Sketch of typical MPM discretization.

constitutive modeling. The fixed Eulerian grid provides the means for determining spatial gradient and divergence terms, and carries no permanent information.

The MPM computational cycle can be divided into two phases, a Lagrangian phase followed by an Eulerian/convective phase. In the Lagrangian phase, the particles are attached to the grid and deform with it so that the grid provides a Lagrangian finite element discretization of the material domain. As a result, the momentum equation of the material domain can be solved by using the standard finite element formulation with the grid. In this phase, the grid serves as an updated Lagrangian frame, and the usual convection term associated with Eulerian formulations does not appear. In the Eulerian phase, the grid is simply reset to its original position to take care of the convection term, while the particles remain in their updated positions within a time step, as detailed below.

3.1.1 Lagrangian Phase

Because the material domain is discretized into particles, as shown in Fig. 3.1, the material density can be approximated with

$$\rho(\boldsymbol{x}) = \sum_{p=1}^{n_p} m_p \delta(\boldsymbol{x} - \boldsymbol{x}_p) \tag{3.1}$$

where n_p is the total number of the particles, m_p is the mass of particle p, δ is the Dirac delta function with dimension of the inverse of volume, and $\boldsymbol{x}_p$ is the spatial coordinates of particle p. Substituting Eq. (3.1) into the weak form

(2.70) gives

$$\sum_{p=1}^{n_p} m_p \ddot{u}_{ip} \delta u_{ip} + \sum_{p=1}^{n_p} m_p \sigma_{ijp}^{s} \delta u_{ip,j} - \sum_{p=1}^{n_p} m_p b_{ip} \delta u_{ip} - \sum_{p=1}^{n_p} m_p \bar{t}_{ip}^{s} h^{-1} \delta u_{ip} = 0 \tag{3.2}$$

where $u_{ip} = u_i(\boldsymbol{x}_p)$, $\delta u_{ip,j} = \delta u_{i,j}(\boldsymbol{x}_p)$, $\sigma_{ijp}^s = \sigma_{ij}^s(\boldsymbol{x}_p)$, $b_{ip} = b_i(\boldsymbol{x}_p)$, $\bar{t}_{ip}^s = \bar{t}_i^s(\boldsymbol{x}_p)$, and h is the thickness of the fictitious layer used to convert the surface integral in the last term on the left side of Eq. (2.70) into a volume integral. Eq. (3.2) shows that the volume integrals in the weak form are evaluated in the MPM as the sum of values of the integrand at each particle multiplied by the particle's volume, namely

$$\int g(x, y, z) \mathrm{d}x \mathrm{d}y \mathrm{d}z = \sum_{p=1}^{n_p} g(\xi_p, \eta_p, \zeta_p) V_p. \tag{3.3}$$

As compared with the Gauss quadrature (5.31) in the FEM, Eq. (3.3) can be viewed as a particle quadrature.

In the Lagrangian phase, the particles are rigidly attached to the grid so that the background grid serves as the finite element discretization of the material domain. Therefore, the spatial coordinates x_{ip} of a particle p can be interpolated from the grid nodal spatial coordinates x_{iI} as

$$x_{ip} = N_{Ip} x_{iI}, \quad I = 1, 2, \ldots, n_g \tag{3.4}$$

where subscript I denotes the variables associated with the grid nodal point I, subscript p denotes the variables associated with the particle p, and n_g is the total number of grid nodal points related to the particle p. The repeated index I denotes the summation over its values. $N_{Ip} = N_I(\boldsymbol{x}_p)$ is the shape function $N_I(\boldsymbol{x})$ associated with grid point I evaluated at the position of the particle p. For a background grid consisting of hexahedron cells for 3D problems, the shape function is given by

$$N_I(\xi, \eta, \zeta) = \frac{1}{8}(1 + \xi_I \xi)(1 + \eta_I \eta)(1 + \zeta_I \zeta)$$

where ξ, η, and ζ are the natural coordinates of the particle, and ξ_I, η_I, and ζ_I are the natural coordinates of node I, which take values of $(\pm 1, \pm 1)$.

Similarly, the displacement u_{ip} of the particle p and its derivatives $u_{ip,j}$ can also be interpolated from the grid nodal displacement u_{iI}, i.e.,

$$u_{ip} = N_{Ip} u_{iI}, \tag{3.5}$$

$$u_{ip,j} = N_{Ip,j} u_{iI}. \tag{3.6}$$

Other kinematic quantities of a particle, such as the velocity $\dot{u}_{ip}$ and acceleration $\ddot{u}_{ip}$, can be interpolated from their grid nodal quantities in the same way. For example, the virtual displacement δu_{ip} of a particle p can be approximated by

$$\delta u_{ip} = N_{Ip}\delta u_{iI}. \tag{3.7}$$

Substituting Eqs. (3.5)–(3.7) into the weak form (3.2) and invoking the arbitrariness of δu_{iI} and $\delta u_{iI}|_{\Gamma_u} = 0$ leads to the discrete momentum equation at each grid point, as follows:

$$\dot{p}_{iI} = f_{iI}^{\text{int}} + f_{iI}^{\text{ext}}, \quad x_I \notin \Gamma_u \tag{3.8}$$

in which

$$p_{iI} = m_{IJ}\dot{u}_{iJ} \tag{3.9}$$

is the ith component of momentum at grid point I,

$$m_{IJ} = \sum_{p=1}^{n_p} m_p N_{Ip} N_{Jp} \tag{3.10}$$

is the consistent mass matrix associated with the background grid, and

$$f_{iI}^{\text{int}} = -\sum_{p=1}^{n_p} N_{Ip,j}\sigma_{ijp}\frac{m_p}{\rho_p} \tag{3.11}$$

and

$$f_{iI}^{\text{ext}} = \sum_{p=1}^{n_p} m_p N_{Ip} b_{ip} + \sum_{p=1}^{n_p} N_{Ip}\bar{t}_{ip}h^{-1}\frac{m_p}{\rho_p} \tag{3.12}$$

are the internal and external nodal forces, respectively, with $\sigma_{ijp} = \sigma_{ij}(\boldsymbol{x}_p)$ being the stress of a particle p which can be obtained from its rate of deformation and spin tensors with the use of a constitutive model (refer to Chapter 6 for more details).

To implement a nonlinear constitutive model into the MPM code, an incremental strain–stress relation is used in this book. The incremental strain $\Delta\varepsilon_{ij}$ and incremental vorticity $\Delta\Omega_{ij}$ of a particle p can be obtained from Eqs. (3.5), (2.19), and (2.18), as follows:

$$\Delta\varepsilon_{ijp} = \frac{1}{2}(N_{Ip,j}v_{iI} + N_{Ip,i}v_{jI})\Delta t, \tag{3.13}$$

$$\Delta\Omega_{ijp} = \frac{1}{2}(N_{Ip,j}v_{iI} - N_{Ip,i}v_{jI})\Delta t. \tag{3.14}$$

To improve the computational efficiency, the lumped grid mass matrix

$$m_I = \sum_{J=1}^{n_g} m_{IJ} = \sum_{p=1}^{n_p} m_p N_{Ip} \tag{3.15}$$

can be used such that the grid nodal momentum p_{iI} can be simplified to

$$p_{iI} = m_I \dot{u}_{iI}, \tag{3.16}$$

and the grid nodal momentum equation (3.8) can then be rewritten as

$$m_I \ddot{u}_{iI} = f_{iI}^{\text{int}} + f_{iI}^{\text{ext}}, \quad x_I \notin \Gamma_u. \tag{3.17}$$

Note that in Eqs. (3.16) and (3.17), the index I is a free index so that the repeated index I in these two equations does not represent the summation over its values.

Once the accelerations $\ddot{u}_{iI}$ at the grid points are determined from Eq. (3.17), the use of an explicit time integration gives the grid nodal velocity $\dot{u}_{iI}^L$ at the end of the Lagrangian phase such that the velocity v_{ip}^L and position x_{ip}^L of each particle at the end of the Lagrangian phase can be updated with the use of

$$x_{ip}^L = x_{ip} + \Delta t \sum_{I=1}^{8} \dot{u}_{iI}^L N_{Ip}, \tag{3.18}$$

$$v_{ip}^L = v_{ip} + \Delta t \sum_{I=1}^{8} \ddot{u}_{iI} N_{Ip} \tag{3.19}$$

where x_{ip} and v_{ip} are the position and velocity of the particle at the beginning of the Lagrangian phase.

3.1.2 Convective Phase

If we keep the grid attached to the material during the whole solution process, the grid will be severely distorted as in the Lagrangian finite element methods so that the computation may be terminated abnormally. In the MPM, all material properties are carried by the particles. Thus, the grid is solely used as a scratch pad for determining the gradient and divergence terms, and carries no permanent information. A great advantage of the MPM is that the background grid can be chosen freely for convenience. For example, the grid can be fixed in space when the particles move through the grid, and transport the material properties assigned to them without introducing an error.

In the convective phase following the Lagrangian phase, the grid is simply reset from its deformed position to its original position, while the particles remain in their current positions. The solution on the new grid can be reconstructed from the information carried by the particles. The consistent and lumped grid mass matrices can be reconstructed using Eqs. (3.10) and (3.15) such that only the reconstruction of velocity on the new grid is specified below.

If the grid nodal velocities are known, the velocity of all particles can be easily determined using the grid nodal shape functions, as follows:

$$v_{ip} = N_{Ip} v_{iI}, \quad p = 1, 2, \ldots, n_p; \quad I = 1, 2, \ldots, n_g. \tag{3.20}$$

However, the construction of grid nodal velocities from the updated particle velocities are not straightforward due to $n_p \neq n_g$. The FLIP determines the grid nodal velocities from the particle velocities using the weighted least squares. Multiplying both sides of Eq. (3.20) by $m_p N_{Jp}$ and summing over all particles yields

$$\sum_{p=1}^{n_p} m_p N_{Jp} v_{ip} = \sum_{p=1}^{n_p} m_p N_{Jp} N_{Ip} v_{iI}. \tag{3.21}$$

Eq. (3.21) can be rewritten using Eq. (3.10) as

$$m_{IJ} v_{iJ} = \sum_{p=1}^{n_p} N_{Ip} m_p v_{ip}. \tag{3.22}$$

If the lumped mass matrix Eq. (3.15) is used, Eq. (3.22) can be further simplified to

$$v_{iI} = \frac{1}{m_I} \sum_{p=1}^{n_p} N_{Ip} m_p v_{ip}. \tag{3.23}$$

Burgess [57] has shown that the consistent mass matrix formulation (3.22) conserves kinetic energy, linear and angular momenta. However, the lumped mass matrix formulation (3.23) results in some numerical dissipation of kinetic energy [34,35,57], although its use reduces the computational cost compared with the consistent mass matrix.

3.2 EXPLICIT MATERIAL POINT METHOD

The equation of motion (3.8) is a second-order ordinary differential equation with respect to time and can be solved by using an explicit integration scheme or an implicit integration scheme. The explicit integration scheme finds the state variable $y(t + \Delta t)$ at next time step based on the state variable $y(t)$ at current time step, while the implicit integration scheme finds the state variable $y(t + \Delta t)$ by solving an equation $G(y(t), y(t + \Delta t)) = 0$ that involves both the current and later state variables.

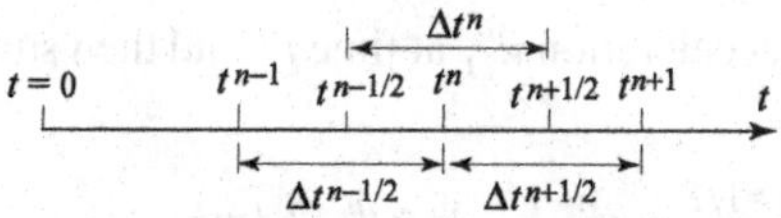

FIGURE 3.2 Central difference method.

3.2.1 Explicit Time Integration

In transient problems such as impact and blast, the duration of load is very small (on the order of microseconds) so that a large number of high frequencies are excited in the system. To accurately capture the high-frequency transient response, a very small time step is required which is about the same as the time step size required by the stability limit of explicit methods. Hence, an explicit method with a lumped (diagonal) mass matrix is usually used because its computational cost per time step is much lower than that required for an implicit method. The central difference method is a common choice for explicit time integration.

3.2.1.1 Central Difference Method

Assume that the displacement, velocity and acceleration at time 0, $t^1, t^2, \ldots, t^n$ are known, and the solution at time t^{n+1} is needed. In the variable-step central difference method, the velocity $\dot{u}_{iI}^{n+1/2}$ at time $t^{n+1/2}$ and acceleration $\ddot{u}_{iI}^{n}$ at time t^n are approximated as

$$\dot{u}_{iI}^{n+1/2} = \frac{u_{iI}^{n+1} - u_{iI}^{n}}{t^{n+1} - t^{n}} = \frac{1}{\Delta t^{n+1/2}}\left(u_{iI}^{n+1} - u_{iI}^{n}\right), \tag{3.24}$$

$$\ddot{u}_{iI}^{n} = \frac{\dot{u}_{iI}^{n+1/2} - \dot{u}_{iI}^{n-1/2}}{t^{n+1/2} - t^{n-1/2}} = \frac{1}{\Delta t^{n}}\left(\dot{u}_{iI}^{n+1/2} - \dot{u}_{iI}^{n-1/2}\right) \tag{3.25}$$

where $\Delta t^{n+1/2} = t^{n+1} - t^n$ and $\Delta t^n = t^{n+1/2} - t^{n-1/2} = \frac{1}{2}(\Delta t^{n-1/2} + \Delta t^{n+1/2})$, as shown in Fig. 3.2. u_{iI}^{n+1} and u_{iI}^{n} denote the displacement vectors at time t^{n+1} and t^n, respectively, and $\dot{u}_{iI}^{n-1/2}$ denotes the velocity vector at time $t^{n-1/2}$.

Eqs. (3.24) and (3.25) can be rewritten as

$$u_{iI}^{n+1} = u_{iI}^{n} + \Delta t^{n+1/2}\dot{u}_{iI}^{n+1/2}, \tag{3.26}$$

$$\dot{u}_{iI}^{n+1/2} = \dot{u}_{iI}^{n-1/2} + \Delta t^{n}\ddot{u}_{iI}^{n}. \tag{3.27}$$

The equation of motion at time t^n is given by

$$m_I \ddot{u}_{iI}^{n} = f_{iI}^{n}. \tag{3.28}$$

Solving Eq. (3.28) for acceleration $\ddot{u}_{iI}^n$ at time t^n and then substituting it into Eq. (3.27) leads to

$$\dot{u}_{iI}^{n+1/2} = \dot{u}_{iI}^{n-1/2} + \Delta t^n f_{iI}^n / m_I. \tag{3.29}$$

Solving velocity $\dot{u}_{iI}^{n+1/2}$ at time $t^{n+1/2}$ from Eq. (3.29) and then substituting it into Eq. (3.26) gives the displacement u_{iI}^{n+1} at time t^{n+1}.

This version of central difference method is known as **leapfrog integration** which updates the position at integer time steps, but updates the velocity at integer-plus-a-half time steps. Positions and velocities are updated at interleaved time points, staggered in such a way that they 'leapfrog' over each other. To solve the velocity $\dot{u}_{iI}^{n+1}$ at time t^{n+1}, Eq. (3.29) can be reformulated into two steps as follows:

$$\dot{u}_{iI}^n = \dot{u}_{iI}^{n-1/2} + \frac{1}{2}\Delta t^n f_{iI}^n / m_I, \tag{3.30}$$

$$\dot{u}_{iI}^{n+1/2} = \dot{u}_{iI}^n + \frac{1}{2}\Delta t^n f_{iI}^n / m_I. \tag{3.31}$$

Replacing n by $n+1$ in Eq. (3.30) results in the velocity $\dot{u}_{iI}^{n+1}$ at time t^{n+1}, namely

$$\dot{u}_{iI}^{n+1} = \dot{u}_{iI}^{n+1/2} + \frac{1}{2}\Delta t^{n+1} f_{iI}^{n+1} / m_I. \tag{3.32}$$

The above scheme is called **leapfrog Verlet**, or **velocity Verlet**. The time step size in any explicit time integration scheme is small due to the stability requirement. Hence, the velocity $\dot{u}_{iI}^{n+1/2}$ at time $t^{n+1/2}$ can be used approximately to calculate the kinetic energy of the system at time t^{n+1}.

In summary, the numerical implementation of the leapfrog integration scheme has the following steps:

1. Calculate the nodal force f_{iI}^n at time t^n;
2. Calculate the nodal velocity $\ddot{u}_{iI}^n$ from the equation of motion at time t^n with the use of

$$\ddot{u}_{iI}^n = f_{iI}^n / m_I; \tag{3.33}$$

3. Impose essential boundary conditions;
4. Update the nodal velocity $\dot{u}_{iI}^{n+1/2}$ at time $t^{n+\frac{1}{2}}$ using Eq. (3.27);
5. Update the nodal position u_{iI}^{n+1} at time t^{n+1} using Eq. (3.26);
6. Let $t^{n+1} = t^n + \Delta t^{n+1/2}$, and $n = n + 1$; and
7. Output the solutions at current time step if required.

The central difference method with a diagonal mass matrix is highly efficient because it does not require the factorization of an 'effective stiffness' matrix, and requires only vector calculations. However, the numerical solutions obtained using the central difference method could exhibit spurious oscillations associated with high frequency modes that cannot be represented by the chosen mesh. Noh and Bathe presented an alternative explicit time integration scheme [58], which automatically suppresses spurious high frequency responses without using any unphysical parameters, at the expense of additional computational costs.

3.2.1.2 Stability Requirement

The central difference method is conditionally stable, whose time step Δt must be less than a critical time step Δt_{cr}, i.e., $\Delta t \leqslant \Delta t_{\mathrm{cr}}$. For an undamped linear system, the critical time step is given by [59]

$$\Delta t_{\mathrm{cr}} = \frac{T_n}{\pi} \tag{3.34}$$

where T_n is the smallest natural period of the system. For a damped linear system, the critical time step is given by [14,60]

$$\Delta t_{\mathrm{cr}} = \frac{T_n}{\pi}\left(\sqrt{1+\xi^2}-\xi\right) \tag{3.35}$$

where ξ is the fraction of critical damping. Note that the critical time steps given in Eqs. (3.34) and (3.35) are obtained from a linear system such that they are only valid for linear systems. For a nonlinear system, the time step can be chosen as

$$\Delta t = \alpha \Delta t_{\mathrm{cr}} \tag{3.36}$$

where α is a constant, and is usually taken to be $0.8 \leqslant \alpha \leqslant 0.98$, depending on the nonlinearity of the system.

In the FEM, it has been proved that the smallest period of a mesh is always greater than or equal to the smallest period of any element in the mesh [61,62]. Therefore, the critical time step of the central difference method can be chosen as

$$\Delta t_{\mathrm{cr}} = \min_e \frac{T^e_{\min}}{\pi} = \min_e \frac{l^e}{c} \tag{3.37}$$

where $T^e_{\min}$ is the smallest period of element e, l_e is the characteristic length of element e, and

$$c = \left[\frac{4G}{3\rho} + \left.\frac{\partial p}{\partial \rho}\right|_S\right]^{1/2} \tag{3.38}$$

is the adiabatic sound speed. Condition (3.37) implies that the time step has to be limited such that a disturbance (stress wave) cannot travel across the smallest characteristic element length in the mesh within a single time step. This condition is usually known as the Courant–Friedrichs–Lewy or **CFL condition** [63].

The pressure p in material is a function of its density ρ and internal energy E per unit initial volume, i.e., $p = p(\rho, E)$. Hence, it follows that

$$\left.\frac{\partial p}{\partial \rho}\right|_S = \left.\frac{\partial p}{\partial \rho}\right|_E + \left.\frac{\partial p}{\partial E}\right|_\rho \left.\frac{\partial E}{\partial \rho}\right|_S . \tag{3.39}$$

Along an isentrope, the differential energy $\mathrm{d}E$ is the product of pressure p and differential volume $\mathrm{d}V$, namely, $\mathrm{d}E = -p\mathrm{d}V$. Thus, we can get

$$\left.\frac{\partial E}{\partial V}\right|_S = -p. \tag{3.40}$$

Taking the first order derivative of the relation $\rho V = \rho_0$ results in

$$\frac{\mathrm{d}V}{\mathrm{d}\rho} = -\frac{V}{\rho} = -\frac{V^2}{\rho_0}. \tag{3.41}$$

Using Eqs. (3.40) and (3.41) yields

$$\left.\frac{\partial E}{\partial \rho}\right|_S = \left.\frac{\partial E}{\partial V}\right|_S \frac{\mathrm{d}V}{\mathrm{d}\rho} = \frac{pV^2}{\rho_0}. \tag{3.42}$$

Substituting Eqs. (3.39) and (3.42) into Eq. (3.38) gives the sound speed as

$$c = \left[\frac{4G}{3\rho} + \left.\frac{\partial p}{\partial \rho}\right|_E + \frac{pV^2}{\rho_0}\left.\frac{\partial p}{\partial E}\right|_\rho\right]^{1/2} . \tag{3.43}$$

For linear elasticity, $p = -K \ln V$ so that

$$\left.\frac{\partial p}{\partial \rho}\right|_S = \left.\frac{\partial p}{\partial V}\right|_S \frac{\mathrm{d}V}{\mathrm{d}\rho} = \frac{K}{\rho} \tag{3.44}$$

where

$$K = \frac{E}{3(1-2\nu)}$$

is the bulk modulus with E and ν being Young's modulus and Poisson's ratio, respectively. Therefore, the sound speed of linearly elastic material is obtained by substituting Eq. (3.44) into Eq. (3.43) as

$$c = \sqrt{\frac{E(1-\nu)}{(1+\nu)(1-2\nu)\rho}}. \tag{3.45}$$

For other material models, the sound speed can be calculated from Eq. (3.43) by invoking an EOS of the material. Please refer to Sect. 6.3 for detailed information.

The computational grid in the MPM is fixed in space and particles move relatively to the grid such that the particle velocity should be taken into consideration in determining the critical time step [64], especially for hypervelocity impact problems in which the particle velocity is comparable to the sound speed. Consequently, if a uniform background grid is used in the MPM, Eq. (3.37) should be rewritten as

$$\Delta t_{\mathrm{cr}} = \frac{d_c}{\max\limits_p (c_p + |u_p|)} \tag{3.46}$$

where d_c is the grid cell size, and c_p and u_p are the sound speed and velocity of particle p, respectively.

3.2.2 Explicit MPM Scheme

In the MPM, all material properties are carried by particles, and no permanent information is stored on the grid nodes. When solving the grid nodal momentum equations at time t^k, hence, the mass, momentum, and stress of each particle are mapped to the corresponding grid nodes by using Eqs. (3.15), (3.23), (3.11), and (3.12). After solving the grid nodal momentum equations with an explicit time integrations scheme, the grid nodal acceleration and velocity values are mapped back to the corresponding particles to update their velocities and positions.

The stress state of a particle can be updated with the use of a constitutive model solver, as presented in Sect. 6.1, which first finds the incremental strain and incremental vorticity of the particle, and then calculates its current stress with the constitutive model. The stress could be updated at the beginning of each time step, or at the end of each time step. The MPM scheme with these two options can be referred to as the update-stress-first (USF) scheme and the update-stress-last (USL) scheme, respectively [65,66]. To update stress, the rates of deformation tensor and vorticity tensor are calculated based on the grid nodal velocity field. Different MPM schemes employ different grid nodal velocity fields to update the stress state as below.

1. In the USF scheme, the grid nodal velocity obtained from the grid nodal momentum $p_{iI}^{k-1/2}$ at the beginning of each time step [65], i.e., $v_{iI}^{k-1/2} = p_{iI}^{k-1/2}/m_I^k = \sum_{p=1}^{n_p} m_p v_{ip}^{k-1/2} N_{Ip}^k / m_I^k$, is used to update the stress state.
2. In the USL scheme, the grid nodal velocity obtained from the updated momentum $p_{iI}^{k+1/2}$ [5], i.e., $v_{iI}^{k+1/2} = p_{iI}^{k+1/2}/m_I^k$, is used to update the stress state.

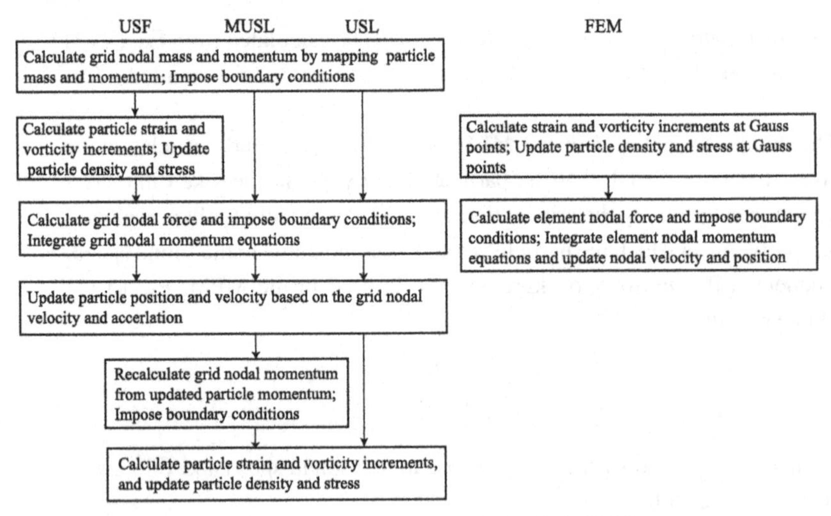
USF
MUSL
USL
FEM
Calculate grid nodal mass and momentum by mapping particle mass and momentum; Impose boundary conditions
Calculate particle strain and vorticity increments; Update particle density and stress
Calculate strain and vorticity increments at Gauss points; Update particle density and stress at Gauss points
Calculate grid nodal force and impose boundary conditions; Integrate grid nodal momentum equations
Calculate element nodal force and impose boundary conditions; Integrate element nodal momentum equations and update nodal velocity and position
Update particle position and velocity based on the grid nodal velocity and accerlation
Recalculate grid nodal momentum from updated particle momentum; Impose boundary conditions
Calculate particle strain and vorticity increments, and update particle density and stress

FIGURE 3.3 Flow chart of the MPM schemes as compared with the FEM.

3. In the MUSL (Modified Update-Stress-Last) scheme, the grid nodal velocity obtained by mapping the updated particle momentum $p_{ip}^{k+1/2} = m_p v_{ip}^{k+1/2}$ back to the grid nodes [67], i.e., $v_{iI}^{k+1/2} = \sum_{p=1}^{n_p} m_p v_{ip}^{k+1/2} N_{Ip}^k / m_I^k$, is used to update the stress state. The MUSL is an improvement over the USL, which does not update the stress state directly based on the updated grid nodal velocity, and instead, updates based on the grid nodal velocity calculated from the updated particle velocity.

The MUSL maps the particle momentum $p_{ip}^{k+1/2}$ at the end of each time step to the grid nodes to calculate their velocities, while the USF maps the particle momentum $p_{ip}^{k+1/2}$ at the beginning of the next time step to the grid nodes to calculate their velocities. Therefore, both the MUSL and USF schemes are quite similar, and the difference between these two schemes is that the MUSL employs N_{Ip}^k while the USF employs N_{Ip}^{k+1} to map the particle velocity to the corresponding grid nodes. Numerical studies [65] show that for the modes resolved on the computational grid both USF and USL schemes result in identical results with a negligible energy error or numerical dissipation. For the unresolved modes, however, the USL is dissipative while the USF is conservative. As compared with the USF, the USL may be a better choice as the damping is consistent with the accuracy of the solution, which damps out the unresolved modes.

The flow chart of the MPM schemes is illustrated and compared with the FEM in Fig. 3.3. In the MPM, the deformed grid is discarded at the end of Lagrangian phase and a new grid will be used in the next time step if necessary. The solution on the new grid is reconstructed from the information carried by

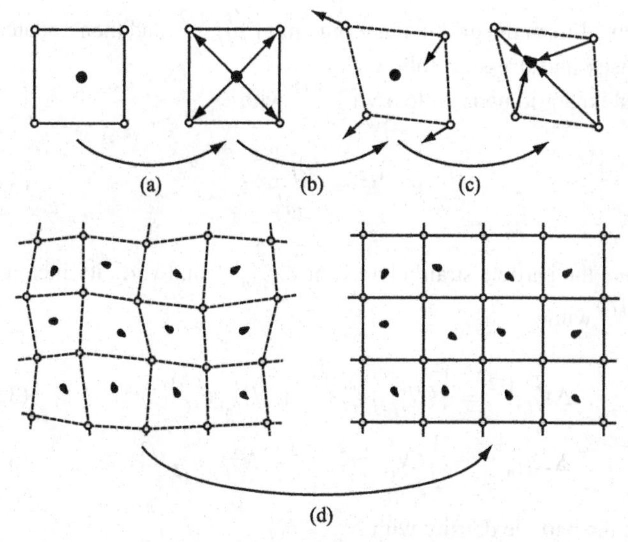

FIGURE 3.4 Schematic illustration of the MPM algorithm.

the particles. Note that the solution reconstruction process can be performed either at the end of the previous time step $t - \Delta t$, or at the beginning of the current time step. In the MPM scheme presented here, the solution reconstruction is performed at the beginning of the current time step using the particle information obtained at time $t - \Delta t$.

The MPM schemes consist of the following steps:

1. Calculate the grid nodal mass and momentum by mapping the particle mass and momentum to the corresponding grid nodes as shown in Fig. 3.4(a), namely

$$m_I^k = \sum_{p=1}^{n_p} m_p N_{Ip}^k, \tag{3.47}$$

$$p_{iI}^{k-1/2} = \sum_{p=1}^{n_p} m_p v_{ip}^{k-1/2} N_{Ip}^k. \tag{3.48}$$

In Fig. 3.4(a), only one particle per cell is used for clarity of illustration.

2. Impose essential boundary conditions on the grid nodal momentum. At the fixed boundary, set $p_{iI}^{k-1/2} = 0$.
3. For the USF only, calculate the particle strain increment $\Delta\varepsilon_{ijp}^{k-1/2}$ and vorticity increment $\Delta\Omega_{ijp}^{k-1/2}$ based on the grid nodal velocity $v_{iI}^{k-1/2}$ which

can be obtained from the grid nodal momentum $p_{iI}^{k-1/2}$, and then update the particle density and stress, as follows:

a. Calculate the grid nodal velocity $v_{iI}^{k-1/2}$ with

$$v_{iI}^{k-1/2} = \frac{p_{iI}^{k-1/2}}{m_I^k}; \tag{3.49}$$

b. Calculate the particle strain increment $\Delta\varepsilon_{ijp}^{k-1/2}$ and vorticity increment $\Delta\Omega_{ijp}^{k-1/2}$ with

$$\Delta\varepsilon_{ijp}^{k-1/2} = \frac{1}{2}(N_{Ip,j}^k v_{iI}^{k-1/2} + N_{Ip,i}^k v_{jI}^{k-1/2})\Delta t, \tag{3.50}$$

$$\Delta\Omega_{ijp}^{k-1/2} = \frac{1}{2}(N_{Ip,j}^k v_{iI}^{k-1/2} - N_{Ip,i}^k v_{jI}^{k-1/2})\Delta t; \tag{3.51}$$

c. Update the particle density with

$$\rho_p^{k+1} = \rho_p^k/(1 + \Delta\varepsilon_{iip}^{k-1/2}); \tag{3.52}$$

d. Update the particle stress state based on $\Delta\varepsilon_{ijp}^{k-1/2}$ and $\Delta\Omega_{ijp}^{k-1/2}$. Please refer to Sect. 6.1 for detailed formulation.

4. Calculate the grid nodal internal force $f_{iI}^{\text{int},k}$, external force $f_{iI}^{\text{ext},k}$, and the total grid nodal force f_{iI}^k with the use of

$$f_{iI}^{\text{int},k} = -\sum_{p=1}^{n_p} N_{Ip,j}^k \sigma_{ijp} \frac{m_p}{\rho_p}, \tag{3.53}$$

$$f_{iI}^{\text{ext},k} = \sum_{p=1}^{n_p} m_p N_{Ip}^k b_{ip}^k + \sum_{p=1}^{n_p} N_{Ip}^k \bar{t}_{ip}^k h^{-1} \frac{m_p}{\rho_p}, \tag{3.54}$$

$$f_{iI}^k = f_{iI}^{\text{int},k} + f_{iI}^{\text{ext},k}. \tag{3.55}$$

For the USF, let $\sigma_{ijp} = \sigma_{ijp}^{k+1}$ and $\rho_p = \rho_p^{k+1}$. Otherwise, let $\sigma_{ijp} = \sigma_{ijp}^k$ and $\rho_p = \rho_p^k$. If the grid node I is fixed in the ith coordinate direction, let $f_{iI}^k = 0$ to make the grid nodal acceleration zero in that direction.

5. Integrate the grid nodal momentum equation as shown in Fig. 3.4(b) by using

$$p_{iI}^{k+1/2} = p_{iI}^{k-1/2} + f_{iI}^k \Delta t^k \tag{3.56}$$

where $\Delta t^k = \frac{1}{2}(\Delta t^{k-1/2} + \Delta t^{k+1/2})$. In the MPM, the deformed position of the background grid does not need to be calculated explicitly so that the grid is shown in Fig. 3.4(b) with dashed line.

6. Update the particle velocity and position based on the grid nodal velocity and acceleration as shown in Fig. 3.4(c) with

$$v_{ip}^{k+1/2} = v_{ip}^{k-1/2} + \sum_{I=1}^{8} \frac{f_{iI}^k N_{Ip}^k}{m_I^k} \Delta t^k, \tag{3.57}$$

$$x_{ip}^{k+1} = x_{ip}^k + \sum_{I=1}^{8} \frac{p_{iI}^{k+1/2} N_{Ip}^k}{m_I^k} \Delta t^{k+1/2}. \tag{3.58}$$

7. For the MUSL only, recalculate the grid nodal momentum based on the updated particle momentum $p_{ip}^{k+1/2}$ with

$$p_{iI}^{k+1/2} = \sum_{p=1}^{n_p} m_p v_{ip}^{k+1/2} N_{Ip}^k \tag{3.59}$$

and impose essential boundary conditions.

8. For the MUSL or USL only, calculate the grid nodal velocity $v_{iI}^{k+1/2}$, particle strain increment $\Delta\varepsilon_{ijp}^{k+1/2}$ and vorticity increment $\Delta\Omega_{ijp}^{k+1/2}$, and then update the particle density and stress, as follows:

   a. Calculate the grid nodal velocity $v_{iI}^{k+1/2}$ with

$$v_{iI}^{k+1/2} = \frac{p_{iI}^{k+1/2}}{m_I^k}; \tag{3.60}$$

   b. Calculate the particle strain increment $\Delta\varepsilon_{ijp}^{k+1/2}$ and vorticity increment $\Delta\Omega_{ijp}^{k+1/2}$ with

$$\Delta\varepsilon_{ijp}^{k+1/2} = \frac{1}{2}(N_{Ip,j}^k v_{iI}^{k+1/2} + N_{Ip,i}^k v_{jI}^{k+1/2})\Delta t^{k+1/2}, \tag{3.61}$$

$$\Delta\Omega_{ijp}^{k+1/2} = \frac{1}{2}(N_{Ip,j}^k v_{iI}^{k+1/2} - N_{Ip,i}^k v_{jI}^{k+1/2})\Delta t^{k+1/2}; \tag{3.62}$$

   c. Update the particle density with

$$\rho_p^{k+1} = \rho_p^k/(1 + \Delta\varepsilon_{iip}^{k+1/2}); \tag{3.63}$$

   d. Update the particle stress state based on $\Delta\varepsilon_{ijp}^{k+1/2}$ and $\Delta\Omega_{ijp}^{k+1/2}$. Please refer to Sect. 6.1 for detailed formulation.

9. Store all material properties in the particles so that the deformed grid can be discarded, if needed, to employ a new grid in the next time step, as shown in Fig. 3.4(d).

Eq. (3.57) shows that particles move according to the velocity field defined by Eq. (3.20) using the grid nodal velocities. In other words, the position of a particle p is updated using the velocity $\hat{v}_{ip}^{k+1/2} = \sum_{I=1}^{8} v_{iI}^{k+1/2} N_{Ip}^{k}$ of the point that coincides with the particle, rather than the velocity $v_{ip}^{k+1/2}$ of the particle itself. Because the velocity field defined by Eq. (3.20) is single-valued, unphysical material interpenetration is not possible. Thus, the non-slip contact condition is satisfied automatically in the MPM without using any special treatment such as the master/slave nodes required for the FEM. For a single-particle problem ($n_p = 1$), the velocity filed $\hat{v}_{ip}^{k+1/2}$ used to move the particle can be obtained from Eqs. (3.56), (3.47), (3.48), and (3.58) as

$$\hat{v}_{ip}^{k+1/2} = \sum_{I=1}^{8} \frac{p_{iI}^{k-1/2} + f_{iI}^{k} \Delta t^{k}}{m_{I}^{k}} N_{Ip}^{k} = v_{ip}^{k+1/2}.$$

Thus, this single particle moves with its own velocity and is not affected by the grid. In other words, the MPM can correctly describe the motion of a single particle.

Next we will compare the three MPM schemes using a special problem. Assume that the grid node I is only related to the particle p. According to Eqs. (3.47) and (3.55), the mass and force of the grid node I are given by

$$m_{I}^{k} = m_{p} N_{Ip}^{k}, \tag{3.64}$$

$$f_{iI}^{k} = -N_{Ip,j}^{k} \sigma_{ijp} \frac{m_{p}}{\rho_{p}} + m_{p} N_{Ip}^{k} b_{ip}^{k}. \tag{3.65}$$

Substituting Eqs. (3.64) and (3.65) into Eqs. (3.60), (3.59), and (3.49), the grid nodal velocities used to update the stress state in the USL, MUSL, and USF are respectively obtained as follows:

$$\text{(USL)} \quad v_{iI}^{k+1/2} = v_{ip}^{k-1/2} + \left(-\frac{N_{Ip,j}^{k} \sigma_{ijp}}{N_{Ip}^{k} \rho_{p}} + b_{ip}^{k} \right) \Delta t^{k}, \tag{3.66}$$

$$\text{(MUSL)} \quad v_{iI}^{k+1/2} = v_{ip}^{k-1/2} + \sum_{J=1}^{8} \frac{f_{iJ}^{k} N_{Jp}^{k}}{m_{J}^{k}} \Delta t^{k}, \tag{3.67}$$

$$\text{(USF)} \quad v_{iI}^{k-1/2} = v_{ip}^{k-1/2}. \tag{3.68}$$

When the particle p moves close to the opposite side of the grid node I, we have $N_{Ip}^{k} \to 0$, but $N_{Ip,j}^{k} \neq 0$. In this case, the second term on the right side of Eq. (3.66) approaches infinity, which makes the USL unstable. However, the m_{J}^{k} and N_{Jp}^{k} in the second term on the right side of Eq. (3.67) are the infinitesimal functions of the same order so that the MUSL remains stable. Therefore, when

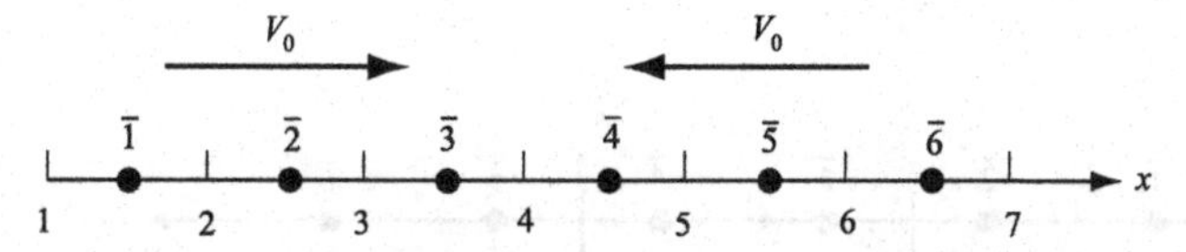

FIGURE 3.5 Impact between two separate elastic bars.

a grid node is only related to a particle, the USL is unstable while the MUSL and USF are stable.

In some material models, such as the Moony–Rivlin model, the particle deformation gradient tensor $F_{ij} = \partial x_i / \partial X_j$ is required to be updated. The deformation gradient tensor at time $t + \Delta t$ can be evaluated by

$$F_{ij}^{t+\Delta t} = \frac{\partial x_i^{t+\Delta t}}{\partial X_j} = \frac{\partial x_i^{t+\Delta t}}{\partial x_k^t} \frac{\partial x_k^t}{\partial X_j}. \tag{3.69}$$

Substituting Eq. (3.4) into Eq. (3.69) gives the deformation gradient of particle p at time $t + \Delta t$, i.e.,

$$F_{ijp}^{t+\Delta t} = x_{iI}^{t+\Delta t} N_{Ip,k} F_{kjp}^t. \tag{3.70}$$

3.2.3 Qualitative Demonstration

To further demonstrate the MPM, consider a one-dimensional example [68] in which the salient feature of the contact/impact scheme in the MPM is qualitatively described. The impact between two separate elastic bars of unit area with an initial velocity V_0 is illustrated in Fig. 3.5. Left and right elastic bars are discretized into three particles (solid dots), respectively.

Recall that each nodal value of any field variable in the MPM is influenced only by those particles within the support domain of the node. Therefore, the velocity gradient is nonzero only if the particles within the support domain of the node have different velocities. While the left and right bars are freely translating in the space with a constant velocity, the velocity gradient in each bar must be zero based on the physics. This physical property is preserved in the MPM because the velocity assigned to the boundary node is determined by the particle in the support domain of that node. As stated before, the boundary nodal velocity is equal to the boundary particle velocity, and the boundary particle therefore experiences no velocity gradient, as long as the nodal force vector is zero. This is the case for a bar moving in the space with a constant velocity, for which the nodal momentum will not change with time. A zero velocity gradient will persist until the support domain of the boundary node contains the particles with different velocities. Consequently, the two bars will not interact with each other

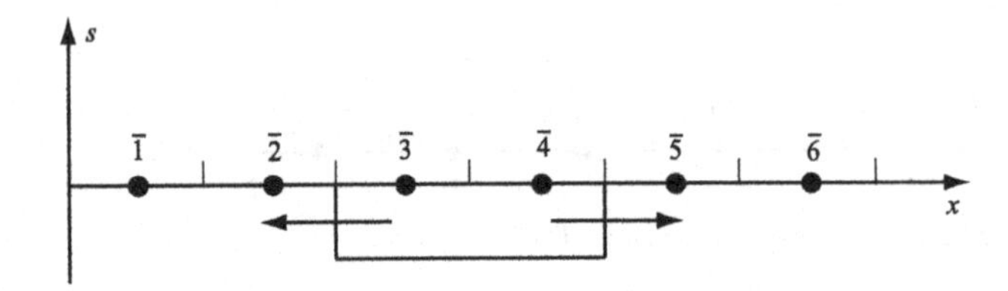

FIGURE 3.6 Stress distribution just after impact.

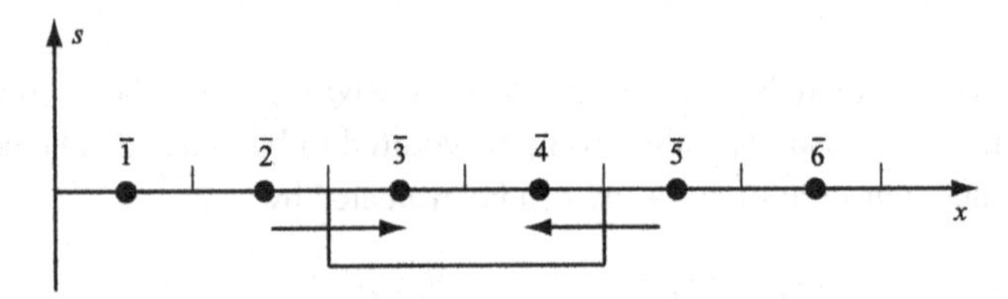

FIGURE 3.7 Stress distribution just before separation.

until the boundary particle of each bar is within the support domain of a single node.

Just before the impact between two bars occurs as shown in Fig. 3.5, the strain increment of particle $\bar{3}$ and $\bar{4}$ can be calculated, from the linear shape functions as defined before, to be

$$\Delta e_{\bar{3}} = \frac{0 - v_3}{h_c} \Delta t = -\frac{v_3}{h_c} \Delta t < 0 \tag{3.71}$$

and

$$\Delta e_{\bar{4}} = \frac{v_5 - 0}{h_c} \Delta t = \frac{v_5}{h_c} \Delta t < 0, \tag{3.72}$$

respectively, because of $v_3 > 0$, $v_4 = 0$, and $v_5 < 0$ in the given coordinate system (positive to the right). In Eqs. (3.71) and (3.72), h_c represents the background cell size.

The corresponding stress increments are given by $\Delta s_{\bar{3}} = E \Delta e_{\bar{3}}$ and $\Delta s_{\bar{4}} = E \Delta e_{\bar{4}}$, which are negative due to the negative increments in strain. The stress distribution just after impact is shown in Fig. 3.6.

In the restitution phase, the reflected tensile wave cancels the compressive wave. Just before separation, the stress distribution is depicted in Fig. 3.7.

Just after the impact and before the separation between the two bars of unit area, the force mapped from particle $\bar{3}$ to node 3 is given by

$$f_3 = -(-1)\, s_{\bar{3}} = s_{\bar{3}} < 0 \tag{3.73}$$

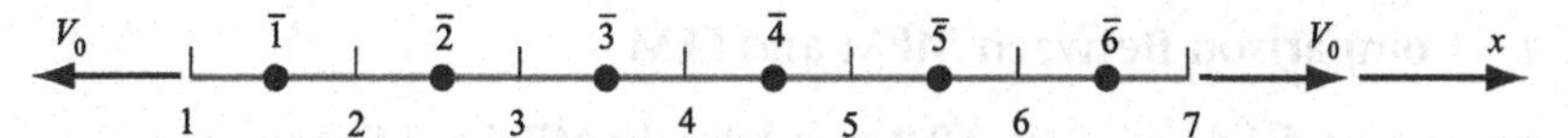

FIGURE 3.8 A single elastic bar under tensile loading at both ends.

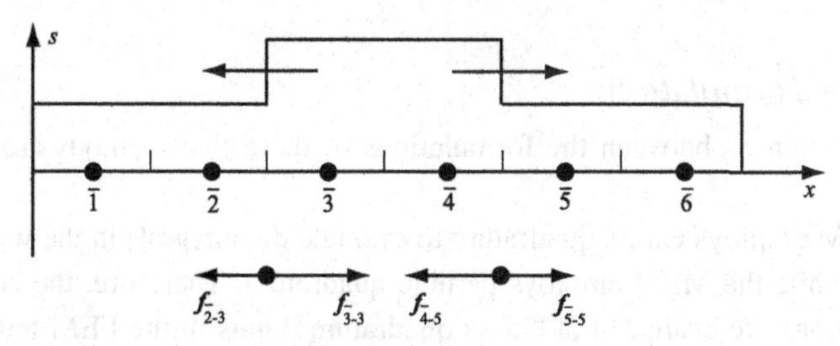

FIGURE 3.9 Stress distribution after the magnitude of stress wave is doubled at the middle point of the bar.

as can be found from Figs. 3.6 and 3.7. The force mapped from particle $\bar{4}$ to node 5 is given by

$$f_5 = -(1)\, s_{\bar{4}} = -s_{\bar{4}} > 0. \tag{3.74}$$

At node 4, the condition of $f_4 = 0$ holds because the forces mapped from particle $\bar{3}$ and $\bar{4}$ cancel each other. Hence, particles $\bar{3}$ and $\bar{4}$ will move away from each other when the reflected tensile wave approaches the contact node. As a result, two bars separate from each other.

To demonstrate the difference between impact and non-impact problems, consider now a single elastic bar of unit area under tensile loading at both ends, as shown in Fig. 3.8. The elastic bar is discretized into 6 particles (solid dots).

Just before two tensile waves meets in the middle point of the bar, the magnitude of the reaction from particle $\bar{5}$ to node 5, $|f_{\bar{5}-5}|$, is larger than that from particle $\bar{4}$ to node 5, $|f_{\bar{4}-5}|$, so that particle $\bar{4}$ moves toward node 5. Similarly, particle $\bar{3}$ moves toward node 3. After the magnitude of stress wave is doubled at the middle point of the bar, the stress distribution is shown in Fig. 3.9.

Now the magnitude of the reaction from particle $\bar{5}$ to node 5, $|f_{\bar{5}}|$, is less than that from particle $\bar{4}$ to node 5, $|f_{\bar{4}}|$, so that particle $\bar{4}$ moves toward node 4. Similarly, particle $\bar{3}$ moves toward node 4. Hence, both particles $\bar{3}$ and $\bar{4}$ would move toward each other. In other words, the bar would not be separated at the center.

As can be seen from the above qualitative analysis, the MPM can simulate the contact/impact phenomena based on the physics involved, without invoking master/slave nodes as required in the conventional mesh-based methods.

3.2.4 Comparison Between MPM and FEM

Comparing Sect. 5.1.4 with Sect. 3.2.2 shows that the MPM and FEM are very similar in each time step. The efficiency and accuracy of both methods are compared below.

3.2.4.1 *Basic Formulation*

The major differences between the formulations of these two methods are as follows:

1. The FEM employs Gauss quadrature to evaluate the integrals in the weak formulation, while the MPM employs particle quadrature. Therefore, the constitutive equations are evaluated at Gauss quadrature points in the FEM but at particles in the MPM.

2. The computational mesh of a Lagrangian FEM is attached to the material during the whole solution process, while a specific background grid of the MPM is only attached to the material in each time step. At the end of each time step, the deformed grid could be discarded to employ a new regular grid in the next time step. As a result, no fixed mesh connectivity is required in the MPM so that crack propagation could be simulated without changing the mesh connectivity as needed in the FEM. Because all the material properties are carried by the particles, the solution on the grid at next time step must be reconstructed from the particle information. Thus, the mass matrix in the MPM is no longer a constant matrix as that in the FEM, and has to be recalculated in each time step.

Consequently, the MPM can be viewed as a special Lagrangian FEM with particle quadrature and continuous mesh-update. Gauss quadrature can produce accurate results for polynomial integrands, but the particle quadrature usually can not. In addition, an explicit FEM usually employs the one-point Gauss quadrature, but $2 \times 2 \times 2$ particles are usually used in each cell in the MPM for 3D problems such that the number of particles in the MPM is usually much larger than the number of Gauss quadrature points in the FEM. Thus, the accuracy and efficiency of the MPM are lower than those of the FEM for small deformation problems. For large deformation problems, however, the accuracy of the Lagrangian FEM deteriorates rapidly and the computational cost increases dramatically due to mesh distortion and the need for remeshing. Furthermore, the Lagrangian FEM is not suitable for hyper velocity impact problems in which metals behave like fluids and become a debris cloud (see Table 3.1).

3.2.4.2 *Computational Efficiency*

Computational efficiency depends on the computational cost per time step and the time step size. As shown in Fig. 3.3, compared with the FEM, the MPM USL formulation performs two additional steps while the MUSL performs three

TABLE 3.1 Comparison Between the MPM and FEM Formulations

Explicit material point method	Explicit finite element method
Grid nodal mass and momentum: $m_I^k = \sum_{p=1}^{n_p} m_p N_{Ip}^k,$ $p_{iI}^{k-1/2} = \sum_{p=1}^{n_p} m_p v_{ip}^{k-1/2} N_{Ip}^k$	Skipped because the mesh nodes carry mass and momentum
Grid nodal force (particle quadrature): $f_{iI}^{\text{int},k} = -\sum_{p=1}^{n_p} N_{Ip,j}^k \sigma_{ijp} \frac{m_p}{\rho_p},$ $f_{iI}^{\text{ext},k} = \sum_{p=1}^{n_p} N_{Ip}^k b_{ip}^k m_p$	Nodal force (one-point Gauss quadrature): $f_{iI}^{\text{int},k} = -\sum_e N_{Ie,j}^k \sigma_{ije} V_e,$ $f_{iI}^{\text{ext}} = \sum_e N_{Ie} b_{ie} m_e$
Grid nodal momentum: $p_{iI}^{k+1/2} = p_{iI}^{k-1/2} + f_{iI}^k \Delta t^k$. The positions of deformed grid nodes are not required to be calculated.	Nodal velocity and position: $v_{iI}^{k+1/2} = v_{iI}^{k-1/2} + \Delta t^k f_{iI}^k / M_I,$ $x_{iI}^{k+1} = x_{iI}^k + \Delta t^{k+1/2} v_{iI}^{k+1/2}$
Grid nodal velocity and position: $v_{ip}^{k+1/2} = v_{ip}^{k-1/2} + \sum_{I=1}^{8} \frac{f_{iI}^k N_{Ip}^k}{m_I^k} \Delta t^k,$ $x_{ip}^{k+1} = x_{ip}^k + \sum_{I=1}^{8} \frac{p_{iI}^{k+1/2} N_{Ip}^k}{m_I^k} \Delta t^{k+1/2}$	Skipped because Gauss points are fixed in an element such that it is unnecessary to recalculate their coordinates

additional steps. At the beginning of each time step, the mass and momentum of particles are mapped to grid nodes. At the end of each time step, the updated grid nodal velocity and acceleration are mapped to the particles to update their positions and velocities. In the FEM, the mass and momentum are carried by the mesh nodes so that they are not recalculated at the beginning of each time step. Furthermore, the Gauss points do not move relative to each element during the whole solution process such that it is unnecessary to update their positions and velocities.

In addition, the FEM has only one Gauss point in each element, while the MPM usually employs 1 particle in 1D problems, 4 particles in 2D problems, and 8 particles in 3D problems in each grid cell. Stress update and nodal internal force calculation will loop over the Gauss points (FEM) or particles (MPM) so that the computational cost of the MPM is much higher than that in the FEM in this part. As a result, the computational cost per time step in the MPM is higher than that in the FEM.

Both the explicit MPM and explicit FEM employ the central difference method, whose critical time step size depends on the characteristic element length. In the MPM, the characteristic element length is the cell size which is constant during the whole solution process. However, the characteristic element length in the FEM decreases with the element deformation. Because the sound speeds in both MPM and FEM are almost the same, the time step of the FEM is smaller than that of the MPM so that the total number of steps required in the FEM is larger than that in the MPM. For small deformation problems, the characteristic element length in the FEM will not decrease significantly so that the computational efficiency of the FEM is higher than that of the MPM. For large deformation problems, however, the characteristic element length in the FEM decreases rapidly, which results in a significant decrease in the time step size and significant increase in the total number of required time steps. Thus, the computational efficiency of the FEM is lower than that of the MPM for large deformation problems. For example, in the slope failure simulation given in Sect. 4.7, due to the severe element distortion near the failure surface, the time step size in the FEM simulation decreases from its initial value of 261 μs to 38 μs so that the total computer time used is 5632 s that is about 10 times of that required in the MPM simulation.

3.2.4.3 Computational Accuracy

When the background grid cell size in the MPM is comparable to the element size in the FEM, the difference in accuracy between these two methods mainly depends on the quadrature scheme used and the technique to deal with large deformations. Both the MPM and FEM employ polynomial-based shape functions. Hence, the Gauss quadrature used in the FEM can integrate accurately the weak form, but the particle quadrature used in the MPM cannot. As a result, the accuracy of the weak form integration in the MPM is lower than that in the FEM. Furthermore, the original MPM also suffers from a cell crossing instability, although several improvements have been proposed to eliminate the instability with additional computational expenses, as discussed later. For large deformation problems, however, the Jacobian of an element in the FEM will decrease to zero, even become negative due to element distortion, which leads to a significant error, even abnormal termination of the simulation. To continue an FEM simulation, an erosion technique is usually used, which simply deletes the distorted or failed elements from the system. However, erosion is not physical, which not only makes mass, momentum, and energy nonconservative, but also unable to simulate hypervelocity impact problems. For example, Fig. 3.10 illustrates the FEM results of a hypervelocity impact simulation using LS-DYNA. Whether the erosion technique is used or not, the FEM simulation cannot obtain the debris cloud. Although remeshing can alleviate the element distortion, remapping material properties of history-dependent materials will result in a significant error. Furthermore, designing an efficient remeshing technique for complicated 3D problems remains a challenging task for the FEM. Therefore,

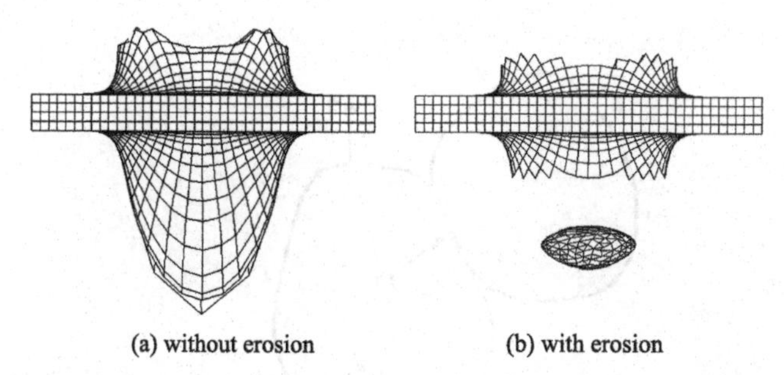

FIGURE 3.10 High velocity impact simulation using LS-DYNA: (a) without erosion, and (b) with erosion.

the accuracy of the MPM is lower than that of the FEM for small deformation problems, but could be much higher than that of the FEM for large deformation problems.

To sum up, the computational efficiency and accuracy of the MPM are lower than those of the FEM in small deformation cases, but could become much higher than those of the FEM in large deformation cases. Hence, an optimized combination of the FEM with MPM could greatly enhance both efficiency and accuracy in large-scale computer modeling and simulation, as illustrated in Chapter 8 for practical applications.

3.3 CONTACT METHOD

In the MPM, all particles move in a single-valued velocity field defined by the grid nodal velocity so that unphysical material interpenetration is not possible and the non-slip contact condition is satisfied automatically. In other words, the non-slip contact constraint is inherent in the MPM without requiring any additional treatment. In many engineering problems, however, bodies often contact and slide against each other such that a contact method with a physics-based criterion is required for the MPM. A simple contact algorithm was proposed by York et al. [69] to allow the release of the no-slip contact constraint in the original MPM. If two bodies are coming into contact with each other, the original MPM is used to impose the impenetrability condition. If the bodies are moving away from one another, they move with their own velocity fields to allow separation. To avoid interpenetration and allow separation in the gear contact process, Hu and Chen [70] presented a contact/sliding/separation algorithm in a multigrid environment. The normal velocity of each particle at the contact surface is calculated in the common background grid, whereas the tangential velocity is found based on the respective individual grid. Although the aforementioned contact algorithms are efficient to handle separation, the friction between contact bodies is not considered. Bardenhagen et al. [71,72] proposed a contact/fric-

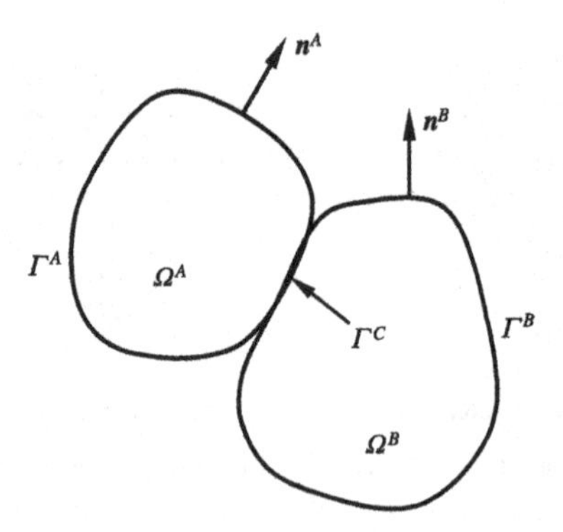

FIGURE 3.11 Two bodies in contact.

tion/separation algorithm in multi-velocity fields. The impenetrability condition and Coulomb friction between bodies are incorporated into the MPM when the contact occurs. The contact force between bodies is obtained from the relative nodal velocity at the contact surface. This approach has been demonstrated with the use of a sphere rolling on an inclined plane and the granular shearing simulation. The contact force in Bardenhagen's contact method is not along a common normal, which results in errors in momentum conservation [71,72], especially for large deformation problems. Huang et al. [73] proposed three methods for impact and penetration simulations to determine the surface normal vectors that satisfy the collinearity condition at the contact surface. Different methods are compared below.

3.3.1 Boundary Conditions at Contact Surface

Consider the contact between two bodies A and B, denoted by Ω_A and Ω_B in the current configuration. The boundaries of these two bodies are denoted by Γ_A and Γ_B, while their contact surface is denoted by $\Gamma_C (= \Gamma_A \cap \Gamma_B)$, as shown in Fig. 3.11. The boundary of each body is composed of displacement boundary, traction boundary and contact surface, namely, $\Gamma^A = \Gamma_t^A \cup \Gamma_u^A \cup \Gamma_c$, $\Gamma^B = \Gamma_t^B \cup \Gamma_u^B \cup \Gamma_c$, $\Gamma_t^A \cap \Gamma_u^A = 0$, $\Gamma_t^A \cap \Gamma_c = 0$, $\Gamma_u^A \cap \Gamma_c = 0$, $\Gamma_t^B \cap \Gamma_u^B = 0$, $\Gamma_t^B \cap \Gamma_c = 0$, and $\Gamma_u^B \cap \Gamma_c = 0$.

It is usually convenient to express the contact surface equation in the local coordinate system of the contact surface. Take body A as the master body whose potential contact surface is called master surface, and take body B as the slave body whose potential contact surface is called slave surface. The local coordinate system at any point on the boundary of a body is established by taking the boundary's tangential unit vectors $\boldsymbol{e}_1^A$, $\boldsymbol{e}_2^A$ and normal unit vector $\boldsymbol{n}^A = \boldsymbol{e}_1^A \times \boldsymbol{e}_2^A$ as the base vectors. At a contact point, the normal vectors of body A and body B are in the opposite direction, i.e., $\boldsymbol{n}^B = -\boldsymbol{n}^A$. The velocity at a contact point of each body can then be expressed in the local coordinate system as

$$\boldsymbol{v}^A = v_N^A \boldsymbol{n}^A + v_\alpha^A \boldsymbol{e}_\alpha^A = v_N^A \boldsymbol{n}^A + \boldsymbol{v}_T^A, \tag{3.75}$$

$$\boldsymbol{v}^B = v_N^B \boldsymbol{n}^A + v_\alpha^B \boldsymbol{e}_\alpha^A = -v_N^B \boldsymbol{n}^B + \boldsymbol{v}_T^B \tag{3.76}$$

where the repeated index α represents the summation over its values (2 in 3D problems while 1 in 2D problems), $v_N^A = \boldsymbol{v}^A \cdot \boldsymbol{n}^A$ and $v_N^B = \boldsymbol{v}^B \cdot \boldsymbol{n}^A$ denote the normal velocity of body A and body B at the contact point, respectively, and $\boldsymbol{v}_T^A = v_\alpha^A \boldsymbol{e}_\alpha^A$ and $\boldsymbol{v}_T^B = v_\alpha^B \boldsymbol{e}_\alpha^A$ denote their tangential velocities.

3.3.1.1 Impenetrability Condition

The impenetrability condition for a pair of bodies can be expressed as

$$\Omega^A \cap \Omega^B = 0. \tag{3.77}$$

Eq. (3.77) states that two bodies cannot occupy the same position at the same time. It is usually unable to express the impenetrability condition as an algebraic or differential equation in terms of displacement in large deformation problems. Therefore, it is convenient to express the impenetrability condition at any point on the contact surface Γ_c in the rate form as follows:

$$\gamma_N = v_N^A - v_N^B = (\boldsymbol{v}^A - \boldsymbol{v}^B) \cdot \boldsymbol{n}^A = \boldsymbol{v}^A \cdot \boldsymbol{n}^A + \boldsymbol{v}^B \cdot \boldsymbol{n}^B \leqslant 0. \tag{3.78}$$

Eq. (3.78) states that two contacted bodies will either remain in contact ($\gamma_N = 0$) or separate from each other ($\gamma_N < 0$). The relative tangential velocity between the two bodies is given by

$$\boldsymbol{\gamma}_T = \boldsymbol{v}_T^A - \boldsymbol{v}_T^B = v_\alpha^A \boldsymbol{e}_\alpha^A - v_\alpha^B \boldsymbol{e}_\alpha^A. \tag{3.79}$$

3.3.1.2 Traction Condition

According to the Newton's third law of motion, the sum of the tractions across the contact surface must be zero, i.e.,

$$\boldsymbol{t}^A + \boldsymbol{t}^B = 0, \tag{3.80}$$

which can be decomposed into normal and tangential components as

$$t_N^A + t_N^B = 0, \tag{3.81}$$

$$\boldsymbol{t}_T^A + \boldsymbol{t}_T^B = 0 \tag{3.82}$$

where $t_N^A = \boldsymbol{t}^A \cdot \boldsymbol{n}^A$ and $t_N^B = \boldsymbol{t}^B \cdot \boldsymbol{n}^A$ are the normal tractions of body A and body B, and $\boldsymbol{t}_T^A = \boldsymbol{t}^A - t_N^A \boldsymbol{n}^A$ and $\boldsymbol{t}_T^B = \boldsymbol{t}^B - t_N^B \boldsymbol{n}^B$ are the corresponding tangential tractions.

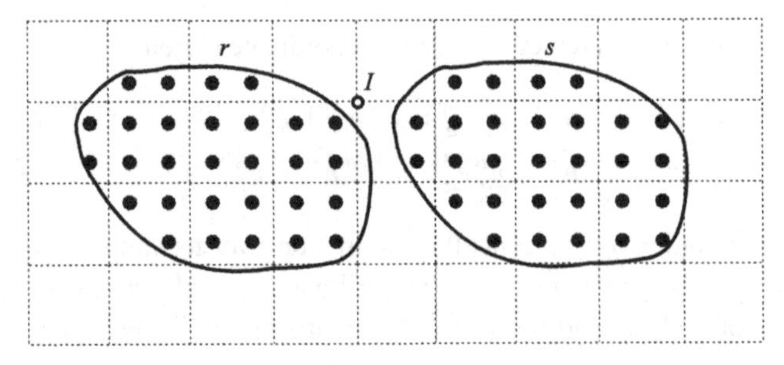

FIGURE 3.12 Two bodies in contact.

If two bodies are not stuck together at a contact point, the normal traction cannot be tensile, i.e.,

$$t_N = t_N^A = -t_N^B \leqslant 0. \tag{3.83}$$

The tangential traction must be limited by the maximum static friction as

$$\|\boldsymbol{t}_T(x,t)\| \leqslant \mu |t_N(x,t)|$$

where μ is the coefficient of static friction.

3.3.2 Contact Detection

Consider two bodies r and s discretized by particles, as shown in Fig. 3.12. If the momenta of both bodies are mapped to the same grid node I, i.e., $p_{iI}^r \neq 0$ and $p_{iI}^s \neq 0$, the two bodies are identified to be in contact at the grid node I. When the normal velocities at the contact grid node I of the two bodies satisfy

$$(v_{iI}^r - v_{iI}^s) n_{iI}^r > 0, \tag{3.84}$$

the two bodies are approaching to each other, and may penetrate into each other. In Eq. (3.84), n_{iI}^r is the unit normal of body r at the contact grid nod I, and the repeated index i represents summation.

Bardenhagen employed a center-of-mass velocity of the grid node I [71,72]

$$v_{iI}^{\text{cm}} = \frac{m_I^r v_{iI}^r + m_I^s v_{iI}^s}{m_I^r + m_I^s} \tag{3.85}$$

to detect the contact state of the two bodies at the grid node I. In Eq. (3.85),

$$m_I^b = \sum_{p=1}^{n_p^b} m_p N_{Ip} \tag{3.86}$$

is the mass at grid node I of body b, n_p^b is the total number of particles in body b. When the two bodies are not in contact, only one body contributes to the momentum of grid node I so that the center-of-mass velocity of grid node I only depends on one body, namely

$$(v_{iI}^b - v_{iI}^{\text{cm}})n_{iI}^b = 0 \quad (b = r, s). \tag{3.87}$$

If the two bodies are in contact and may penetrate into each other at the grid node I, the condition (3.84) can be rewritten in terms of the center-of-mass velocity by using Eq. (3.85) as follows:

$$(v_{iI}^b - v_{iI}^{\text{cm}})n_{iI}^b > 0. \tag{3.88}$$

If the two bodies contacted at the grid node I will separate from each other, it follows that

$$(v_{iI}^b - v_{iI}^{\text{cm}})n_{iI}^b < 0. \tag{3.89}$$

The contact state of the two bodies at the grid node I can be identified by using Eqs. (3.87)–(3.89). If two bodies are in contact at a grid node and may penetrate into each other, a contact force must be applied at the grid node to resist the penetration.

The unit normal $\hat{n}_{iI}^b$ to the surface of body b at grid node I can be calculated from the mass gradient as

$$\hat{n}_{iI}^b = \frac{\sum_{p=1}^{n_p^b} m_p N_{Ip,i}}{\left|\sum_{p=1}^{n_p^b} m_p N_{Ip,i}\right|}, \qquad b = r, s. \tag{3.90}$$

The unit normal obtained from Eq. (3.90) does not satisfy the collinearity condition $\hat{n}_{iI}^r = -\hat{n}_{iI}^s$, which leads to non-conservation of momentum, and even penetration.

A collinear unit normal can be obtained by averaging the two unit normals [73] given by Eq. (3.90), i.e.,

$$n_{iI}^r = -n_{iI}^s = \frac{\hat{n}_{iI}^r - \hat{n}_{iI}^s}{|\hat{n}_{iI}^r - \hat{n}_{iI}^s|}. \tag{3.91}$$

If body r is stiffer than body s, or if the surface of body r is flat/convex but the surface of body s is concave, choose the unit normal of body r as the collinear unit normal, i.e.,

$$n_{iI}^r = -n_{iI}^s = \hat{n}_{iI}^r. \tag{3.92}$$

3.3.3 Contact Force

A contact algorithm can be implemented to find the contact force in a trial-correction fashion. The momentum equation of each body is first integrated independently to obtain the trial solution as if both bodies were not in contact. If the trial solution satisfies the impenetrability condition, take the trial solution as the final true solution. If not, the contact force is applied at the contact grid nodes to prevent penetration.

Integrating the momentum equation of each body independently gives the trial grid nodal momentum

$$\bar{p}_{iI}^{b,k+1/2} = p_{iI}^{b,k-1/2} + \Delta t^k f_{iI}^{b,k} \tag{3.93}$$

and trial grid nodal velocity

$$\bar{v}_{iI}^{b,k+1/2} = v_{iI}^{b,k-1/2} + \Delta t^k \frac{f_{iI}^{b,k}}{m_I^{b,k}} \tag{3.94}$$

of each body b at grid node I.

If the trial grid nodal velocity $\bar{v}_{iI}^{b,k+1/2}$ does not satisfy the condition

$$(\bar{v}_{iI}^{r,k+1/2} - \bar{v}_{iI}^{s,k+1/2})n_{iI}^{r,k} > 0, \tag{3.95}$$

or

$$(\bar{v}_{iI}^{b,k+1/2} - \bar{v}_{iI}^{\text{cm},k+1/2})n_{iI}^{b,k} > 0, \tag{3.96}$$

the two bodies are not in contact such that the trial solution is the true solution and let $v_{iI}^{b,k+1/2} = \bar{v}_{iI}^{b,k+1/2}$. If the condition (3.95) or (3.96) is satisfied, the two bodies will penetrate into each other so that a contact force should be applied at the grid node to prevent penetration. In Eq. (3.96),

$$\bar{v}_{iI}^{\text{cm},k+1/2} = \frac{\bar{p}_{iI}^{r,k+1/2} + \bar{p}_{iI}^{s,k+1/2}}{m_I^{r,k} + m_I^{s,k}} \tag{3.97}$$

is the trial center-of-mass velocity of grid node I. Eq. (3.95) can be rewritten in the momentum form by multiplying $m_I^{r,k} m_I^{s,k}$ as

$$(m_I^{s,k} \bar{p}_{iI}^{r,k+1/2} - m_I^{r,k} \bar{p}_{iI}^{s,k+1/2})n_{iI}^{r,k} > 0. \tag{3.98}$$

Eq. (3.96) can also be given in the momentum form as

$$\left(\bar{p}_{iI}^{b,k+1/2} - m_I^{b,k} \bar{v}_{iI}^{\text{cm},k+1/2}\right) n_{iI}^{b,k} > 0. \tag{3.99}$$

When the trial solution violates the impenetrability condition at grid node I, a contact force $f_{iI}^{b,c,k}$ is applied to correct the trial solution, with the use of the above momentum form, as follows:

$$p_{iI}^{b,k+1/2} = \bar{p}_{iI}^{b,k+1/2} + \Delta t^k f_{iI}^{b,c,k}, \tag{3.100}$$

$$v_{iI}^{b,k+1/2} = \bar{v}_{iI}^{b,k+1/2} + \Delta t^k \frac{f_{iI}^{b,c,k}}{m_I^{b,k}}. \tag{3.101}$$

For a sticking contact, the corrected grid nodal velocity $v_{iI}^{b,k+1/2}$ must satisfy the velocity continuity condition

$$v_{iI}^{r,k+1/2} - v_{iI}^{s,k+1/2} = 0. \tag{3.102}$$

Substituting Eq. (3.101) into (3.102) yields the contact force for the sticking contact as

$$f_{iI}^{r,c,k} = -f_{iI}^{s,c,k} = \frac{m_I^{r,k} m_I^{s,k}}{(m_I^{r,k} + m_I^{s,k})\Delta t^k}(\bar{v}_{iI}^{s,k+1/2} - \bar{v}_{iI}^{r,k+1/2}). \tag{3.103}$$

Eq. (3.103) can be further rewritten in the momentum form as

$$f_{iI}^{r,c,k} = \frac{1}{(m_I^{r,k} + m_I^{s,k})\Delta t^k}(m_I^{r,k} \bar{p}_{iI}^{s,k+1/2} - m_I^{s,k} \bar{p}_{iI}^{r,k+1/2}). \tag{3.104}$$

Eq. (3.103) can be simplified, by using Eq. (3.97), to

$$f_{iI}^{b,c,k} = \frac{m_I^{b,k}}{\Delta t^k}(\bar{v}_{iI}^{\mathrm{cm},k+1/2} - \bar{v}_{iI}^{b,k+1/2}). \tag{3.105}$$

The normal and tangential contact forces for the sticking contact can be found to be

$$f_{iI}^{b,\mathrm{nor},k} = f_{jI}^{b,c,k} n_{jI}^{b,k} n_{iI}^{b,k}, \tag{3.106}$$

$$f_{iI}^{b,\tan,k} = f_{iI}^{b,c,k} - f_{iI}^{b,\mathrm{nor},k}. \tag{3.107}$$

The magnitude of the tangential contact force $\| f_{iI}^{b,\tan,k} \|$ is limited by the maximum static friction force $\mu \| f_{iI}^{b,\mathrm{nor},k} \|$. Hence, the contact force can be finally expressed as

$$f_{iI}^{b,c,k} = f_{iI}^{b,\mathrm{nor},k} + \min(\| f_{iI}^{b,\tan,k} \|, \mu \| f_{iI}^{b,\mathrm{nor},k} \|) \frac{f_{iI}^{b,\tan,k}}{\| f_{iI}^{b,\tan,k} \|}. \tag{3.108}$$

The condition (3.98) may result in a spurious contact. For example, when the space between two bodies approaching each other is less than 2 times the

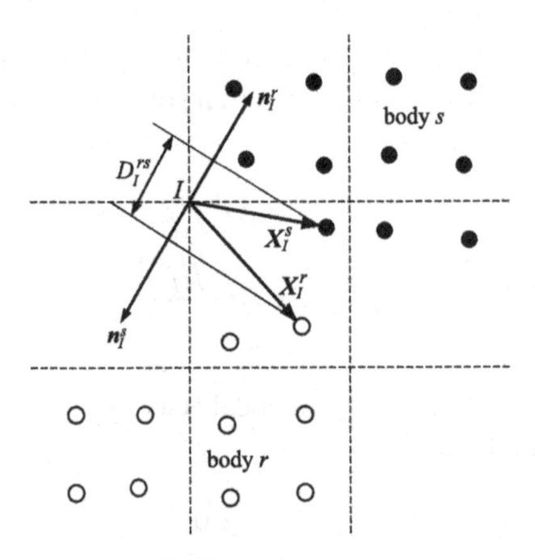

FIGURE 3.13 Distance between two bodies.

cell size (as shown in Fig. 3.11), condition (3.98) is satisfied at grid node I, and identifies gird node I as a contacted grid node, but the two bodies are not actually in contact at this time. To avoid the spurious contact, Ma et al. proposed an improved contact detection condition [74] by calculating the real distance between two bodies. As shown in Fig. 3.13, let $\boldsymbol{X}_I^r$ and $\boldsymbol{X}_I^s$ denote the position vector emanating from grid node I to its closest particle in body r and body s, respectively. The distance between the two bodies can be calculated as the sum of the projections of these two position vectors onto the normal vectors of the two bodies at the grid node I, namely

$$D_I^{rs} = -\boldsymbol{X}_I^r \cdot \boldsymbol{n}_I^r - \boldsymbol{X}_I^s \cdot \boldsymbol{n}_I^s. \tag{3.109}$$

If $D_I^{rs} \leqslant 0$, two bodies overlap each other, and penetration occurs. Thus, the contact detection condition can be modified as

$$\begin{cases} (\bar{v}_{iI}^{r,n+1/2} - \bar{v}_{iI}^{s,n+1/2})n_{iI}^{r,n} > 0, \\ D_I^{rs} \leqslant \lambda d_c \end{cases} \tag{3.110}$$

where d_c is the cell size. In Eq. (3.110), λd_c is used to take the particle size into account, and λ can be set to 0.5 if 2 particles are used initially in each direction of a cell.

In the above contact methods, the momentum equation of each body is solved in its own background grid, which requires significant storage [70,71]. Ma et al. proposed a local multi-mesh technique [74], in which the multi-mesh is only employed at contact nodes, as shown in Fig. 3.14. The local multi-mesh technique reduces the storage requirement and computational cost significantly.

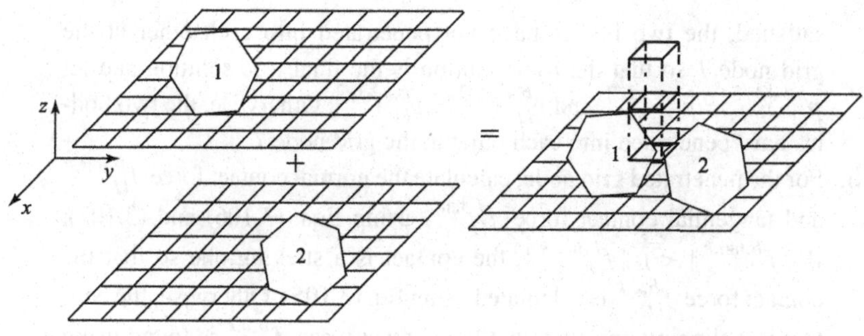

FIGURE 3.14 Local multi-mesh.

3.3.4 Numerical Algorithm for Contact Method

The numerical algorithm of the explicit MPM, as presented in Sect. 3.2.2, can be revised to include the contact method as follows:

1. Loop over all bodies to reconstruct the grid nodal mass and momentum of each body b by mapping their particle mass and momentum to the corresponding grid nodes, namely,

$$m_I^{b,k} = \sum_{p=1}^{n_p^b} m_p N_{Ip}^k,$$

$$p_{iI}^{b,k-1/2} = \sum_{p=1}^{n_p^b} m_p v_{ip}^{b,k-1/2} N_{Ip}^k.$$

2. Impose essential boundary conditions on the grid nodal momentum $p_{iI}^{b,k-1/2}$ of each body b. At the fixed boundary nodes, set $p_{iI}^{b,k-1/2} = 0$.
3. For the USF only, calculate the particle strain increment $\Delta\varepsilon_{ijp}^{b,k-1/2}$ and vorticity increment $\Delta\Omega_{ijp}^{b,k-1/2}$ of each body b based on its grid nodal velocity $v_{iI}^{b,k-1/2}$ that can be obtained from the grid nodal momentum $p_{iI}^{b,k-1/2}$, and then update the particle density $\rho_p^{b,k+1}$ and stress $\sigma_{ijp}^{b,k+1}$.
4. Calculate the gird nodal internal force $f_{iI}^{b,\text{int},k}$, external force $f_{iI}^{b,\text{ext},k}$, and total force $f_{iI}^{b,k} = f_{iI}^{b,\text{int},k} + f_{iI}^{b,\text{ext},k}$ of each body.
5. Integrate the momentum equation of each body independently as if two bodies were not in contact, namely, calculate the grid nodal trial momentum $\bar{p}_{iI}^{b,k+1/2}$ using Eq. (3.93).
6. Contact detecting and contact force calculation:
   a. Calculate the unit normal $n_{iI}^{b,k}$ to each body surface at grid nodes using Eq. (3.90), (3.91), or (3.92). If condition (3.98) or (3.99) is not

satisfied, the two bodies have not penetrated into each other at the grid node I so that the trial solution is the final true solution and let $p_{iI}^{b,k+1/2} = \bar{p}_{iI}^{b,k+1/2}$, and $v_{iI}^{b,k+1/2} = \bar{v}_{iI}^{b,k+1/2}$. Otherwise, the two bodies have penetrated into each other at the grid node I.

b. For the penetrated grid node, calculate the normal contact force $f_{iI}^{b,\mathrm{nor},k}$ and tangential contact force $f_{iI}^{b,\tan,k}$ using Eqs. (3.106) and (3.107). If $\|f_{iI}^{b,\tan,k}\| < \mu \|f_{iI}^{b,\mathrm{nor},k}\|$, the contact is a sticking one so that the contact force $f_{iI}^{b,c,k}$ is calculated using Eq. (3.105). Otherwise, the contact is a slipping one such that the contact force $f_{iI}^{b,c,k}$ is found using Eq. (3.108).

c. The corrected grid nodal momentum $p_{iI}^{b,k+1/2}$ of each body which satisfies the contact surface condition is finally obtained using Eq. (3.100).

7. Update the position and velocity of each body by mapping the grid nodal velocity and acceleration of each body back to the particles as follows:

$$x_{ip}^{b,k+1} = x_{ip}^{b,k} + \Delta t^{k+1/2} \sum_{I=1}^{8} \frac{p_{iI}^{b,k+1/2}}{m_I^b} N_{Ip}^k, \tag{3.111}$$

$$v_{ip}^{b,k+1/2} = v_{ip}^{b,k-1/2} + \Delta t^k \sum_{I=1}^{8} \frac{f_{iI}^{b,k} + f_{iI}^{b,c,k}}{m_I^b} N_{Ip}^k. \tag{3.112}$$

8. For the MUSL only, recalculate the grid nodal momentum of each body based on its updated particle momenta, and impose essential boundary conditions.

9. For the MUSL and USL only, calculate the grid nodal velocity, particle strain increment and vorticity increment, and then update particle density and stress.

10. All material properties are stored with the particles so that the deformed grid could be discarded, if needed, to employ a new grid in the next time step.

Compared with the numerical algorithm of the explicit MPM without contact treatment as presented in Sect. 3.2.2, the revised algorithm only employs one additional step, i.e., step 6, which calculates the contact force and then corrects the trial solution.

3.4 GENERALIZED INTERPOLATION MPM AND OTHER IMPROVEMENTS

The original MPM discretizes a material domain into a set of particles, with the use of linear grid nodal shape functions for computational efficiency, which results in a discontinuous gradient of shape functions. In addition, the shape functions are local in the sense that they are defined within their own cells. As a result, a particle on the cell boundary would not be covered by the local shape functions defined within the respective cells around the particle. This issue would produce noise in the numerical solution, called **cell crossing noise**, which results in unsatisfactory and even unphysical results.

To alleviate the cell crossing noise, Bardenhagen et al. developed the Generalized Interpolation Material Point (GIMP) method [75] using a Petrov–Galerkin discretization scheme and discretizing a continuum as a collection of particles defined by particle characteristic functions that are nonzero (nonlocal) over a representative volume. A particle characteristic function $\chi_p(x)$ defines the spatial volume occupied by the particle. In the initial configuration, the particle characteristic functions should be a partition of unity, i.e.,

$$\sum_p \chi_p(\boldsymbol{x}) = 1. \tag{3.113}$$

The current volume of particle p is given by

$$V_p = \int_{\Omega_p \bigcap \Omega} \chi_p(\boldsymbol{x}) \mathrm{d}V \tag{3.114}$$

where Ω is the current domain occupied by the continuum, and Ω_p is the current support of the characteristic function of particle p, which can also be considered as the current domain occupied by the particle.

Any material property $f(\boldsymbol{x})$ can be approximated by its particle value f_p as

$$f(\boldsymbol{x}) = \sum_p f_p \chi_p(\boldsymbol{x}). \tag{3.115}$$

Eq. (3.115) shows that the particle characteristic functions smooth out the particle properties over the entire computational domain, and determine the smoothness of the spatial variation. The density ρ, stress σ_{ij}, and acceleration $\ddot{u}_i$ can all be approximated using Eq. (3.115). The weak form (2.70) can then be expressed using the approximation (3.115) as

$$\sum_p \int_{\Omega_p \bigcap \Omega} \frac{\dot{p}_{ip}}{V_p} \chi_p \delta u_i \mathrm{d}V + \sum_p \int_{\Omega_p \bigcap \Omega} \sigma_{ijp} \chi_p \delta u_{i,j} \mathrm{d}V$$
$$- \sum_p \int_{\Omega_p \bigcap \Omega} \rho_p b_{ip} \chi_p \delta u_i \mathrm{d}V - \int_{\Gamma_t} \bar{t}_i \delta u_i \mathrm{d}A = 0. \tag{3.116}$$

The GIMP method employs the Petrov–Galerkin method to solve Eq. (3.116). The particle characteristic functions $\chi_p(\boldsymbol{x})$ are used in the trial functions (see Eq. (3.115)), and the grid nodal shape functions $N_I(\boldsymbol{x})$ are used in the test functions. Thus, the virtual displacement δu_i in Eq. (3.116) can be expressed as

$$\delta u_i = \sum_I \delta u_{iI} N_I(\boldsymbol{x}) \tag{3.117}$$

where the grid nodal shape functions are also a partition of unity, i.e.,

$$\sum_I N_I(\boldsymbol{x}) = 1. \tag{3.118}$$

Substituting Eq. (3.117) into Eq. (3.116) and invoking the arbitrariness of δu_{iI} and $\delta u_{iI}|_{\Gamma_u} = 0$ yields the grid nodal momentum equations of the GIMP similar to Eq. (3.8) as follows:

$$\dot{p}_{iI} = f_{iI}^{\text{int}} + f_{iI}^{\text{ext}}, \quad x_I \notin \Gamma_u \tag{3.119}$$

where

$$p_{iI} = \sum_p S_{Ip} p_{ip}, \tag{3.120}$$

$$f_{iI}^{\text{int}} = -\sum_p \sigma_{ijp} S_{Ip,j} V_p, \tag{3.121}$$

$$f_{iI}^{\text{ext}} = \sum_p m_p S_{Ip} b_{ip} + \int_{\Gamma_t} N_I(\boldsymbol{x}) \bar{t}_i \mathrm{d}\Gamma, \tag{3.122}$$

and

$$S_{Ip} = \frac{1}{V_p} \int_{\Omega_p \cap \Omega} \chi_p(\boldsymbol{x}) N_I(\boldsymbol{x}) \mathrm{d}\Omega, \tag{3.123}$$

$$S_{Ip,j} = \frac{1}{V_p} \int_{\Omega_p \cap \Omega} \chi_p(\boldsymbol{x}) N_{I,j}(\boldsymbol{x}) \mathrm{d}\Omega. \tag{3.124}$$

Eqs. (3.123) and (3.124) show that the GIMP shape function S_{Ip} (also referred to as weighting function) and its gradients $S_{Ip,j}$ are implicitly functions of grid nodal position $\boldsymbol{x}_I$, particle position $\boldsymbol{x}_p$, and current particle volume Ω_p. Compared with the shape function N_I of the computational grid, the shape function S_I of the GIMP is smooth with a nonlocal support. It can be proved, using Eqs. (3.118) and (3.114), that the shape function S_{Ip} of the GIMP is also a partition of unity, i.e.,

$$\sum_I S_{Ip} = 1 \qquad \forall x_{iI}, \, x_{ip}. \tag{3.125}$$

The grid nodal mass m_I and momentum p_{iI} are then given by

$$m_I = \sum_p m_p S_{Ip}, \tag{3.126}$$

$$p_{iI} = \sum_p p_{ip} S_{Ip}. \tag{3.127}$$

In the GIMP, the mass and momentum are also conserved, i.e.,

$$\sum_I m_I = \sum_I \sum_p m_p S_{Ip} = \sum_p m_p, \tag{3.128}$$

$$\sum_I p_{iI} = \sum_I \sum_p p_{ip} S_{Ip} = \sum_p p_{ip}. \tag{3.129}$$

The original MPM formulation can be recovered from the GIMP by choosing the Dirac delta function as the particle characteristic function $\chi_p(\boldsymbol{x})$, i.e.,

$$\chi_p(\boldsymbol{x}) = \delta(\boldsymbol{x} - \boldsymbol{x}_p) V_p \tag{3.130}$$

because of $S_{Ip} = N_I(\boldsymbol{x}_p)$ and $S_{Ip,j} = N_{I,j}(\boldsymbol{x}_p)$.

Choosing different particle characteristic functions and grid nodal shape functions results in different weighting functions. The grid nodal shape function is usually chosen to be a linear function for computational efficiency. In 1D problems, it follows that

$$N_I(x) = \begin{cases} 0 & |x - x_I| \geqslant L, \\ 1 + (x - x_I)/L & -L < x - x_I \leqslant 0, \\ 1 - (x - x_I)/L & 0 < x - x_I < L \end{cases} \tag{3.131}$$

where L is the grid cell spacing size.

3.4.1 Contiguous Particle GIMP

The simplest characteristic function in 1D cases is

$$\chi_p(x) = H(x - (x_p - l_p)) - H(x - (x_p + l_p)) \tag{3.132}$$

where $2l_p$ is the current particle size, and

$$H(x) = \begin{cases} 0 & x < 0, \\ 1 & x > 0 \end{cases}$$

is the step function. The particle characteristic function (3.132) defines "contiguous particles", i.e., contiguous regions of non-overlapping support Ω_p [75].

The corresponding GIMP is denoted as the contiguous particle GIMP (cpGIMP) [75,76], in which the initial particle size is determined by dividing the cell spacing L by the number of particles per cell.

Eq. (3.132) can also be written as

$$\chi_p(x) = \begin{cases} 1 & x \in \Omega_p, \\ 0 & x \notin \Omega_p \end{cases} \tag{3.133}$$

so that Eqs. (3.123) and (3.124) can be simplified to

$$S_{Ip} = \frac{1}{2l_p} \int_{x_p-l_p}^{x_p+l_p} N_I(x)\mathrm{d}x, \tag{3.134}$$

$$S_{Ip,j} = \frac{1}{2l_p} \int_{x_p-l_p}^{x_p+l_p} N_{I,j}(x)\mathrm{d}x. \tag{3.135}$$

Substituting Eq. (3.131) into Eqs. (3.134) and (3.135) results in the cpGIMP shape function with C^1 continuity as follows:

$$S_{Ip} = \begin{cases} 0 & |x_p - x_I| \geqslant L + l_p, \\ \frac{(L+l_p+(x_p-x_I))^2}{4Ll_p} & -L - l_p < x_p - x_I \leqslant -L + l_p, \\ 1 + \frac{x_p-x_I}{L} & -L + l_p < x_p - x_I \leqslant -l_p, \\ 1 - \frac{(x_p-x_I)^2+l_p^2}{2Ll_p} & -l_p < x_p - x_I \leqslant l_p, \\ 1 - \frac{x_p-x_I}{L} & l_p < x_p - x_I \leqslant L - l_p, \\ \frac{(L+l_p-(x_p-x_I))^2}{4Ll_p} & L - l_p < x_p - x_I \leqslant L + l_p. \end{cases} \tag{3.136}$$

The derivative of the cpGIMP shape function then takes the form of

$$\nabla S_{Ip} = \begin{cases} 0 & |x_p - x_I| \geqslant L + l_p, \\ \frac{L+l_p+(x_p-x_I)}{2Ll_p} & -L - l_p < x_p - x_I \leqslant -L + l_p, \\ \frac{1}{L} & -L + l_p < x_p - x_I \leqslant -l_p, \\ -\frac{x_p-x_I}{Ll_p} & -l_p < x_p - x_I \leqslant l_p, \\ -\frac{1}{L} & l_p < x_p - x_I \leqslant L - l_p, \\ -\frac{L+l_p-(x_p-x_I)}{2Ll_p} & L - l_p < x_p - x_I \leqslant L + l_p. \end{cases} \tag{3.137}$$

3.4.2 Uniform GIMP

As shown in Eqs. (3.123) and (3.124), the GIMP requires integration over the current support of the particle characteristic functions that deform and rotate

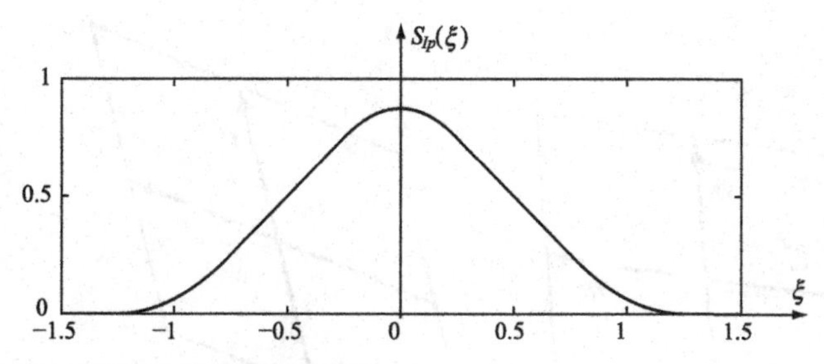

FIGURE 3.15 uGIMP shape function in 1D.

relative to the computational grid. In other words, the full version of the GIMP requires tracking the support of the particle characteristic function, or tracking the particle shape, which is very difficult for multi-dimensional problems. For 1D cases, the current length l_p^n of particle p at time t^n can be obtained from its initial length l_p^0 and deformation gradient F_p^n as

$$l_p^n = F_p^n l_p^0 \tag{3.138}$$

where F_p^n is the deformation gradient of particle p at time t^n.

To avoid the difficulty of tracking the particle shape for multi-dimensional problems, several approximations have been proposed. For example, particles can be assumed rectangular or cuboid particles initially and remain rectangular or cuboid during deformations, i.e., neglecting the shear deformation and rigid-body rotation of a particle. Thus, their current size in each direction can be obtained from [76]

$$l_{ip}^n = F_{iip}^n l_{ip}^0 \tag{3.139}$$

where F_{iip}^n is the ith diagonal element of the deformation gradient tensor $\boldsymbol{F}$ of particle p at time t^n.

Assuming that two particles initially occupy each cell and the size of the particles is fixed, the current length of each particle is $2l_p = L/2$, i.e., $l_p = L/4$. Letting $\xi = |(x_p - x_I)/L|$, Eq. (3.136) can be simplified to [48]

$$S_{Ip}(\xi) = \begin{cases} \frac{7-16\xi^2}{8} & \xi \leqslant 0.25, \\ 1-\xi & 0.25 < \xi \leqslant 0.75, \\ \frac{(5-4\xi)^2}{16} & 0.75 < \xi \leqslant 1.25, \\ 0 & \xi > 1.25. \end{cases} \tag{3.140}$$

The GIMP shape function given by Eq. (3.140) is shown in Fig. 3.15. The three-dimensional shape function can be chosen as the product of three one-

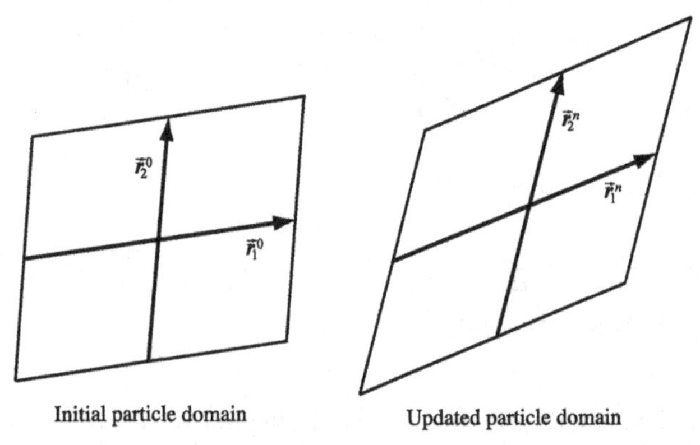

FIGURE 3.16 Initial and updated particle domains [77].

dimensional shape functions (3.140), namely,

$$S_{Ip}(\boldsymbol{x}) = S_{Ip}(\xi)S_{Ip}(\eta)S_{Ip}(\zeta) \tag{3.141}$$

where $\eta = |(y_p - y_I)/L|$ and $\zeta = |(z_p - z_I)/L|$. This version of the GIMP is called the uniform GIMP (uGIMP) that assumes that all particles are cuboid particles of fixed size.

The uGIMP assumes that the sizes of particles are fixed during material deformations. The particle characteristic functions, whose supports may overlap or leave gaps for large deformations, are no longer a partition of unity. Therefore, the uGIMP is unable to completely eliminate the cell crossing noise.

3.4.3 Convected Particle Domain Interpolation

Sadeghirad et al. proposed a convected particle domain interpolation (CPDI) method [77] which employs initially parallelogram-shaped particles and assumes the deformation gradient to be constant over the particle domain. The convected particle domain is a reshaped parallelogram in the deformed configuration, as shown in Fig. 3.16. The vectors defining the parallelogram domain in the current configuration can be obtained from

$$\begin{aligned} \boldsymbol{r}_1^n &= \boldsymbol{F}_p^n \boldsymbol{r}_1^0, \\ \boldsymbol{r}_2^n &= \boldsymbol{F}_p^n \boldsymbol{r}_2^0, \\ \boldsymbol{r}_3^n &= \boldsymbol{F}_p^n \boldsymbol{r}_3^0 \end{aligned} \tag{3.142}$$

where $(\boldsymbol{r}_1^0, \boldsymbol{r}_2^0, \boldsymbol{r}_3^0)$ and $(\boldsymbol{r}_1^n, \boldsymbol{r}_2^n, \boldsymbol{r}_3^n)$ define the parallelogram particle in the initial and current configuration, respectively, as shown in Fig. 3.16.

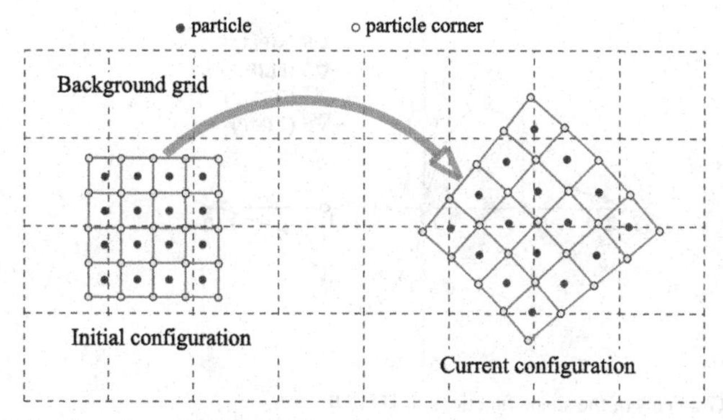

FIGURE 3.17 The enhanced CPDI method [78].

To reduce the computational expense of dividing the particle domains along the cell boundaries in calculating the integrals in Eqs. (3.123) and (3.137), Sadeghirad et al. proposed alternative grid shape functions by interpolating the standard grid shape functions at the four corners of each particle domain as

$$N_I^{\text{app}}(\boldsymbol{x}) = \sum_{\alpha=1}^{4} Q_\alpha^p(\boldsymbol{x}) N_I(\boldsymbol{x}_\alpha^p) \quad \text{on } \Omega_p \tag{3.143}$$

where Q_α^p is the standard FE 4-node shape function related to the αth corner of the domain corresponding to particle p, and $\boldsymbol{x}_\alpha^p$ is the position vector of this corner.

The parallelogram domain defined by Eq. (3.142) is a first-order accurate approximation of the actual particle domain Ω_p. Sadeghirad et al. further enhanced the CPDI method to more accurately track the particle domains as quadrilaterals in 2D (hexahedrons in 3D) [78], as shown in Fig. 3.17. The field values are saved at the particle centroid $\boldsymbol{x}_p$. The corner positions of particle p are updated at the end of each time step by

$$\boldsymbol{x}_\alpha^{p(n+1)} = \boldsymbol{x}_\alpha^{p(n)} + \sum_I N_I(\boldsymbol{x}_\alpha^{p(n)}) \boldsymbol{v}_I^{n+1} \Delta t. \tag{3.144}$$

The enhanced CPDI method is a second-order accurate approximation of the actual particle domain Ω_p, which not only removes the overlaps or gaps between particle domains, but also provides the flexibility in choosing the particle domain shape in the initial configuration.

3.4.4 Dual Domain Material Point Method

Instead of modifying the shape functions from the original MPM, Zhang et al. proposed to define the gradient of the shape function as a weighted average of

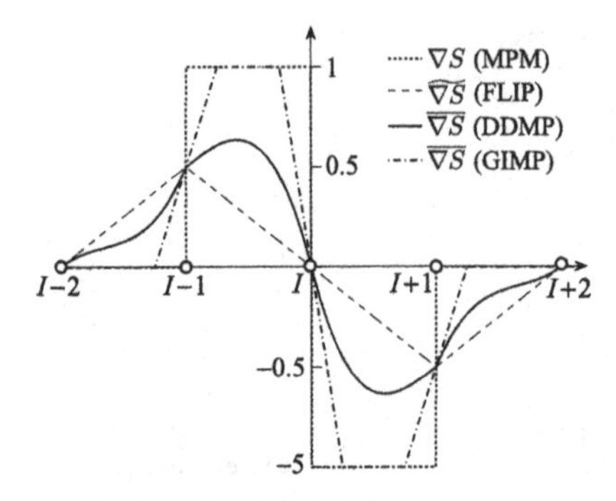

FIGURE 3.18 Gradients of the shape functions in 1D [79].

the gradient calculated in the original MPM and the gradient from the FLIP node-based calculation [79]. The modified gradient is continuous as particles move across cell boundaries such that the cell crossing noise is eliminated. Zhang et al. named this method as the dual domain material point (DDMP) method [79] because the support of the modified gradient is greater than that of the shape function itself. Unlike the GIMP, the DDMP does not make use of particle characteristic functions so that the difficulty of tracking the support of a characteristic function is avoided for large deformation problems. The modified gradient of the DDMP is given by [79]

$$\overline{\nabla S_I}(\boldsymbol{x}) = \alpha(\boldsymbol{x})\nabla S_I(\boldsymbol{x}) + (1 - \alpha(\boldsymbol{x}))\widetilde{\nabla S_I}(\boldsymbol{x}) \tag{3.145}$$

where $\alpha(\boldsymbol{x})$ is a bounded function, $\nabla S_I(\boldsymbol{x})$ is the gradient of the shape function calculated using the original MPM, and

$$\widetilde{\nabla S_I}(\boldsymbol{x}) = \sum_{J=1}^{n_n} \frac{S_J(\boldsymbol{x})}{V_J} \int_{\Omega} S_J(\boldsymbol{x})\nabla S_I(\boldsymbol{x})\mathrm{d}\Omega \tag{3.146}$$

is the gradient from the node-based calculation as used in FLIP [57], with n_n being the number of nodes in the computational grid. Fig. 3.18 compares the gradients of the shape functions in 1D as used in the MPM, FLIP, GIMP, and DDMP.

3.4.5 Spline Grid Shape Function

The particle quadrature used in the original MPM is equivalent to a midpoint quadrature with an uneven spacing during deformation. The piecewise linear grid shape functions are usually used such that the integrals in the calculation of internal forces are discontinuous across cell boundaries. Integrating discontinuous functions with the midpoint rule is valid only when the particle domain

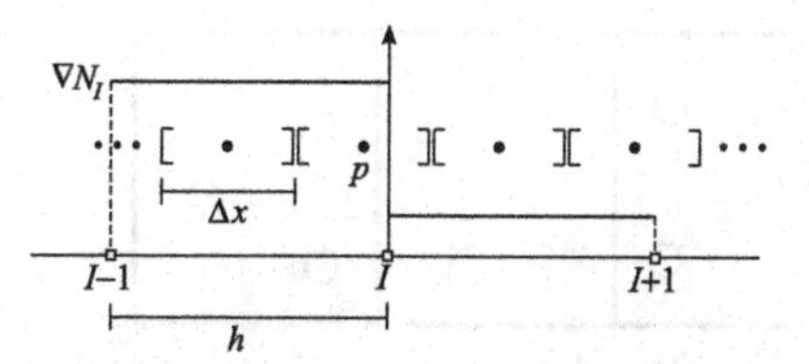

FIGURE 3.19 The vertex of particle p does not coincide with the discontinuity [80]. Δx is the particle length, h is the cell width, and ∇N_I is the gradient of shape function associated with grid node I.

vertices coincide with these discontinuities. However, the particle domain vertices usually do not coincide with these discontinuities all the time as particles advect through the background grid, which will lead to quadrature errors when integrating across the discontinuities [80], as shown in Fig. 3.19.

Steffen et al. [80] demonstrated that the integration error in finding the internal force for a uniformly stressed body with the use of piecewise linear shape functions is $O(\Delta x/h)$. They showed that simply using smoother basis functions could drastically reduce the integration error. For example, using the quadratic B-spline

$$N(x)=\begin{cases}\frac{1}{2h^2}x^2+\frac{3}{2h}x+\frac{9}{8}, & -\frac{3}{2}h\leqslant x\leqslant -\frac{1}{2}h,\\ -\frac{1}{h^2}x^2+\frac{3}{4}, & -\frac{1}{2}h\leqslant x\leqslant \frac{1}{2}h,\\ \frac{1}{2h^2}x^2-\frac{3}{2h}x+\frac{9}{8}, & \frac{1}{2}h\leqslant x\leqslant \frac{3}{2}h,\\ 0, & \text{otherwise}\end{cases} \tag{3.147}$$

reduces the integration error to $O((\Delta x/h)^2)$, whereas using the cubic B-spline

$$N(x)=\begin{cases}\frac{1}{6h^3}x^3+\frac{1}{h^2}x^2+\frac{2}{h}x+\frac{4}{3}, & -2h\leqslant x\leqslant -h,\\ -\frac{1}{2h^3}x^3-\frac{1}{h^2}x^2+\frac{4}{3}, & -h\leqslant x\leqslant 0,\\ \frac{1}{2h^3}x^3-\frac{1}{h^2}x^2+\frac{4}{3}, & 0\leqslant x\leqslant h,\\ -\frac{1}{6h^3}x^3+\frac{1}{h^2}x^2-\frac{2}{h}x+\frac{4}{3}, & h\leqslant x\leqslant 2h,\\ 0, & \text{otherwise}\end{cases} \tag{3.148}$$

reduces the integration error to $O((\Delta x/h)^3)$. Multi-dimensional B-spline shape functions can be obtained by taking the product of these 1D shape functions.

3.5 ADAPTIVE MATERIAL POINT METHOD

3.5.1 Particle Adaptive Split

In the MPM, the interactions between particles are carried out via the computational grid. There will be no interaction between two particles when they are

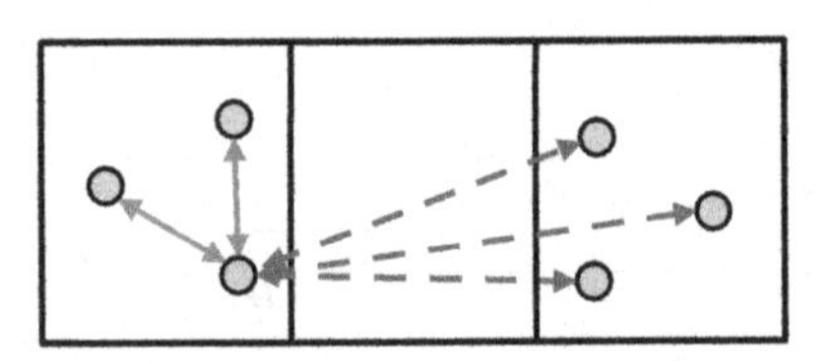

FIGURE 3.20 Numerical fracture [81].

separated by a grid cell, as shown in Fig. 3.20, which may lead to numerical fracture. Therefore, in the cases with extremely large deformations such as the expansion of explosion product and the extreme stretch of the material in shaped charge, the original MPM may be unable to yield the reasonable solution without using a particle rearrangement or adding scheme.

Ma et al. proposed an adaptive particle splitting scheme [81] to eliminate the numerical fracture. A particle is split into two particles when its accumulative strain in one direction exceeds a specified value. The current accumulative strain of a particle in the ith direction is given by

$$\varepsilon_i = \sum_k \Delta\varepsilon_i^k = \sum_k \dot{\varepsilon}_{ii}^k \Delta t \tag{3.149}$$

where $\Delta\varepsilon_i^k$ is the strain increment in the ith direction at time step k. The current equivalent length of the particle in the ith direction can be obtained as

$$L_i = L_0(1 + \varepsilon_i) \tag{3.150}$$

where $L_0 = \sqrt[3]{m_p/\rho_0}$ is the initial equivalent length of the particle. When the current equivalent length of the particle exceeds a given value in any direction, i.e.,

$$L_i > \alpha d_c, \tag{3.151}$$

the particle is divided into two particles in that direction, where d_c is the grid cell spacing and $\alpha < 1$ is a user-specified constant. The spacing between the two new particles is chosen as $0.5\alpha d_c$, as shown in Fig. 3.21.

When a particle is split into two particles, its mass, volume, and internal energy are halved to each new particle, while other variables such as stress, strain, and temperature are copied to the new particles directly. Accumulated strain ε_i of the new particles is determined as follows:

$$L_i' = \begin{cases} 0.5L_i & \text{in split direction,} \\ L_i & \text{in other direction} \end{cases} \tag{3.152}$$

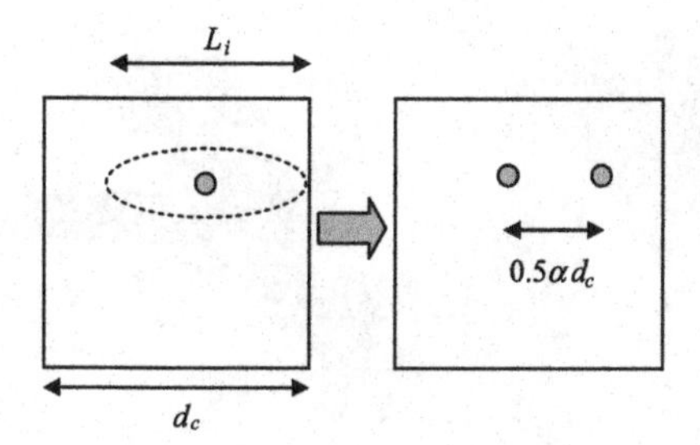

FIGURE 3.21 Particle adaptive splitting [81].

where L_i' denotes the equivalent length of the new particles. Note that the mass of the new particles is a half of the original mass m_p. Substituting the equivalent length defined in Eq. (3.150) into Eq. (3.152), the adjusted accumulated strain can be found to be

$$\varepsilon_i' = \begin{cases} 0.63\varepsilon_i - 0.37 & \text{in split direction,} \\ 1.26\varepsilon_2 + 0.26 & \text{in the other two directions.} \end{cases} \tag{3.153}$$

3.5.2 Adaptive Computational Grid

The MPM integrates the momentum equations on the computational grid so that the grid has a significant effect on the accuracy and efficiency of the MPM calculation. A regular grid with uniform cells is commonly used in the MPM, which is inefficient for solving the problems involving discontinuities of different degrees, such as evolution of localization and crack propagation. Tan et al. developed a hierarchically adaptive MPM for dynamic energy release rate calculations, which automatically refines the mesh around the crack tip to meet the local resolution requirement as judged by tracking the gradients in the solution process [82]. Based on the Structured Adaptive Mesh Refinement Application Infrastructure (SAMRAI), Ma et al. presented a parallel GIMP computational method [83] which uses the nested computational grid levels (with successive spatial and temporal refinements) to improve the computational accuracy and to reduce the overall computational cost in simulating the problems with multiple length scales. Ma et al. [84] also proposed a spatial refinement scheme of the structured grid for the GIMP in the simulations with highly localized stress gradients by modifying the grid shape function for the transitional node. Several representative methods are discussed below.

3.5.2.1 Dynamic Grid

The MPM usually employs a static computational grid, which must cover all the particles during the simulation. In most problems, however, particles only oc-

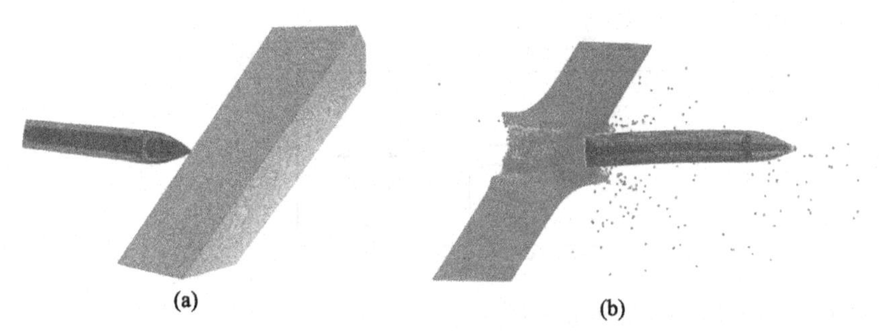

FIGURE 3.22 MPM models for simulating oblique penetration.

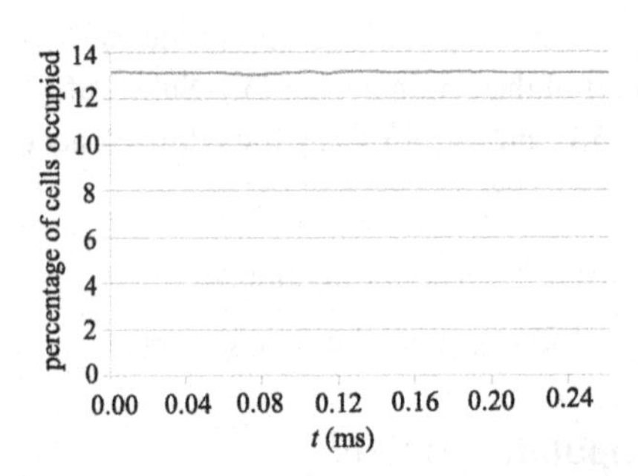

FIGURE 3.23 Time history of the percentage of used cells.

cupy a part of the grid. For example, in the penetration simulation as shown in Fig. 3.22, only about 13% of cells are occupied by the particles, and the remaining 87% of cells are not occupied by any particle, as shown in Fig. 3.23. Hence, the static grid wastes a huge amount of computer storage and computational time.

To minimize the computer storage and computational time required by the grid, Ma et al. proposed the dynamic grid technique [74]. At the beginning of a simulation, two arrays of pointers to node and cell objects are created without instantiating these objects. During the simulation, a node object or a cell object is instantiated only when there are particles around the node or in the cell, as shown in Fig. 3.24. The objects which have not been instantiated will not consume any memory and will not take part in the computation. Thus, the dynamic grid method will decrease the memory allocated and improve the efficiency significantly.

3.5.2.2 Moving Grid

In the problems with a high particle velocity, the initial computational grid may be unable to cover all the particles during the whole simulation. To overcome

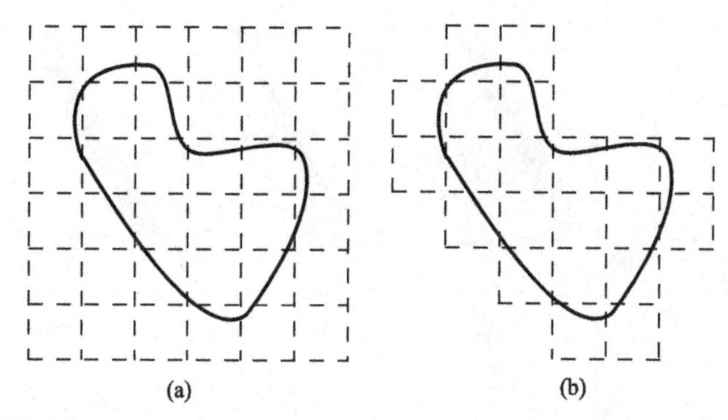

FIGURE 3.24 Dynamic grid technique: (a) static grid, (b) dynamic grid.

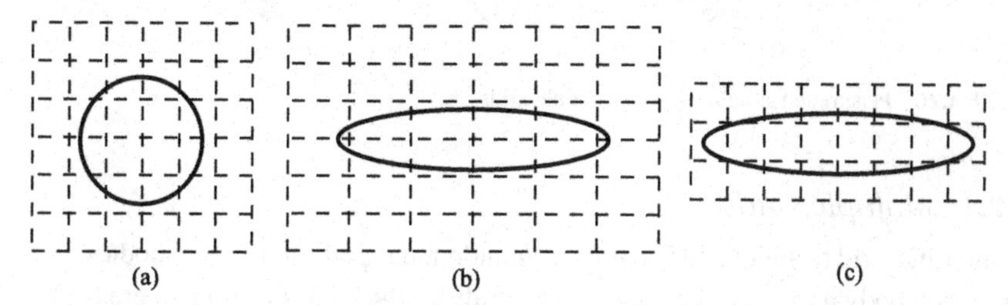

FIGURE 3.25 Moving grid technique: (a) initial grid, (b) original moving grid technique, and (c) improved moving grid technique.

this difficulty, Ma et al. proposed a moving grid technique [74], which adjusts the computational grid at each time step based on the current positions of particles to make the grid just cover all the particles. This technique can not only reduce the computational cost, but also eliminate the numerical fracture to some extent. When the particle spacing increases, the material domain also enlarges which triggers the expansion of the grid. The grid expansion enlarges the influence domains of particles to alleviate the numerical fracture.

The moving grid technique does not change the grid topology, but changes the cell size, which may deteriorate the solution accuracy that depends on the cell size. Yang et al. improved the moving grid technique [85] by keeping the cell size fixed, changing the grid topology, and deleting automatically the void cells. For example, consider that a sphere expands in the horizontal direction, as shown in Fig. 3.25(a). The original moving grid technique expands the grid by enlarging the cells in the horizontal direction to cover all the particles, as shown in Fig. 3.25(b), while the improved moving grid technique expands the grid by appending more cells to the grid and deleting the void cells, as shown in Fig. 3.25(c).

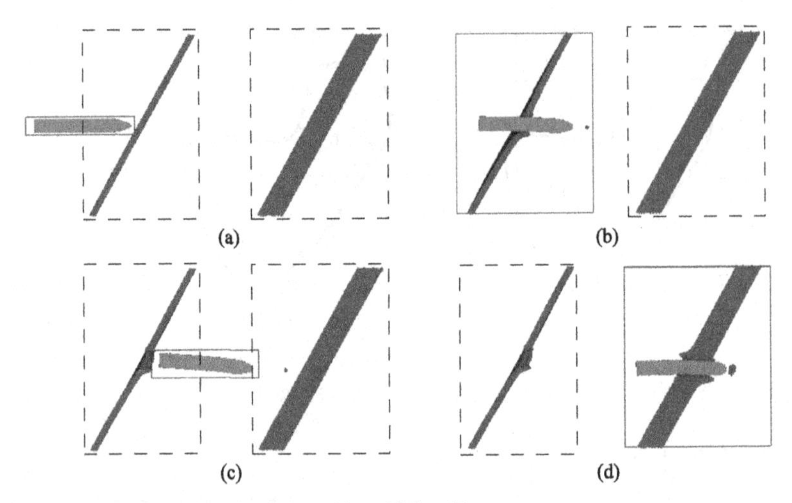

FIGURE 3.26 Penetration simulation with multiple grids.

3.5.2.3 Multiple Grids

The multiple-grid technique [85] employs an individual grid for different bodies so that each body moves in its own grid. The grids for the bodies without interaction with other bodies can be inactivated to reduce the computational cost. When bodies interact with each other, they move in a common grid. This technique is very efficient for penetration to multi-targets, as shown in Fig. 3.26, in which the dashed lines denote the boundary of an inactivated grid while the solid lines denote the boundary of an activated grid. Before the projectile contacts the target, both the projectile and target move in their own grids such that only the grid of projectile is activated, as shown in Fig. 3.26(a). After the projectile contacts the first target, the projectile and the first target move in a common grid so that only the common grid is activated, as shown in Fig. 3.26(b). After penetration into the first target is complete, three bodies move in their own grids again, and only the grid for the projectile is activated, as shown in Fig. 3.26(c). As compared with the original MPM grid, the multiple grid technique saves 50% of the CPU time and 60% of computer storage for the example as shown in Fig. 3.26.

3.5.2.4 Multi-level Grid

To improve the accuracy and efficiency of simulating localized failure evolution, Yang et al. proposed a multi-level grid [85,86] to gradually refine the computational grid. The grid can be gradually refined from level 0 (the original MPM grid) to level n in a localized region. The cell spacing of level n is chosen as a half of that of the level $n-1$ so that $h_n = h_0/2^n$ with h_0 being the cell size of level 0 that corresponds to the coarsest level. Fig. 3.27 illustrates an example of

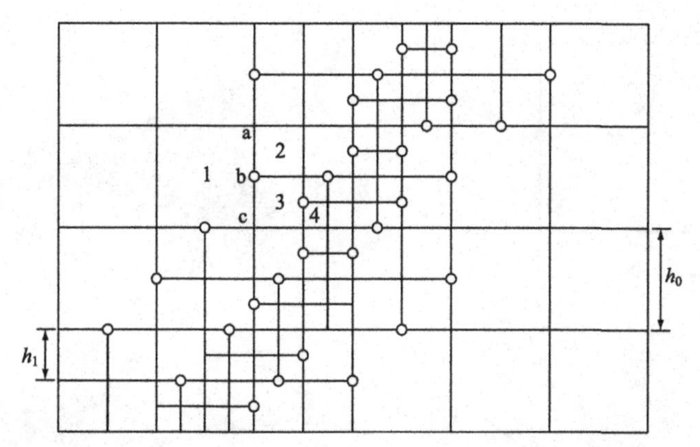

FIGURE 3.27 Three-level grid in 2D. The *hollow dots* denote the hanging nodes [85].

the 3-level grid in 2D. This technique results in additional nodes called hanging nodes in lower level cells at the interface between cells of different levels. In this two-dimensional case, the highest level cells have 4 nodes, but other level cells may have 4 to 8 nodes. For example, in Fig. 3.27, the 0-level cell 1 has 6 nodes, the 1-level cell 3 has 5 nodes, and the 2-level cell 4 only has 4 nodes.

In the above multi-level grid, 4-node cells are connected to 5- to 8-node cells so that the grid nodal shape functions must be modified to ensure the continuity across the interface between different-level cells. For example, the approximation function of cell 1 must conform to the approximation function of cell 2 along the *a–b* side, and conform to the approximation function of cell 3 along the *b–c* side. Therefore, the approximation function of cell 1 must be a piecewise linear function along the *a–b–c* side.

Take a 5-node cell shown in Fig. 3.28(a) as an example, in which the hanging node b is located at the middle of the side *a–c*. The cell is divided into two subcells by the dashed line passing through node b, and the shape function in each subdomain must be linear. Since any shape function must satisfy the requirement of

$$N_I(\boldsymbol{x}_J) = \delta_{IJ}, \tag{3.154}$$

the shape function of node b can be chosen as

$$N_b(\xi, \eta) = \frac{1}{2}(1+\xi)(1-|\eta|)$$

which is linear along the *a–b–c* side. Let N_I^L $(I = a, c, d, e)$ denote the 4-node cell shape function of node I, which is given by

$$N_I^L(\xi, \eta) = \frac{1}{4}(1+\xi_I\xi)(1+\eta_I\eta)$$

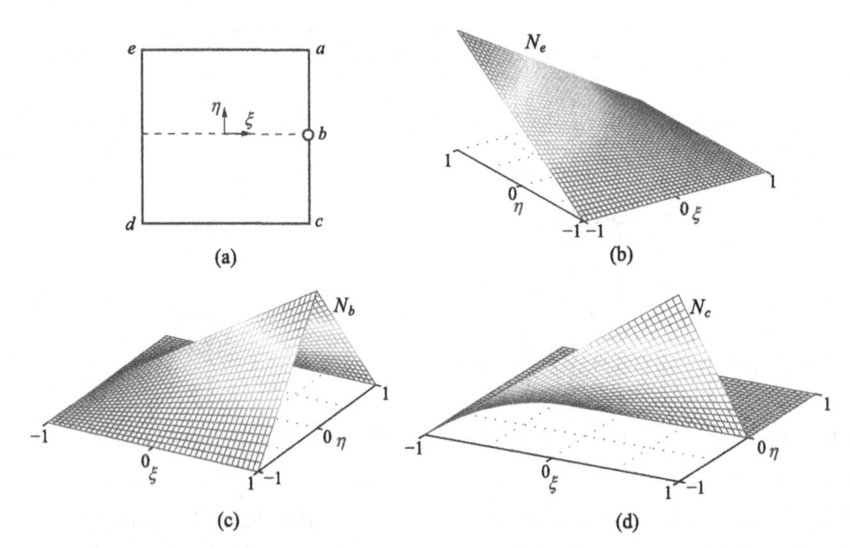

FIGURE 3.28 A 5-node cell with one hanging node. (a) The cell is divided into two subcells by the *dashed line* passing through node b, (b) shape function of node e, (c) shape function of node b, and (d) shape function of node c [85,86].

where $\xi_I = \pm 1$, $\eta_I = \pm 1$ are the natural coordinates of node I. The 4-node cell shape functions $N_d^L(\xi, \eta)$ and $N_e^L(\xi, \eta)$ already satisfy the requirement (3.154) such that they can be used to construct the corresponding 5-node cell shape functions, i.e.,

$$N_I(\xi, \eta) = N_I^L(\xi, \eta), \quad I = d, e.$$

However, the 4-node shape functions equal $1/2$ at node b so that the above functions must be modified before they can be taken as the corresponding 5-node shape functions, namely,

$$N_a(\xi, \eta) = N_a^L(\xi, \eta) - \frac{1}{2} N_b(\xi, \eta), \tag{3.155}$$

$$N_c(\xi, \eta) = N_c^L(\xi, \eta) - \frac{1}{2} N_b(\xi, \eta). \tag{3.156}$$

Figs. 3.28(b)–(d) plot the 5-node shape functions of nodes e, b, and c, respectively.

Note that the particle size should also conform to the cell size such that refined particles should be used in refined cells. For example, the equivalent length of particles in level 0 cells as shown in Fig. 3.29(a) should be 2 times of that in level 1 cells. When a particle in level 0 cells moves into a level 1 cell, it should be split into 4 particles in 2D and 8 particles in 3D, as shown in Fig. 3.29(b). Please refer to Sect. 3.5.1 for the particle splitting scheme.

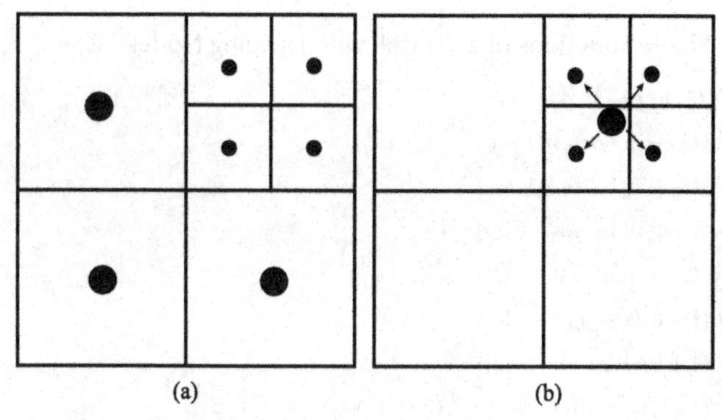

FIGURE 3.29 Two-level grid.

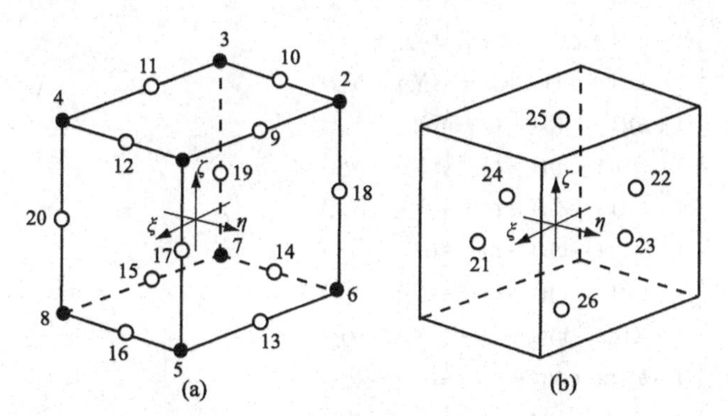

FIGURE 3.30 8- to 26-node linear isoparametric cells.

In three-dimensional cases, a multi-level grid may have 8- to 26-node cells, as shown in Fig. 3.30. In a 3D cell, node 1 to node 8 are mandatory, but other nodes are optional. The shape function of each node is listed in Table 3.2 [87].

3.5.2.5 Adaptive Grid

The above-mentioned multi-level grid technique needs users to identify the refined zone and generate the multi-level grid before starting a simulation. Yang et al. further developed an adaptive grid technique [85] based on the multi-level grid. At each time step, the cells with a sharp gradient of solution are refined automatically. For example, the cell 2 of level 0 is divided into 4 cells of level 1 when its solution gradient reaches a prescribed value. The adaptive grid technique can divide a cell into several levels, and would result in hanging nodes in all the cells except for the level 0 cells (see Fig. 3.31). The shape functions for cells with hanging nodes are given in Sect. 3.5.2.4.

The strain gradient or energy norm can be used to identify the cells which need to be refined. The energy norm criterion calculates the strain energy norm

TABLE 3.2 Shape Functions of a 3D Cell with Hanging Nodes

Node	$N(\xi,\eta,\zeta)$
26	$\frac{1}{2}(1-\lvert\xi\rvert)(1-\lvert\eta\rvert)(1-\zeta)$
25	$\frac{1}{2}(1-\lvert\xi\rvert)(1-\lvert\eta\rvert)(1+\zeta)$
24	$\frac{1}{2}(1-\lvert\xi\rvert)(1-\eta)(1-\lvert\zeta\rvert)$
23	$\frac{1}{2}(1-\lvert\xi\rvert)(1+\eta)(1-\lvert\zeta\rvert)$
22	$\frac{1}{2}(1-\xi)(1-\lvert\eta\rvert)(1-\lvert\zeta\rvert)$
21	$\frac{1}{2}(1+\xi)(1-\lvert\eta\rvert)(1-\lvert\zeta\rvert)$
20	$\frac{1}{4}(1+\xi)(1-\eta)(1-\lvert\zeta\rvert)-\frac{1}{2}(N_{21}+N_{24})$
19	$\frac{1}{4}(1-\xi)(1-\eta)(1-\lvert\zeta\rvert)-\frac{1}{2}(N_{22}+N_{24})$
18	$\frac{1}{4}(1-\xi)(1+\eta)(1-\lvert\zeta\rvert)-\frac{1}{2}(N_{22}+N_{23})$
17	$\frac{1}{4}(1+\xi)(1+\eta)(1-\lvert\zeta\rvert)-\frac{1}{2}(N_{21}+N_{23})$
16	$\frac{1}{4}(1+\xi)(1-\lvert\eta\rvert)(1-\zeta)-\frac{1}{2}(N_{21}+N_{26})$
15	$\frac{1}{4}(1-\lvert\xi\rvert)(1-\eta)(1-\zeta)-\frac{1}{2}(N_{24}+N_{26})$
14	$\frac{1}{4}(1-\xi)(1-\lvert\eta\rvert)(1-\zeta)-\frac{1}{2}(N_{22}+N_{26})$
13	$\frac{1}{4}(1-\lvert\xi\rvert)(1+\eta)(1-\zeta)-\frac{1}{2}(N_{23}+N_{26})$
12	$\frac{1}{4}(1+\xi)(1-\lvert\eta\rvert)(1+\zeta)-\frac{1}{2}(N_{21}+N_{25})$
11	$\frac{1}{4}(1-\lvert\xi\rvert)(1-\eta)(1+\zeta)-\frac{1}{2}(N_{24}+N_{25})$
10	$\frac{1}{4}(1-\xi)(1-\lvert\eta\rvert)(1+\zeta)-\frac{1}{2}(N_{22}+N_{25})$
9	$\frac{1}{4}(1-\lvert\xi\rvert)(1+\eta)(1+\zeta)-\frac{1}{2}(N_{23}+N_{25})$
8	$\frac{1}{8}(1+\xi)(1-\eta)(1-\zeta)-\frac{1}{2}(N_{15}+N_{16}+N_{20})-\frac{1}{4}(N_{21}+N_{24}+N_{26})$
7	$\frac{1}{8}(1-\xi)(1-\eta)(1-\zeta)-\frac{1}{2}(N_{14}+N_{15}+N_{19})-\frac{1}{4}(N_{22}+N_{24}+N_{26})$
6	$\frac{1}{8}(1-\xi)(1+\eta)(1-\zeta)-\frac{1}{2}(N_{13}+N_{14}+N_{18})-\frac{1}{4}(N_{22}+N_{23}+N_{26})$
5	$\frac{1}{8}(1+\xi)(1+\eta)(1-\zeta)-\frac{1}{2}(N_{13}+N_{16}+N_{17})-\frac{1}{4}(N_{21}+N_{23}+N_{26})$
4	$\frac{1}{8}(1+\xi)(1-\eta)(1+\zeta)-\frac{1}{2}(N_{11}+N_{12}+N_{20})-\frac{1}{4}(N_{21}+N_{24}+N_{25})$
3	$\frac{1}{8}(1-\xi)(1-\eta)(1+\zeta)-\frac{1}{2}(N_{10}+N_{11}+N_{19})-\frac{1}{4}(N_{22}+N_{24}+N_{25})$
2	$\frac{1}{8}(1-\xi)(1+\eta)(1+\zeta)-\frac{1}{2}(N_{9}+N_{10}+N_{18})-\frac{1}{4}(N_{22}+N_{23}+N_{25})$
1	$\frac{1}{8}(1+\xi)(1+\eta)(1+\zeta)-\frac{1}{2}(N_{9}+N_{12}+N_{17})-\frac{1}{4}(N_{21}+N_{23}+N_{25})$

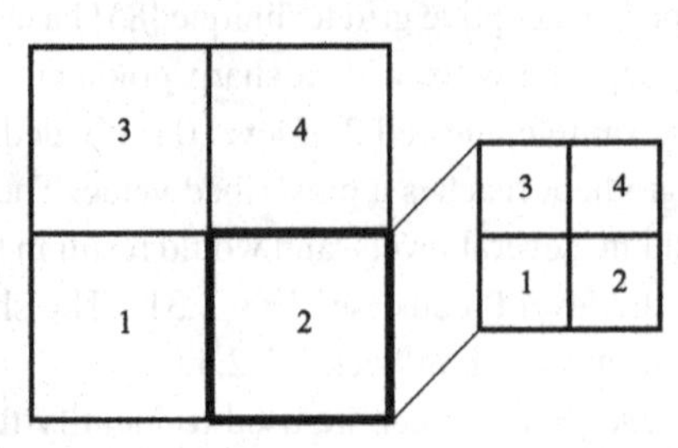

FIGURE 3.31 Two-dimensional adaptive grid.

of each cell at each time step. When the ratio of strain energy norm of a cell to strain energy norm of the system reaches a prescribed threshold, the cell is divided. The current strain energy norm of the system is given by

$$\|U\| = \left(\frac{1}{2} \int_{\Omega} \boldsymbol{\sigma}^T \boldsymbol{D}^{-1} \boldsymbol{\sigma} \mathrm{d}\Omega \right)^{\frac{1}{2}}. \tag{3.157}$$

The current strain energy norm of cell i is defined by

$$\|E_i\| = \left(\frac{1}{2} \int_{\Omega_i} \boldsymbol{\sigma}^T \boldsymbol{D}^{-1} \boldsymbol{\sigma} \mathrm{d}\Omega \right)^{\frac{1}{2}}. \tag{3.158}$$

Thus, the ratio of energy norm of cell i to that of the system can be obtained as

$$\eta_i = \frac{n \|E_i\|}{\|U\|} \tag{3.159}$$

where n is the number of cells that are occupied by any particles. When $\eta_i \geqslant$ TOL (where TOL is a prescribed threshold), cell i is divided into 4 in 2D or 8 subcells in 3D. All the particles in the cell will also be split to match the subcells.

3.6 NON-REFLECTING BOUNDARY

In the cases such as wave propagation through geologic media, the solution domain is a semi-infinite one, but numerical solutions are usually needed only in a finite domain. Isolating a finite domain from the semi-infinite domain results in an artificial boundary that does not exist in the original semi-infinite domain. To eliminate the wave reflection from the artificial boundary, some kind of non-reflecting boundary conditions must be applied on the artificial boundary. With the use of viscous damping forces along the artificial boundary, Lysemer and Kuhlemeyer proposed an approach [88] that can effectively absorb the radiant energy. The normal and tangential viscous damping tractions are given by

$$\sigma_n = \boldsymbol{n} \cdot \boldsymbol{\sigma} \cdot \boldsymbol{n} = -\rho c_d \boldsymbol{v} \cdot \boldsymbol{n} = -\rho c_d v_n, \tag{3.160}$$

$$\boldsymbol{\tau} = (\boldsymbol{\sigma} \cdot \boldsymbol{n} - \sigma_n \boldsymbol{n}) = -\rho c_s (\boldsymbol{v} - v_n \boldsymbol{n}) = -\rho c_s \boldsymbol{v}_\tau \tag{3.161}$$

where $\boldsymbol{n}$ is the unit normal of the artificial boundary, c_d is the velocity of dilatational wave, and c_s is the velocity of shear wave. For an isotropic elastic

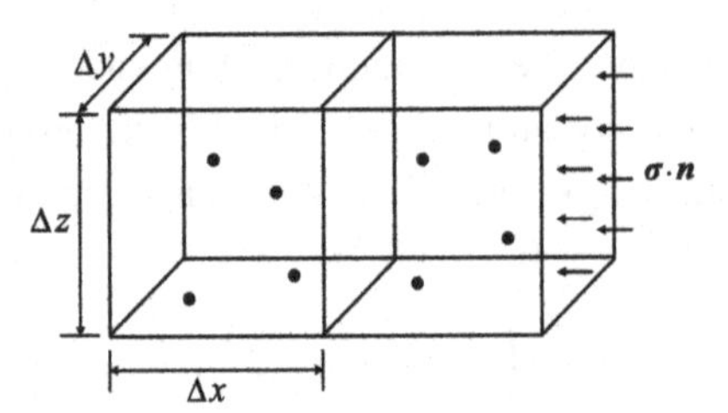

FIGURE 3.32 Converting tractions into equivalent specific forces.

material, it follows that

$$c_d = \sqrt{\frac{\lambda + 2G}{\rho}}, \tag{3.162}$$

$$c_s = \sqrt{\frac{G}{\rho}} \tag{3.163}$$

where λ and G are the Lame constants. This approach is not only relatively easy to be implemented, but also quite reasonable in treating both dilatational and shear waves in many applications. As a result, it has been widely used, e.g., in LS-DYNA [14] and ABAQUS [89].

Shen et al. [90] introduced the viscous damping traction (3.160) and (3.161) into the MPM by converting the viscous damping traction applied on the artificial boundary into the equivalent specific force applied on the boundary cells according to the Saint-Venant's principle, as shown in Fig. 3.32, i.e.,

$$\rho \cdot \Delta x \cdot \Delta y \cdot \Delta z \cdot \boldsymbol{b} = \boldsymbol{\sigma} \cdot \boldsymbol{n} \cdot \Delta y \cdot \Delta z. \tag{3.164}$$

The equivalent specific force can then be obtained from Eq. (3.164) as

$$\boldsymbol{b} = \frac{\boldsymbol{\sigma} \cdot \boldsymbol{n}}{\rho \cdot \Delta x}. \tag{3.165}$$

The equivalent specific force (3.165) will be added into Eq. (3.54).

3.7 INCOMPRESSIBLE MATERIAL POINT METHOD

The MPM has also been applied in fluid mechanics to solve the compressible gas dynamics [81,91–96] and weakly compressible flow [37,97,98]. When investigating the dynamic behavior of sloshing liquids in a moving container, Li et al. [97] proposed a weakly compressible material point method (WCMPM) by employing a nearly incompressible EOS $p = c^2\rho$ in the MPM to compute the particle pressure. The artificial sound speed c is normally taken as 10 times higher than the maximum fluid velocity in order to reduce the density fluctuation down to 1% [99]. As a result, the critical time step size of the explicit time integration is very small and the computational cost is considerably high. A small variation in density ρ will lead to a significant change in pressure p such that the pressure field obtained by the WCMPM exhibits significant oscillations.

To overcome the shortcomings of the WCMPM in solving the free surface flow problems, Zhang et al. proposed an incompressible material point method (iMPM) [100] based on an operator splitting scheme. The momentum equation is split into two sequentially solved subequations connected via the initial conditions.

3.7.1 Momentum Equation of Fluid

In fluid mechanics, the pressure and deviatoric stress are updated independently. Decomposing the stress $\boldsymbol{\sigma}$ into the sum of deviatoric stress $\boldsymbol{s}$ and hydrostatic pressure p, i.e., $\boldsymbol{\sigma} = -p\mathbf{1} + \boldsymbol{s}$, Eq. (2.56) can be written as

$$\rho \dot{\boldsymbol{v}} = -\nabla p + \nabla \cdot \boldsymbol{s} + \rho \boldsymbol{b} \tag{3.166}$$

where $\mathbf{1}$ is the second-order identity tensor.

For fully incompressible fluid, the velocity field must satisfy the divergence-free condition

$$\nabla \cdot \boldsymbol{v} = 0. \tag{3.167}$$

3.7.2 Operator Splitting

Using the operator splitting method, the momentum equation (3.166) is split into the following two sequentially solved subequations connected via the initial conditions:

$$\rho \dot{\boldsymbol{v}}^* = \nabla \cdot \boldsymbol{s} + \rho \boldsymbol{b} \quad \text{with } t \in [t^n, t^{n+1}] \text{ and } \boldsymbol{v}^*(t^n) = \boldsymbol{v}^n, \tag{3.168}$$

$$\rho \dot{\boldsymbol{v}} = -\nabla p \quad \text{with } t \in [t^n, t^{n+1}] \text{ and } \boldsymbol{v}(t^n) = \boldsymbol{v}^{*(n+1)}. \tag{3.169}$$

Eq. (3.168) is almost identical to Eq. (2.56) so that it can be solved by using the explicit MPM as presented in Sect. 3.2.2. Thus, the intermediate grid nodal velocity $v_{iI}^{*(n+1)}$ can be obtained as

$$v_{iI}^{*(n+1)} = v_{iI}^n + \frac{\Delta t}{m_I}(f_{iI}^{\text{ext},n} + f_{iI}^{\text{int,s},n}) \tag{3.170}$$

where the external grid nodal force $f_{iI}^{\text{ext},n}$ is given by Eq. (3.54), and

$$f_{iI}^{\text{int,s},n} = -\sum_{p=1}^{n_p} N_{Ip,j}^{n} s_{ijp} \frac{m_p}{\rho_p} \tag{3.171}$$

is the internal grid nodal force contributed by the deviatoric stress only.

Note that the critical time step size depends only on the speed of shear wave when integrating Eq. (3.168) explicitly. The speed of compressive wave in a fluid is usually several orders of magnitude higher than that of the shear wave. Thus, the time step size in the iMPM can be much larger than that in the WCMPM.

Solving the second subequation (3.169) results in the final solution

$$\boldsymbol{v}^{n+1} = \boldsymbol{v}^{*(n+1)} - \frac{\Delta t}{\rho}\nabla p^{n+1}, \quad \boldsymbol{v}^{n+1}\Big|_{\Gamma_u} = \bar{\boldsymbol{v}} \tag{3.172}$$

which must satisfy the divergence-free condition (3.167).

3.7.3 Pressure Poisson Equations

Substituting Eq. (3.172) into the divergence-free condition (3.167) results in the pressure Poisson equation as

$$\nabla \cdot \boldsymbol{v}^* = \frac{\Delta t}{\rho}\nabla^2 p^{n+1}. \tag{3.173}$$

In the iMPM, the velocity is interpolated from the grid nodal velocities, while the pressure is approximated as a constant in each cell. The pressure Poisson equation (3.173) is solved approximately by collocating at cell centers.

The divergence of the intermediate velocity field $\boldsymbol{v}^*$ at cell center (i, j, k) can be approximated by using Eq. (3.5) as

$$\nabla \cdot \boldsymbol{v}^*(\boldsymbol{x}_{i,j,k}) = \sum_I N_{I,i}(\boldsymbol{x}_{i,j,k}) u_{iI}^*. \tag{3.174}$$

The second-order derivative of pressure p with respect to x can be approximated at the cell center (i, j, k) by using the central difference method as

$$\left(\frac{\mathrm{d}^2 p}{\mathrm{d}x^2}\right)_{i,j,k} = \frac{p_{i+1,j,k} + p_{i-1,j,k} - 2p_{i,j,k}}{\Delta x^2} \tag{3.175}$$

where Δx is the length of side x in the cell. If cubic cells are used, i.e., $\Delta x = \Delta y = \Delta z = h$, the Laplacian $\nabla^2 p$ can be approximated at the cell center (i, j, k)

by using the seven-point stencil finite difference method as follows:

$$\left(\nabla^2 p\right)_{i,j,k} = \frac{p_{i+1,j,k} + p_{i-1,j,k} + p_{i,j+1,k} + p_{i,j-1,k} + p_{i,j,k+1} + p_{i,j,k-1} - 6p_{i,j,k}}{h^2}. \tag{3.176}$$

Substituting Eqs. (3.176) and (3.174) into Eq. (3.173) leads to a large system of linear equations as

$$\boldsymbol{A}\boldsymbol{p} = \boldsymbol{b} \tag{3.177}$$

where $\boldsymbol{A}$ is the coefficient matrix, $\boldsymbol{p}$ is a vector consisting of pressure at all cell centers, and $\boldsymbol{b}$ is a vector consisting of the negative divergences of the intermediate velocity at each cell center. This symmetric positive semi-definite linear systems can be solved efficiently by the preconditioned conjugate gradient (PCG) solver.

3.7.4 Pressure Boundary Conditions

When solving the pressure Poisson equation (3.173), two kinds of boundary conditions need to be imposed. The pressure is prescribed on the free surface, while the zero normal pressure gradient is imposed on the solid boundary to guarantee the normal velocity to be continuous on the interface between the solid and fluid.

The prescribed pressure boundary condition can be imposed on the free surface using the ghost fluid method [101–103]. Consider the two-dimensional Poisson equation as an example. If the free surface is located between the centers of fluid cell (i, j) and air cell $(i+1, j)$, a ghost pressure $p_{i+1,j}^{\mathrm{G}x}$ is defined at the center of cell $(i+1, j)$ so that

$$\left(\frac{\partial^2 p}{\partial x^2}\right)_{i,j} = \frac{p_{i+1,j}^{\mathrm{G}x} + p_{i-1,j}^{\mathrm{f}} - 2p_{i,j}^{\mathrm{f}}}{\Delta x^2} \tag{3.178}$$

where

$$p_{i+1,j}^{\mathrm{G}x} = \frac{p^{\mathrm{fs}} + (\theta - 1)p_{i,j}^{\mathrm{f}}}{\theta}$$

is the ghost pressure of cell (i, j) in the x-direction, p^{fs} is the surface pressure, and $\theta = |(x_{\mathrm{fs}} - x_i)/\Delta x|$. The surface pressure p^{fs} is equal to the air pressure p_{air} or $p_{\mathrm{air}} + \sigma_\kappa^{\mathrm{fs}}$ if the surface tension is taken into consideration. Substituting

$p^{Gx}_{i+1,j}$ into Eq. (3.178) gives

$$\left(\frac{\partial^2 p}{\partial x^2}\right)_{i,j} = \frac{\frac{1}{\theta} p^{fs} + p^{f}_{i-1,j} - (1 + \frac{1}{\theta}) p^{f}_{i,j}}{\Delta x^2}. \tag{3.179}$$

The material surface is not explicitly tracked in the MPM so that it is difficult to accurately impose the pressure boundary conditions on the free surface. In the iMPM, a level set function which represents the signed distance to free surface is used to track the free surface and apply the pressure boundary conditions (please refer to Zhang et al. [100] for details).

3.7.5 Velocity Update

After obtaining the pressure p^{n+1} by solving Eq. (3.177), the final velocity $\boldsymbol{v}^{n+1}$ can be updated using Eq. (3.172). Note that the pressure gradient ∇p is required to update the velocity $\boldsymbol{v}^{n+1}$ using Eq. (3.172). Because the velocity $\boldsymbol{v}^{n+1}$ is updated at grid nodes, the pressure gradient ∇p should also be evaluated at grid nodes. If all cells connected to a grid node are fluid cells, ∇p can be evaluated at the grid node $(i + \frac{1}{2}, j + \frac{1}{2})$ as

$$\begin{aligned} \left(\frac{\partial p}{\partial x}\right)_{i+\frac{1}{2},j+\frac{1}{2}} &= \frac{1}{2}\left[\left(\frac{\partial p}{\partial x}\right)_{i+\frac{1}{2},j} + \left(\frac{\partial p}{\partial x}\right)_{i+\frac{1}{2},j+1}\right], \\ \left(\frac{\partial p}{\partial y}\right)_{i+\frac{1}{2},j+\frac{1}{2}} &= \frac{1}{2}\left[\left(\frac{\partial p}{\partial y}\right)_{i,j+\frac{1}{2}} + \left(\frac{\partial p}{\partial y}\right)_{i+1,j+\frac{1}{2}}\right] \end{aligned} \tag{3.180}$$

where

$$\begin{aligned} \left(\frac{\partial p}{\partial x}\right)_{i+\frac{1}{2},j} &= \frac{p_{i+1,j} - p_{i,j}}{\Delta x}, \\ \left(\frac{\partial p}{\partial x}\right)_{i+\frac{1}{2},j+1} &= \frac{p_{i+1,j+1} - p_{i,j+1}}{\Delta x}, \\ \left(\frac{\partial p}{\partial y}\right)_{i,j+\frac{1}{2}} &= \frac{p_{i,j+1} - p_{i,j}}{\Delta y}, \\ \left(\frac{\partial p}{\partial y}\right)_{i+1,j+\frac{1}{2}} &= \frac{p_{i+1,j+1} - p_{i+1,j}}{\Delta y} \end{aligned}$$

are the pressure gradient components evaluated at face centers. Note that $(\partial p/\partial x)_{i+\frac{1}{2},j+\frac{1}{2}}$ is the average of $\partial p/\partial x$ at face centers $(i + 1/2, j + 1)$ and $(i + 1/2, j)$, and $(\partial p/\partial y)_{i+\frac{1}{2},j+\frac{1}{2}}$ is the average of $\partial p/\partial y$ at face centers $(i, j + 1/2)$ and $(i + 1, j + 1/2)$.

If a grid node is next to the free surface, the pressure $p_{i+1,j+1}$ in Eq. (3.180) should be replaced with a ghost pressure to impose the pressure boundary condition on the free surface. That is to say, $(\partial p/\partial x)_{i+1/2,j+1}$ is calculated from the pressure $p_{i,j+1}$ and ghost pressure $p^{\mathrm{G}x}_{i+1,j+1}$, and $(\partial p/\partial y)_{i+1,j+1/2}$ is calculated from the pressure $p_{i+1,j}$ and ghost pressure $p^{\mathrm{G}y}_{i+1,j+1}$. The ghost pressure $p^{\mathrm{G}x}_{i+1,j+1}$ is extrapolated from $p_{i,j+1}$ and the free surface pressure p^{fs}_{j+1} in the x-direction, while $p^{\mathrm{G}y}_{i+1,j+1}$ is extrapolated from $p_{i+1,j}$ and the free surface pressure p^{fs}_{i+1} in the y-direction. Please refer to Zhang et al. [100] for details.

Because the Poisson equations are discretized at cell centers, the divergence of velocity is also calculated at cell centers. The velocity modes illustrated in Fig. 5.2 are divergence-free at cell centers so that they do not contribute to the Poisson equations. In other words, these spurious velocity modes are not resisted, and will lead to spurious oscillation in the velocity field. These modes are the same as the hourglass modes in the hexahedron element with a single-point Gauss Quadrature in the FEM, and can be suppressed in the same way as that used in the FEM [100].

Compared with the WCMPM, the iMPM is much more efficient, accurate, and stable in simulating the free surface flow problems. Fig. 3.33 compares a sequence of snapshots of a dam breaking simulation obtained by the WCMPM and iMPM [100]. Although the free surface profiles obtained by both methods are very similar until the second plunging wave impacts on the right wall, the pressure distributions in the fluid domain are quite different. At the beginning, the hydrostatic pressure obtained by both methods is linearly distributed along the vertical direction after flap is removed. However, due to the weakly compressible EOS and crossing-cell noise, the pressure obtained by the WCMPM soon exhibits high-frequency oscillations when water flow develops along the deck. Furthermore, there is a lot of unphysical spray and splash in the WCMPM results at $T = 6.969$. Conversely, the pressure distribution obtained by the iMPM is smooth during the whole process, and the free surface profiles are reasonable.

3.8 IMPLICIT MATERIAL POINT METHOD

An explicit time integration method does not solve a system of simultaneous equations, which results in a very efficient step-by-step solution scheme for large-scale computer simulations. However, the explicit scheme is conditionally stable, whose time step size must be very small, due to the stability requirement, in order to resolve all the wave components in transient analysis. On the other hand, an implicit method is unconditionally stable, whose time step size can be 10^2–10^4 times of that of the explicit method. Therefore, for many problems in

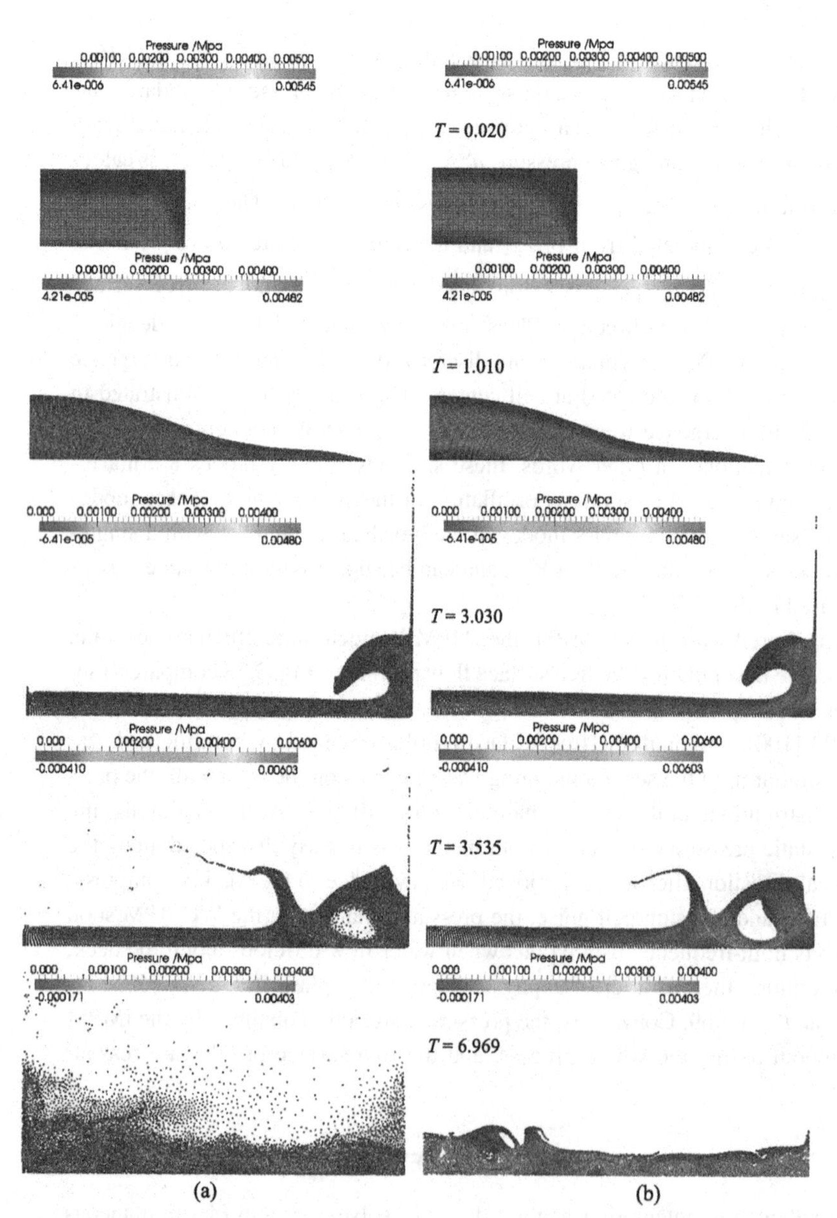

FIGURE 3.33 Dam break configurations obtained by (a) WCMPM and (b) iMPM.

which only the low-frequency motion is of interest, an implicit method may be employed to considerably reduce the computational cost. Research efforts have been made to develop the implicit MPM [104–106] so that only the essential features and numerical implementation are discussed in this section.

3.8.1 Implicit Time Integration

Newmark method is a commonly used implicit integration scheme, in which the velocity and displacement at time t^{n+1} are solved with the use of

$$\dot{u}_{iI}^{n+1} = \dot{u}_{iI}^{n} + \Delta t[(1-\gamma)\ddot{u}_{iI}^{n} + \gamma \ddot{u}_{iI}^{n+1}], \quad 0 \leqslant \gamma \leqslant 1, \tag{3.181}$$

$$u_{iI}^{n+1} = u_{iI}^{n} + \Delta t \dot{u}_{iI}^{n} + \frac{\Delta t^2}{2}[(1-2\beta)\ddot{u}_{iI}^{n} + 2\beta \ddot{u}_{iI}^{n+1}], \quad 0 \leqslant 2\beta \leqslant 1 \tag{3.182}$$

where $\Delta t = t^{n+1} - t^n$. Based on Eq. (3.182), the acceleration at time t^{n+1} can be obtained as

$$\ddot{u}_{iI}^{n+1} = \frac{1}{\beta \Delta t^2}\left(u_{iI}^{n+1} - \tilde{u}_{iI}^{n}\right) \tag{3.183}$$

where

$$\tilde{u}_{iI}^{n} = u_{iI}^{n} + \Delta t \dot{u}_{iI}^{n} + \frac{\Delta t^2}{2}(1-2\beta)\ddot{u}_{iI}^{n}. \tag{3.184}$$

Substituting Eq. (3.183) into the nodal equations of motion (3.8) at time t^{n+1} gives a system of simultaneous equations in terms of the grid nodal displacements u_{iI}^{n+1} at time t^{n+1}, namely,

$$\boldsymbol{r}(\boldsymbol{a}^{n+1}, t^{n+1}) = \frac{1}{\beta \Delta t^2}\boldsymbol{M}\left(\boldsymbol{a}^{n+1} - \tilde{\boldsymbol{a}}^{n}\right) - \boldsymbol{f}(\boldsymbol{a}^{n+1}, t^{n+1}) = 0 \tag{3.185}$$

where the vector $\boldsymbol{a}^{n+1} = [\, u_{11}^{n+1} \quad u_{21}^{n+1} \quad u_{31}^{n+1} \quad \cdots \quad u_{1n_g}^{n+1} \quad u_{2n_g}^{n+1} \quad u_{3n_g}^{n+1} \,]^{\mathrm{T}}$ contains all the displacement components at all the nodes related to the spatial discretization, and has $3n_g$ components in 3D cases. $\boldsymbol{M}$ is the consistent mass matrix, and $\boldsymbol{f}(\boldsymbol{a}^{n+1}, t^{n+1}) = \boldsymbol{f}^{\mathrm{ext}}(\boldsymbol{a}^{n+1}, t^{n+1}) - \boldsymbol{f}^{\mathrm{int}}(\boldsymbol{a}^{n+1}, t^{n+1})$. By neglecting the inertial term, Eq. (3.185) is reduced to the static equilibrium equation as follows:

$$\boldsymbol{r}(\boldsymbol{a}^{n+1}) = \boldsymbol{f}(\boldsymbol{a}^{n+1}) = 0. \tag{3.186}$$

3.8.2 Solution of a System of Nonlinear Equations

The system of simultaneous equations (3.185) is nonlinear, in general, and can be solved using Newton's method (also known as Newton–Raphson method). Let subscript (k) denote the iteration step, $\boldsymbol{a}_{(k)}^{n+1}$ denote the grid nodal displacement vector in the kth iteration step, and $\boldsymbol{a}_{(k)} \equiv \boldsymbol{a}_{(k)}^{n+1}$. At the beginning of the iteration loop, the first guess $\boldsymbol{a}_{(0)}$ for a root of the function $\boldsymbol{r}(\boldsymbol{a}^{n+1}, t^{n+1})$ is required, which can usually be chosen as the solution at time step t^n, i.e.,

$\boldsymbol{a}_{(0)} \equiv \boldsymbol{a}^n$. In the dynamic simulation with Newmark method, the estimation $\tilde{\boldsymbol{a}}^n$ given in Eq. (3.184) can be taken as the first guess $\boldsymbol{a}_{(0)}$.

The solution of the kth iteration, $\boldsymbol{a}_{(k)}^{n+1}$, might not satisfy the nonlinear equation (3.185), i.e.,

$$\boldsymbol{r}(\boldsymbol{a}_{(k)}, t^{n+1}) = \frac{1}{\beta \Delta t^2} \boldsymbol{M} \left(\boldsymbol{a}_{(k)} - \tilde{\boldsymbol{a}}^n \right) - \boldsymbol{f}(\boldsymbol{a}_{(k)}, t^{n+1}) \neq 0. \tag{3.187}$$

To obtain a better solution $\boldsymbol{a}_{(k+1)}$ of Eq. (3.185), which approaches $\boldsymbol{r}(\boldsymbol{a}_{(k+1)}, t^{n+1}) = 0$, let

$$\boldsymbol{a}^{n+1} = \boldsymbol{a}_{(k+1)} = \boldsymbol{a}_{(k)} + \Delta \boldsymbol{a}_{(k)}. \tag{3.188}$$

Substituting Eq. (3.188) into Eq. (3.185) and expanding it in a Taylor series around the current value $\boldsymbol{a}_{(k)}$ of $\boldsymbol{a}^{n+1}$ yields

$$\boldsymbol{r}(\boldsymbol{a}_{(k+1)}, t^{n+1}) = \boldsymbol{r}(\boldsymbol{a}_{(k)}, t^{n+1}) + \frac{\partial \boldsymbol{r}(\boldsymbol{a}_{(k)}, t^{n+1})}{\partial \boldsymbol{a}} \Delta \boldsymbol{a}_{(k)} + O(\Delta \boldsymbol{a}_{(k)}^2) = 0. \tag{3.189}$$

Ignoring higher-order terms in Eq. (3.189) produces a system of linear equations in $\Delta \boldsymbol{a}_{(k)}$ as follows:

$$\boldsymbol{K}^{\text{eff}} \Delta \boldsymbol{a}_{(k)} = -\boldsymbol{r}(\boldsymbol{a}_{(k)}, t^{n+1}) \tag{3.190}$$

where the Jacobian

$$\boldsymbol{K}^{\text{eff}} = \frac{\partial \boldsymbol{r}(\boldsymbol{a}_{(k)}, t^{n+1})}{\partial \boldsymbol{a}} = \frac{1}{\beta \Delta t^2} \boldsymbol{M} + \boldsymbol{K}^{\text{int}} - \boldsymbol{K}^{\text{ext}} \tag{3.191}$$

is called the effective tangent stiffness matrix,

$$\boldsymbol{K}^{\text{int}} = \frac{\partial \boldsymbol{f}^{\text{int}}}{\partial \boldsymbol{a}} \tag{3.192}$$

is the Jacobian of grid nodal internal forces, also known as the tangent stiffness matrix, and

$$\boldsymbol{K}^{\text{ext}} = \frac{\partial \boldsymbol{f}^{\text{ext}}}{\partial \boldsymbol{a}} \tag{3.193}$$

is the Jacobian of grid nodal external forces.

Eq. (3.190) is often called a linear model of the nonlinear equations (3.185). The above process of obtaining a linear model is called **linearization**.

Three types of convergence criteria are usually used to control the iterations [107], namely,

1. The criterion based on the magnitude of the residual $\boldsymbol{r}$, i.e.,

$$\|\boldsymbol{r}\|_{L_2} = \left(\sum_{i=1}^{n_{\text{dof}}} r_i^2\right)^{\frac{1}{2}} \leqslant \varepsilon \max\left(\left\|\boldsymbol{f}^{\text{ext}}\right\|_{L_2}, \left\|\boldsymbol{f}^{\text{int}}\right\|_{L_2}, \|\boldsymbol{M}\ddot{\boldsymbol{a}}\|_{L_2}\right) \tag{3.194}$$

where n_{dof} is the number of degrees of freedom.

2. The criterion based on the magnitude of the displacement increments $\Delta \boldsymbol{a}_{(k)}$, i.e.,

$$\|\Delta \boldsymbol{a}\|_{L_2} = \left(\sum_{i=1}^{n_{\text{dof}}} \Delta a_i^2\right)^{\frac{1}{2}} \leqslant \varepsilon \, \|\boldsymbol{a}\|_{L_2}. \tag{3.195}$$

3. The energy error criterion, i.e.,

$$\left|\Delta \boldsymbol{a}^{\mathrm{T}} \boldsymbol{r}\right| \leqslant \varepsilon \max\left(W^{\text{ext}}, W^{\text{int}}, W^{\text{kin}}\right). \tag{3.196}$$

3.8.3 The Jacobian of Grid Nodal Internal Force

$\boldsymbol{K}^{\text{int}}$ can be directly obtained by taking the time derivative of the internal force vector in the current configuration. Another way to obtain the tangent stiffness matrix is to take the time derivative in the original configuration and then transform $\boldsymbol{K}^{\text{int}}$ into the current configuration, which could be much easier. Hence, the second way is chosen here to find the time derivative. The internal force vector f_{iI}^{int} described in the original configuration is given by

$$f_{iI}^{\text{int}} = \sum_p V_p^0 \frac{\partial N_{Ip}}{\partial X_k} F_{ilp} S_{klp} \tag{3.197}$$

so that the material derivative of internal force can be obtained by

$$\dot{f}_{iI}^{\text{int}} = \sum_p V_p^0 \frac{\partial N_{Ip}}{\partial X_k} \left(\dot{F}_{ilp} S_{klp} + F_{ilp} \dot{S}_{klp}\right) \tag{3.198}$$

where $\boldsymbol{F}_p$ and $\boldsymbol{S}_p$ signify the deformation gradient and PK2 stress tensors of material point p. Eq. (3.198) can be divided into two parts, material rate and geometric rate, as follows:

$$\dot{f}_{iI}^{\text{int}} = \dot{f}_{iI}^{\text{mat}} + \dot{f}_{iI}^{\text{geo}} \tag{3.199}$$

where

$$\dot{f}_{iI}^{\text{mat}} = \sum_p V_p^0 \frac{\partial N_{Ip}}{\partial X_k} F_{ilp} \dot{S}_{klp}, \tag{3.200}$$

$$\dot{f}_{iI}^{\text{geo}} = \sum_p V_p^0 \frac{\partial N_{Ip}}{\partial X_k} \dot{F}_{ilp} S_{klp}. \tag{3.201}$$

In order to simplify the deviation of the tangent stiffness matrix, rewrite the material rate of nodal internal force in Voigt format, i.e.,

$$\dot{\boldsymbol{f}}^{\text{mat}} = \sum_p V_p^0 \boldsymbol{B}_0^{\text{T}} \{\dot{\boldsymbol{S}}\} \tag{3.202}$$

where $\{\dot{\boldsymbol{S}}\}$ represents the PK2 stress rate in Voigt format. The rate-form of a constitutive model takes the form of

$$\{\dot{\boldsymbol{S}}\} = \left[\boldsymbol{C}^{SE}\right] \{\dot{\boldsymbol{E}}\} \tag{3.203}$$

in which $\left[\boldsymbol{C}^{SE}\right]$ is the constitutive tangent stiffness matrix related to the PK2 stress rate, and $\{\dot{\boldsymbol{E}}\} = \boldsymbol{B}_0 \dot{\boldsymbol{a}}$ is the Green strain rate. Substituting Eq. (3.203) into Eq. (3.202) yields

$$\dot{\boldsymbol{f}}^{\text{mat}} = \boldsymbol{K}^{\text{mat}} \dot{\boldsymbol{a}} \tag{3.204}$$

where

$$\boldsymbol{K}^{\text{mat}} = \sum_p V_p^0 \boldsymbol{B}_0^{\text{T}} \left[\boldsymbol{C}^{SE}\right] \boldsymbol{B}_0 \tag{3.205}$$

is the material tangent stiffness matrix. Eq. (3.205) can be rewritten in the current configuration as

$$\boldsymbol{K}^{\text{mat}} = \sum_p V_p^0 \boldsymbol{B}^{\text{T}} \left[\boldsymbol{C}^{\sigma T}\right] \boldsymbol{B} \tag{3.206}$$

where $\left[\boldsymbol{C}^{\sigma T}\right]$ is the constitutive tangent stiffness matrix related to the Truesdell stress rate. The Truesdell and Jaumann moduli are related by the 4th order tensor relationship as follows:

$$\boldsymbol{C}^{\sigma T} = \boldsymbol{C}^{\sigma J} - \boldsymbol{C}^{*} \tag{3.207}$$

with

$$C_{ijkl}^{*} = \frac{1}{2} \left(\delta_{ik}\sigma_{jl} + \delta_{il}\sigma_{jk} + \delta_{jk}\sigma_{il} + \delta_{jl}\sigma_{ik}\right) - \sigma_{ij}\delta_{kl}. \tag{3.208}$$

The geometric tangent stiffness matrix can be derived as follows:

$$
\begin{aligned}
\dot{f}_{iI}^{\text{geo}} &= \sum_p V_p^0 \frac{\partial N_{Ip}}{\partial X_k} \dot{F}_{ilp} S_{klp} \\
&= \sum_p V_p^0 \frac{\partial N_{Ip}}{\partial X_k} S_{klp} \frac{\partial N_{Jp}}{\partial X_l} \dot{u}_{iJ} \\
&= \sum_p V_p^0 \frac{\partial N_{Ip}}{\partial X_k} S_{klp} \frac{\partial N_{Jp}}{\partial X_l} \delta_{ir} \dot{u}_{rJ}.
\end{aligned} \tag{3.209}
$$

Eq. (3.209) can be reorganized in the matrix form, namely,

$$
\dot{\boldsymbol{f}}_I^{\text{geo}} = \boldsymbol{K}_{IJ}^{\text{geo}} \dot{\boldsymbol{a}}_J \tag{3.210}
$$

where

$$
\boldsymbol{K}_{IJ}^{\text{geo}} = \boldsymbol{I} \left(\sum_p V_p^0 \frac{\partial N_{Ip}}{\partial X_k} S_{klp} \frac{\partial N_{Jp}}{\partial X_l} \right) = \boldsymbol{I} H_{IJ}, \tag{3.211}
$$

and

$$
H_{IJ} = \sum_p V_p^0 \frac{\partial N_{Ip}}{\partial X_k} S_{klp} \frac{\partial N_{Jp}}{\partial X_l} \tag{3.212}
$$

is a scalar, and $\boldsymbol{I}$ is an identity matrix. The geometric tangent stiffness matrix can be rewritten in the current configuration as

$$
\boldsymbol{K}_{IJ}^{\text{geo}} = \boldsymbol{I} \left(\sum_p V_p \frac{\partial N_{Ip}}{\partial x_k} \sigma_{klp} \frac{\partial N_{Jp}}{\partial x_l} \right). \tag{3.213}
$$

Notice that

$$
\dot{\boldsymbol{f}}^{\text{int}} = \frac{\mathrm{d}\boldsymbol{f}^{\text{int}}}{\mathrm{d}t} = \frac{\partial \boldsymbol{f}^{\text{int}}}{\partial \boldsymbol{a}} \frac{\mathrm{d}\boldsymbol{a}}{\mathrm{d}t} = \frac{\partial \boldsymbol{f}^{\text{int}}}{\partial \boldsymbol{a}} \cdot \dot{\boldsymbol{a}}, \tag{3.214}
$$

so

$$
\frac{\partial \boldsymbol{f}^{\text{int}}}{\partial \boldsymbol{a}} = \boldsymbol{K}^{\text{int}} + \boldsymbol{K}^{\text{geo}}. \tag{3.215}
$$

3.8.4 Solution of a Linearized System of Equations

The linearized system of Eqs. (3.190) can be solved using a direct solver that requires calculating and factoring the Jacobian matrix $\boldsymbol{K}^{\text{eff}}$. To avoid the calculation and factorization of Jacobian, Eqs. (3.190) can also be solved in a matrix-free fashion by using a Krylov method [108], such as the conjugate gradient (CG) method and generalized minimum residual (GMRES) method. In a Krylov method, only the product of the Jacobian $\boldsymbol{K}^{\text{eff}} = \partial \boldsymbol{r}(\boldsymbol{a}_{(k)}, t^{n+1})/\partial \boldsymbol{a}$ and the conjugate vector or Krylov vector $\boldsymbol{p}$, $\boldsymbol{K}^{\text{eff}} \cdot \boldsymbol{p}$, needs to be evaluated, and the Jacobian matrix $\boldsymbol{K}^{\text{eff}}$ does not need to be formed or stored. Because the Jacobian matrix $\partial \boldsymbol{r}(\boldsymbol{a}_{(k)}, t^{n+1})/\partial \boldsymbol{a}$ is the gradient of residual $\boldsymbol{r}(\boldsymbol{a}_{(k)}, t^{n+1})$, the matrix–vector product $\boldsymbol{K}^{\text{eff}} \cdot \boldsymbol{p}$ is the directional derivative of $\boldsymbol{r}(\boldsymbol{a}_{(k)}, t^{n+1})$ in the direction of the vector $\boldsymbol{a}$, which can be found using a difference approximation as

$$\frac{\partial \boldsymbol{r}(\boldsymbol{a}_{(k)}, t^{n+1})}{\partial \boldsymbol{a}} \cdot \boldsymbol{p} \approx \frac{\boldsymbol{r}(\boldsymbol{a}_{(k)} + h\boldsymbol{p}, t^{n+1}) - \boldsymbol{r}(\boldsymbol{a}_{(k)}, t^{n+1})}{h} \tag{3.216}$$

where h is a perturbation parameter that affects the solution convergence considerably. A very large h could yield an inaccurate approximation to the local gradient, but a very small h may lead to severe local numerical noise.

To improve its overall efficiency, the Krylov method usually requires a preconditioner to reduce the condition number of the problem. There are many types of preconditioner [108], such as Jacobi (or diagonal) preconditioner, sparse approximate inverse preconditioner (SPAI), incomplete Cholesky factorization or incomplete LU factorization, successive over-relaxation (SOR) or symmetric successive over-relaxation (SSOR), and multigrid preconditioning.

Cummins et al. proposed an implicit MPM implementation for quasi-static granular problems [104]. They employed the implicit theta method in which the momentum equation (3.8) was discretized by

$$\boldsymbol{M}\dot{\boldsymbol{a}}^{n+1} = \boldsymbol{M}\dot{\boldsymbol{a}}^{n} + \Delta t[\theta \boldsymbol{f}(\boldsymbol{a}^{n+1}, t^{n+1}) + (1-\theta)\boldsymbol{f}(\boldsymbol{a}^{n}, t^{n})], \quad \frac{1}{2} \leqslant \theta \leqslant 1 \tag{3.217}$$

which was solved using the Jacobian-free Newton–Krylov method [109]. To improve the efficiency of the Krylov iterative method, the multigrid preconditioning [110] was used, but the results were somewhat disappointing because little improvement could be observed for the preconditioners with more than two levels [104].

Guilkey et al. proposed an implicit MPM implementation [105] using the Newmark method with Newton's type. The linearized system of equations was solved using a direct solver instead of an iterative solver. Compared with the implementation presented by Cummins et al., the procedure proposed by Guilkey et al. can employ much larger time steps due to the fact that their procedure employs a consistent tangent stiffness matrix that provides the best convergence behavior possible with Newton's methods [105].

Sulsky et al. also presented an implicit MPM implementation [106] in a matrix-free fashion. The momentum equation (3.8) was also discretized by the implicit theta method as Eq. (3.217), which was solved using the Newton's method combined with either the conjugate gradient method or the generalized minimum residual method. This implicit scheme for the MPM was extended to the GIMP method by Nair et al. [111].

For material nonlinear problems involving failure evolution, the implicit methods might not yield reasonable solutions when bifurcation occurs due to the use of a large time step. Hence, the explicit methods are more effective in simulating multiphase interactions under extreme loading conditions.

Chapter 4

Computer Implementation of the MPM

Contents

MPM3D is a 3D explicit and parallel MPM code developed since 2004 in the Computational Dynamics Laboratory led by Professor Xiong Zhang at Tsinghua University in China. The MPM3D can be used to simulate the transient responses of structures to extreme loadings such as impact and blast. For the sake of easy update and maintenance, the development of the MPM3D was shifted from FORTRAN to C++ in 2007. The simplified FORTRAN 90 version of the MPM3D, MPM3D-F90, has been made available as the companion open source code for the Chinese book about the MPM [6] since 2013, and can be downloaded from our website, http://mpm3d.comdyn.cn.

The C++ version of the MPM3D is still under development, which has implemented the USL, USF, and MUSL schemes of the MPM, GIMP, contact algorithm, adaptive scheme (with both the particle [81] and grid [74, 112,113]), explicit FEM (with bar element, hexahedron element, membrane element, shell element and beam element), hybrid finite element material point (HFEMP) method [39], coupled finite element material point (CFEMP) method [37], and adaptive finite element material point (AFEMP) method [41]. The material model library consists of elasticity, elasto-perfect plasticity, elasto-

The Material Point Method. http://dx.doi.org/10.1016/B978-0-12-407716-4.00004-1

plasticity with isotropic hardening and/or pressure-dependence, Johnson–Cook plasticity, high explosive, Newtonian fluid, Holmquist–Johnson–Cook concrete, RHT concrete, Taylor–Chen–Kuszmau concrete, Holmquist–Johnson ceramic, Deshpande–Fleck model and Mooney–Rivlin hyperelasticity, together with several equations of state (EOS) such as polynomial, JWL, Grüneisen, and P–α, as discussed in Chapter 6. To predict the evolution of failure, several types of failure models have also been implemented based on the maximum equivalent plastic strain, maximum hydrostatic pressure, maximum principal stress/shear stress, maximum principal strain/shear strain, and instantaneous geometric strain. The MPM3D has been parallelized using both OpenMP and MPI so that it can be run on both SMP (symmetric multi-processing) and MPP (massively parallel processing) architectures. The graphical user interface (GUI) of the MPM3D was developed using the cross-platform application and UI development framework Qt [114], the visualization toolkit VTK [115], and the cross-platform, open-source build system CMake [116]. The MPM3D can be run on Windows, Linux, or Mac OS X. The open-source, multi-platform data analysis and visualization code, ParaView [7], is used as its post processor.

This chapter uses the MPM3D-F90 as an example to describe the computer implementation of the MPM. The MPM3D-F90 has implemented the USF, USL, and MUSL versions of the MPM with the variable-step central difference method, as well as the GIMP and contact algorithm. The material model library consists of elasticity, perfect plasticity, linear isotropic-hardening plasticity, Johnson–Cook plasticity with failure and its simplified version, null material, high explosive, and Drucker–Prager model. The linear polynomial EOS, Mie–Grüneisen EOS and JWL EOS are also implemented to describe the relations among state variables of fluid, solid, and high explosive. The essential or natural boundary conditions can be applied on the six faces of the 3D background grid, whereas the external force and initial velocity are applied on the particles. The simulation results can be exported to TecPlot [117] and ParaView [7] for data analysis and visualization.

4.1 EXECUTION OF THE MPM3D-F90

The MPM3D-F90 can be executed from the command line (i.e., the Terminal in Linux or Mac OS X or the Command Prompt in Windows) by typing

```
mpm3d-f90 JobName
```

where "mpm3d-f90" is the base name of the MPM3D-F90 executable file, "JobName" is the base name of the input data file whose extension must be "mpm", namely, MPM3D-F90 will read the input data from the file "JobName.mpm". Under Linux or Mac OS X, the executable file "mpm3d-f90" can be generated by GNU FORTRAN compiler using the makefile provided in the subdirectory "make/gnu". Under Windows, the executable file "mpm3d-f90.exe" can be generated by Intel Visual Fortran compiler using the solution file mpm3d-f90.sln provided in the subdirectory "make/ivf".

The MPM3D-F90 creates the following output files:

1. JobName.out – Log file, which records the basic output message during the execution of the MPM3D-F90;
2. JobName_anim.dat – TecPlot [117] data format file storing the simulation results, which can be used by TecPlot for data analysis and visualization;
3. JobName_anim_nn.vtu – A series of ParaView VTK Unstructured Grid files that stores simulation results for given time steps, where nn is an integer number;
4. JobName_anim.pvd – ParaView data format file storing pointers to the series of ParaView VTK Unstructured Grid files;
5. JobName_curv.dat – Data file storing the time history of user specified variables, which can be used by Tecplot or Origin to plot the time history of these variables;
6. ContforcPlot.dat – Data file storing the time history of the total contact forces when the contact algorithm is activated;
7. EnergyPlot.dat – Data file storing the time history of the total energy, kinetic energy and internal energy; and
8. MomentumPlot.dat – Data file storing the time history of the total momentum.

4.2 INPUT DATA FILE FORMAT OF THE MPM3D-F90

4.2.1 Unit

The MPM3D-F90 does not have a built-in unit system so that it is the user's responsibility to ensure that the units chosen are self-consistent, i.e., the derived units can be expressed in terms of the fundamental units without conversion factors. The SI system "m–kg–s" and the system "mm–g–ms" are two examples of a self-consistent unit system. In a self-consistent system, the derived units are related to the fundamental units by

1 force unit = 1 mass unit * 1 acceleration unit,

1 acceleration unit = 1 length unit/(1 time unit)^2,

1 density unit = 1 mass unit/(1 length unit)^3.

4.2.2 Keywords

The input data file of the MPM3D-F90 is in a free format. In contrast to the fixed format data, the free format data are not arranged in columns/fields, but separated by SPACE, COMMA, or TAB. Thus, it is much easier for users to prepare a free format data file than a fixed format data file.

In the MPM3D-F90 input data file, each line contains maximum 256 characters and at most 15 data items each of which is composed of not more than 20 characters. Comments begin with the exclamation mark (!) which can start anywhere in a line and continue until the end of the line.

The MPM3D-F90 input data file is organized by a series of keywords. There are seven categories of keywords, namely, global information, material model, computational grid, domain discretization, loading, solution setup, and post processing. The MPM3D-F90 only recognizes the first 4 characters of and ignores the remaining of a keyword. For example, the keyword "particle" will be recognized as "part" by the MPM3D-F90. The keywords are case-insensitive.

In what follows, an underlined word represents a keyword whose parameters are represented by italic words. The data type of a parameter is described by a character placed in a pair of parentheses after the parameter, where 's' means a string, 'i' means an integer number, and 'r' means a real number.

The keywords marked with an asterisk are mandatory keywords, which must appear in an input data file. The remaining keywords are optional.

4.2.3 Global Information

There are six keywords in this category, five of which are located at the beginning of the input data file, and the remaining one is located at the end of the file.

1. mpm3* *title*(s) – Give the title of the problem, which is used to describe the problem to be solved.

2. nbco* *ComponentNumber*(i) – Set the total number of components, which equals to 2 if contact is on, and 1 if contact is off. In the current version, the contact between more than two components is not supported.

3. nbbo* *BodyNumber*(i) – Set the total number of bodies.

4. nbmp* *ParticleNumber*(i) – Set the total number of particles.

5. nmat* *MaterialNumber*(i) – Set the total number of material sets.

6. endi* – End of the input data file. This keyword must be the last keyword in an input data file.

4.2.4 Material Model

There are 3 keywords in this category. They are used to specify the artificial bulk viscosity, constitutive model, and equation of state, respectively.

1. bulk *Q1*(r) *Q2*(r)

Specify the artificial bulk viscosity (refer to Sect. 2.8.2) for solving shock physics problems. *Q1*(r) and *Q2*(r) are the coefficients of quadratic term and linear term, respectively.

2. mate

mid(i) *mtype*(s) ...

mid(i) *mtype*(s) ...

...

Specify the constitutive models that will be used in the simulation. The total number of material models defined here must be equal to the number specified by the keyword nmat. This keyword requires material ID *mid*, material type *mtype*, and other parameters which depend on the material type *mtype*. The material type *mtype* can be one of the following types: elas, pla1, pla2, john, sjc, sjcf, hiex, null, jcf, and dpm. The formats for different material models are listed as follows:

(a) elas ρ(r) E(r) ν(r) – Elasticity for which the parameters are density, Young's modulus and Poisson's ratio, respectively.

(b) pla1 ρ(r) E(r) ν(r) σ_y(r) – Perfect plasticity for which the parameters are density, Young's modulus, Poisson's ratio, and yield strength, respectively.

(c) pla2 ρ(r) E(r) ν(r) σ_y(r) E_t(r) – Linear isotropic-hardening plasticity for which the parameters are density, Young's modulus, Poisson's ratio, initial yield strength, and hardening modulus, respectively.

(d) john ρ(r) E(r) A(r) B(r) n(r) C(r) m(r) *roomt*(r) *meltt*(r) C_v(r) *epso*(r) – Johnson–Cook plasticity for which ρ and E are density and Young's modulus, and the parameters A, B, n, C, and m are the material constants used in the model (refer to Sect. 6.2.4.4). The parameter *roomt* is the room temperate, *meltt* is the melting temperate, C_v is the specific heat at constant volume, *epso* is the effective plastic strain-rate of the quasi-static test used to determine the yield and hardening parameters A, B, and n.

(e) sjc ρ(r) E(r) A(r) B(r) n(r) C(r) *epso*(r) – Simplified Johnson–Cook plasticity which ignores the temperate effect. This model must be used with an EOS that can be set up by using the keyword seos.

(f) sjcf ρ(r) E(r) A(r) B(r) n(r) C(r) *epso*(r) *epf*(r) – Simplified Johnson–Cook plasticity with failure which ignores the temperate effect. The parameter *epf* is the plastic strain at failure. This model must be used with an EOS that can be set up by using the keyword seos.

(g) jcf ρ(r) E(r) A(r) B(r) n(r) C(r) m(r) *roomt*(r) *meltt*(r) C_v(r) *epso*(r) *epf*(r) – Johnson–Cook plasticity with failure, which must be used with an EOS that can be set up by using the keyword seos.

(h) hiex ρ(r) D(r) – High explosive material for which ρ and D are the density and detonation speed, respectively.

(i) null ρ(r) c(r) – Null material for which ρ and c are the density and sound speed, respectively. This model is used to simulate a fluid whose pressure is updated by using an EOS specified by using the keyword seos.

(j) dpm ρ(r) E(r) ν(r) q_ϕ(r) k_ϕ(r) q_ψ(r) σ^t(r) – Drucker–Prager model (refer to Sect. 6.2.5 for a detailed explanation of its parameters).

3. seos *mid*(i) *etype*(i) ...

Specify the EOS to be used with the material set *mid*. The type of EOS is specified by *etype*, namely, *etype* = 1 for polynomial EOS, *etype* = 2 for Mie–Grüneisen EOS, and *etype* = 3 for JWL EOS. The format of keyword seos is described as follows:

seos *mid*(i) 1 C_0(r) C_1(r) C_2(r) C_3(r) C_4(r) C_5(r) C_6(r) E_0(r) – Polynomial EOS (refer to Sect. 6.3.3 for a detailed explanation of its parameters).

seos *mid*(i) 2 c_0(r) λ(r) γ_0(r) E_0(r) – Mie–Grüneisen EOS (refer to Sect. 6.3.5 for a detailed explanation of its parameters).

seos *mid*(i) 3 A(r) B(r) R_1(r) R_2(r) ω(r) E_0(r) – JWL EOS (refer to Sect. 6.3.4 for a detailed explanation of its parameters).

deto x(r) y(r) z(r) – Specify the detonation point. The default detonation point is the origin of the domain. This keyword must be used together with JWL EOS.

4.2.5 Background Grid

In the MPM3D-F90, a 3D uniform regular grid is used as the background grid, which can be defined by the grid spacing, and the minimum and maximum coordinates in the x, y, and z directions. The essential or natural boundary conditions are imposed on the six grid boundary planes.

The background grid can be either specified by keyword grid or by keywords spx, spy and spz:

1. grid *X1*(r) *X2*(r) *Y1*(r) *Y2*(r) *Z1*(r) *Z2*(r) – Specify the minimum and maximum coordinates of the grid in the x, y, and z directions. These six coordinates can also be specified by keywords spx, spy, and spz.

2. spx *X1*(r) *X2*(r) – Specify the minimum and maximum x coordinates of the grid.

3. spy *Y1*(r) *Y2*(r) – Specify the minimum and maximum y coordinates of the grid.

4. spz *Z1*(r) *Z2*(r) – Specify the minimum and maximum z coordinates of the grid.

5. dcel *dcell*(r) – Specify the grid spacing.

6. fixe *c1*(i) *c2*(i) *c1*(i) *c2*(i) *c1*(i) *c2*(i) – Specify the boundary condition types on the six boundary planes; 0 represents a free boundary, 1 represents a fixed boundary, and 2 represents a symmetric boundary which is fixed along

the direction perpendicular to the boundary plane, and free in the other two directions.

4.2.6 Solution Scheme

The keywords in this category specify the solution related parameters.

1. dtsc *TimeStepScale*(r) – Specify the time step safety factor, whose default value is 0.9. The time step size used in an explicit time integration is the critical time step multiplied by the safety factor.

2. gimp – Use the upGIMP shape function rather than the standard trilinear shape function in 3D cases.

3. jaum *switch*(s) – Turn on/off the Jaumann stress rate in stress update.

4. usl *switch*(s) – Turn on/off the USL algorithm.

5. musl *switch*(s) – Turn on/off the MUSL algorithm.

6. usf *switch*(s) – Turn on/off the USF algorithm.

7. cont – Turn on the contact algorithm, which must be followed by a sub-keyword to specify the contact algorithm type, the friction coefficient and unit normal of the contact surface. The current version of the MPM3D-F90 only supports two components in contact, and only provides one sub-keyword as follows:

lagr *FrictionCoefficient*(r) *NormalVectorMethod*(i)

where the sub-keyword lagr denotes the Lagrangian multiplier based contact algorithm, *FrictionCoefficient* specifies the coefficient of friction, and *NormalVectorMethod* specifies the unit normal determination method. If *NormalVectorMethod* equals to 0, the unit normal of the contact surface is determined by the average of the unit normals of both components. If *NormalVectorMethod* equals to 1 or 2, the unit normal of component 1 or 2 is chosen as the unit normal of the contact surface.

8. endt *EndTime*(r) – Specify the termination time of the simulation.

4.2.7 Results Output

Keywords in this category are used to specify how to output simulation results:

1. outt *OutputTimeInterval*(r) – Specify the time interval for outputting simulation results in Techplot or ParaView format.

2. rptt *ReportTimeInterval*(r) – Specify the time interval for simulation status report, which is equal to *OutputTimeInterval* by default.

3. tecp – Output simulation results in Tecplot data file format, which can be used by Tecplot for postprocessing.

4. para – Output simulation results in ParaView VTK Unstructured Grid files (vtu and pvd files), which can be used by ParaView for postprocessing.

TABLE 4.1 The Value of Parameter *var*

seqv	Mises stress	velx	Velocity component v_x
epef	Effective plastic strain	vely	Velocity component v_y
pres	Pressure	velz	Velocity component v_z
engk	Kinetic energy	fail	Failure
engi	Internal energy	damg	Damage
mat	Material set number	cels	Temperature

5. pt2d *x1*(r) *x2*(r) *y1*(r) *y2*(r) *z1*(r) *z2*(r) – Specify a region in space, and only those particles initially located in the region will be outputted into the Tecplot data file. This keyword is used with keyword tecp to reduce the amount of data to be outputted to Tecplot for a large-scale problem. This keyword does not affect the data outputted to ParaView.

6. outr *var*(s) – Specify the variable name whose value will be outputted into the Tecplot data file "JobName_anim.dat". The value of *var* is explained in Table 4.1. This keyword is used with keyword tecp to reduce the amount of data outputted to Tecplot. All variables will be outputted to ParaView.

7. curv *var*(s) [pid(i)] – Specify the variable name of a particle, whose time history will be outputted into the file "JobName_curv.dat". The value of *var* is listed in Table 4.1, and the default particle ID pid is 1.

8. curx *var*(s) *x*(r) *y*(r) *z*(r) – Specify the variable name of the particle closest to the point (*x*, *y*, *z*), whose time history will be outputted into the file "JobName_curv.dat". The value of *var* is listed in Table 4.1.

4.2.8 Bodies

Keywords in this category are used to define bodies in the material domain. The current version of the MPM3D-F90 provides three types of bodies, which are defined directly by particles, blocks, or spheres.

1. Defined directly by particles

A body can be directly defined by particles as

part *point num*(i) *comID*

pid(i) *matid*(i) *pmass*(r) *x*(r) *y*(r) *z*(r)

The first row specifies the total number of particles in this body (*num*) and its component ID (*comID*). After the first row, *num* rows follows with each row defining a particle, including the particle ID *pid*, material set number *matid*, particle mass *pmass*, and its coordinates (*x*, *y*, *z*).

2. Blocks

A body can be defined by a block as

part *block comID*
matid(i) *pmass*(r) *dp*(r) *ox*(r) *oy*(r) *oz*(r) *nx*(i) *ny*(i) *nz*(i)

The first row specifies the component ID (*comID*) of the body. In the second row, (*ox*, *oy*, *oz*) is the coordinate of the lower left corner of the block which is discretized into *nx*, *ny*, and *nz* particles in the x, y, and z directions with particle spacing of *dp*.

3. Spheres

A body can be defined by a sphere as

part *sphe comID*
matid(i) *pmass*(r) *dp*(r) *ox*(r) *oy*(r) *oz*(r) *nx*(i)

The first row specifies the component ID (*comID*) of the body. In the second row, (*ox*, *oy*, *oz*) is the coordinate of the center of the sphere which is discretized into *nx* particles along the radial direction.

4.2.9 Load

Load can be applied on an individual particle or a component.

1. Apply external forces or prescribe accelerations as follows:

load
ltype(s) *bid*/*nid*(i) *fx*(r) *fy*(r) *fz*(r)

ltype is the load type, which can be "body", "node", or "grav". If *ltype* is "body" or "node", the force is applied on the component *bid* or on the particle *nid*. *fx*, *fy*, and *fz* are the x-, y-, and z-components of the force. If *ltype* is "grav", *fx*, *fy*, and *fz* are the x-, y-, and z-components of the gravitational acceleration.

The keyword load can be followed by multiple *ltype* lines, and ended with a sub-keyword "endl".

2. Prescribe initial velocity as follows:

velo
vtype(s) *bid*/*nid*(i) *vx*(r) *vy*(r) *vz*(r)

vtype can be "body" or "node", which represents prescribing an initial velocity of the component *bid* or particle *nid*. *vx*, *vy*, and *vz* are the x-, y-, and z-components of the initial velocity.

The keyword velo can be followed by multiple *vtype* lines, and ended with a sub-keyword "endv".

4.2.10 An Example of Input Data File

As an example, the detonation process of TNT explosive within a copper hollow block is simulated. Due to the geometrical symmetry, only 1/8 of the model is discretized in space. The TNT explosive is modeled with the high explosive material model and JWL EOS, while the copper is modeled with the simplified Johnson–Cook material model and Mie–Grüneisen EOS. The Jaumann rate is used for stress update, and the MUSL version is employed. The simulation termination time is 0.006 ms.

The input data file is TNT3D.mpm located in the subdirectory Data, whose content is listed as follows:

```
mpm3d *** test detonation simulation
! Unit: mm g N ms MPa
nbco 1              ! Number of components
nbbo 4              ! Number of bodies
nbmp 27000          ! Total number of particles
nmat 2              ! Total number material sets

deto 0.0 0.0 0.0    ! Define detonation point

material            ! Define material sets
! High explosive
! num      Mp    mtype density   D
   1 2.0375d-4 hiex  1.63d-3 6930
! Simplified Johnson-Cook material
! num      Mp    mtype density   E    Poission Yield0 B n C
   2 1.11d-3   sjc   8.9d-3 117.0d3 0.31d0 90 392 0.5 0.0 1e-3

! EOS for material set 1
seos 1 3 3.712e5 3.21e4 4.15 0.95 0.3
! EOS for material set 2
seos 2 2 3.3d3 1.49 1.96

spx 0.0 25.0         ! Define background grid
spy 0.0 25.0
spz 0.0 25.0
dcell 1.0            ! Grid spacing
fixed 2 0 2 0 2 0    ! Boundary condition codes

dtscale 0.2         ! Time step scale factor
endt 0.006          ! Termination time
musl on             ! Turn on MUSL version of MPM
jaum on             ! Use Jaumann rate in stress update

outtime 3.0d-4   ! Time interval for result output
rpttime 5.0d-5   ! Time interval for simulation status report
pt2d 0 5 0 10 0 10   ! Define a region to output selected
                     ! particle
outr pres            ! Output pressure of all particles

Particle block 1   ! Define blocks and their discretization
                   ! parameters
! matID compID dp   ox   oy   oz nx ny nz
   1      1    0.5 0.0 0.0 0.0 20 20 20
Particle block 1
   2      2    0.5 10.0 0.0 0.0 10 30 20
Particle block 1
   2      2    0.5 0.0 10.0 0.0 20 10 20
```

```
Particle block 1
    2      2   0.5 0.0 0.0 10.0 30 30 10
endi
```

4.3 SOURCE FILES OF THE MPM3D-F90

The files of the MPM3D-F90 are organized into the following directories:

- src – Containing all the source codes, including
  - MPM3D.f90, main program of the MPM3D-F90;
  - DataIn.f90, defining data input procedures (encapsulated as DataIn module);
  - DataOut.f90, defining data output procedures (encapsulated as DataOut module);
  - Particle.f90, defining particle-related variables and global variables (encapsulated as ParticleData module);
  - Grid.f90, defining a background grid and related operators (encapsulated as GridData module);
  - update_step.f90, defining the MPM calculation steps;
  - Material.f90, defining material parameters (encapsulated as MaterialData module);
  - Constitution.f90, defining material models (encapsulated as MaterialModel module); and
  - FFI.f90, defining free format input procedures (encapsulated as FFI module).
- make – Containing make file/solution file for different compilers as follows:
  - gnu, make file for GNU Fortran compiler; and
  - ivf, solution file for Intel Visual Fortran compiler.
- Data – Containing input data files for the following numerical examples:
  - Deto1k.mpm, 1D slab TNT detonation;
  - Taylor.mpm, Taylor bar impact;
  - PeneOgive.mpm, penetration of an ogival projectile;
  - Slopefail.mpm, failure of a soil slope;
  - LeadHypervelocityImpact.mpm, hypervelocity impact of a lead sphere on a lead plate; and
  - TNT3D.mpm, detonation process of TNT explosive within a copper hollow block.

4.4 FREE FORMAT INPUT

The module FFI (Free Format Input) provided in the MPM3D-F90 is used to read free format data from disk files. The FFI allows each line to contain maximum 256 characters and at most 15 data items, each of which is composed of no more than 20 characters. Comments begin with the exclamation mark (!) which can start anywhere in a line and continue until the end of the line.

The module FFI defines the following global variables:

- integer:: iord = 11 – Unit number for input data file;
- integer:: iomsg = 13 – Unit number for simulation status report file;
- integer:: iow1 = 14 – Unit number for simulation results output in TechPlot data format;
- integer:: iow2 = 15 – Unit number for time history output;
- integer:: iow03 = 101, iow04 = 102 – Unit numbers for the energy and momentum output;
- integer:: iow05 = 103 – Unit number for contact forces output;
- charater(100) FileInp, FileOut – File names of input data file and simulation status file;
- integer:: line_nb, nb_word, nb_read – The current line number in the input data file, the number of words left in the line and the number of words read in the line;
- character(256) sss – The string storing the current line which is composed of at most 256 characters; and
- character(20) cmd_line(15) – Data items read from the current line. Each line allows maximum 15 data items and each date item is composed of at most 20 characters.

In addition to the global variables, the FFI also provides the following functions to read data from input files:

- FFIOpen() – Open the input data file, and initialize the global variables;
- GetString(mystring) – Read a string to mystring from the input data file;
- GetInt() – Read and return an integer number from the input data file;
- GetReal() – Read and return a real number from the input data file;
- KeyWord(kw, nbkw) – Read a keyword from the input data file, and check if the keyword exists in the keyword list kw(nbkw). If so, return its index to the list; Otherwise, return −1;
- ReadLine() – Read the current line from the input data file into the array sss, and parse the valid data items into array cmd_line. The function returns the total number of data items contained in the line;
- pcomb(a, b, n) – Compare the first n characters of the strings a and b case-insensitively. Return "true" if they are the same. Otherwise, return "false";
- isNumber(num) – Check if the data item num is a valid number. Return "true" if it is. Otherwise, return "false"; and
- ErrorMsg() – If an error occurs when reading the input data file, it shows a message explaining the error and the line number in which the error appears.

4.5 MPM DATA ENCAPSULATION

The MPM requires both particle and grid data. The MPM3D-F90 encapsulates the particle and grid data in modules ParticleData and GridData, respectively.

4.5.1 Particle Data

The module ParticleData, defined in the source code file Particle.f90, provides two derived data types, Body and Particle. The derived data type Body, which is used to define a body, is defined as below.

```
type Body
    integer:: mat        ! material set number (1 ~ nb_mat)
    integer:: comID      ! component set number (1 ~ nb_mat)
    real(8):: Gravp(3)   ! gravity
    integer:: par_begin  ! no. of the first particle in
                         ! particle list
    integer:: par_end    ! no. of the last particle in
                         ! particle list
  end type Body
```

The member variable "mat" defines the material set number of the body, "comId" defines its component set number, and "Gravp" defines its gravity. The numbers of the first particle and the last particle of the body in the particle list are defined by "par_begin" and "par_end", respectively. In the MPM3D-F90, a component is composed of one or several bodies. The particles of each body are numbered consecutively in the particle list.

The derived data type Particle, which is used to define a particle, is defined as below.

```
type Particle
  real(8):: XX(3)      ! particle position at time step t+1
  real(8):: Xp(3)      ! particle position at time step t
  real(8):: VXp(3)     ! particle velocity
  real(8):: FXp(3)     ! particle load

  real(8):: VOL        ! current volume
  real(8):: sig_y      ! yield stress
  real(8):: SM, Seqv   ! mean stress and Mises stress
  real(8):: SDxx, SDyy, SDzz, SDxy, SDyz, SDxz ! deviatoric
                                              ! stress
  real(8):: epeff      ! effective plastic strain
  real(8):: celsius_t  ! celsius temperature

  logical:: SkipThis   ! skip this particle in postprocessing
  logical:: failure    ! failure
  integer:: icell      ! cell number
  real(8):: DMG        ! damage
  real(8):: LT         ! lighting time for explosive simulation
```

```
    real(8):: ie          ! internal energy
    real(8):: mass        ! particle mass
    real(8):: cp          ! sound speed
  end type Particle
```

In addition to these derived data types, the module ParticleData also defines many global variables such as the following:

- integer:: nb_particle – Total number of particles used to discretize the material domain;
- integer:: nb_body – Total number of bodies in the material domain;
- integer:: nb_component – Total number of components in the material domain (In the MPM3D-F90, the contact algorithm is implemented between two components, instead of between two bodies. Each component can consist of many bodies, but all the bodies in a component move with the same velocity field.);
- type(Particle), target, allocatable:: particle_list(:) – Particle list to store all the particles in the material domain;
- type(Body), target, allocatable:: body_list(:) – Body list to store all the bodies in the material domain;
- logical:: MUSL – If true, the MUSL MPM scheme is used;
- logical:: USL – If true, the USL MPM scheme is used;
- logical:: USF – If true, the USF MPM scheme is used;
- logical:: GIMP – If true, the GIMP is used;
- logical:: contact – If true, the contact algorithm is activated;
- integer:: istep – Current time step number;
- real(8):: DT – Current time step size;
- real(8):: CurrentTime – Current time;
- real(8):: EndTime – End-time of simulation;
- real(8):: DTScale – Time step size scale ($\leqslant 1$);
- real(8):: EngInternal – Total internal energy of all bodies;
- real(8):: EngKinetic – Total kinetic energy of all bodies;
- real(8):: Momentum – Total momentum of all bodies;
- real(8):: Mombody1 – Momentum of the first body in a contact pair; and
- real(8):: Mombody2 – Momentum of the second body in a contact pair.

The module ParticleData also defines an InitParticle procedure, which initializes the variables carried by the particles.

4.5.2 Grid Data

The module GridData, defined in the source code file Grid.f90, encapsulates all the variables and operators associated with the background grid. In the current version of the MPM3D-F90, only 8-node hexahedron cells are implemented for 3D cases. To unify the implementation of the grid with or without the contact algorithm being activated, GridData provides derived types GridNode, GridNodeProperty and ContactGridNodeProperty. The derived data type GridNode defines a background grid node, including its position (Xg) and boundary condition type (Fix_x, Fix_y and Fix_z). The type GridNode is defined as below.

```
type GridNode
   real(8):: Xg(3)      ! grid node coordinate
   logical:: Fix_x, Fix_y, Fix_z    ! BC
end type GridNode
```

The derived type GridNodeProperty defines the variables carried by a grid node, including its mass (Mg), momentum (PXg), and force (FXg). The type GridNodeProperty is defined as below.

```
type GridNodeProperty
   real(8):: Mg         ! mass on grid node
   real(8):: PXg(3)     ! momentum on grid node
   real(8):: FXg(3)     ! internal/external force on gride node
end type GridNodeProperty
```

The derived type ContactGridNodeProperty defines the extra variables carried by a grid node which is in contact with other components, including the unit normal and unit tangent vectors. The type ContactGridNodeProperty is defined as blow.

```
type ContactGridNodeProperty
   ! the normal direction of contact grid node
   real(8):: ndir(3)
   ! the tangential unit vectors of contact grid node
   real(8):: sdir(3)
end type ContactGridNodeProperty
```

The module GridData also defines the following global variables:

- type(GridNode), target, allocatable:: node_list(:) – List of grid nodal objects, each of which defines a grid node;
- type(GridNodeProperty), target, allocatable:: grid_list(:,:) – List of grid nodal property sets, each of which defines the properties of a grid node;
- type(ContactGridNodeProperty), target, allocatable:: CP_list(:,:) – List of contacted grid nodal properties, each of which defines the extra properties of a grid node which is in contact with other components;
- real(8):: SpanX(2)=0.0, SpanY(2)=0.0, SpanZ(2)=0.0 – Define the background grid region with the x coordinates of the leftmost and rightmost faces, y coordinates of the front and back faces, and z coordinates of the bottom and top faces of the region;

- real(8):: Dcell – The grid cell length (In the MPM3D-F90, each cell is a cube with length Dcell);
- real(8):: CutOff – The cutoff mass of grid nodes. If the mass of a grid node is less than the CutOff, the node will be skipped during the MPM calculation to avoid a singular acceleration (causing the cell-crossing error in the original MPM);
- integer:: nb_gridnode – Total number of grid nodes;
- integer:: FixS(6)=0 – Boundary types of the six faces of the background grid: 0 – free, 1 – fixed, 2 – symmetric (fixed along the direction perpendicular to the face, but free along the directions parallel to the face);
- integer:: NumCell – Total number of grid cells;
- integer:: NumCellx – Total number of grid cells in the x direction;
- integer:: NumCelly – Total number of grid cells in the y direction;
- integer:: NumCellz – Total number of grid cells in the z direction;
- integer:: NumCellxy – Total number of grid cells in the xy plane;
- integer:: NGx – Total number of grid nodes in the x direction;
- integer:: NGy – Total number of grid nodes in the y direction;
- integer:: NGz – Total number of grid nodes in the z direction;
- integer:: NGxy – Total number of grid nodes in the xy plane, which is equal to NGx*NGy;
- integer, allocatable:: CellsNode(:,:): – Grid node numbers of all corresponding cells;
- integer:: nb_InflNode – Total number of influenced nodes of a particle, which equals to 8 when using the MPM, but equals to 8–27 when using the GIMP (An influenced node of a particle is defined as the grid node whose shape function covers the particle.);
- integer:: InflNode(27) – The nodal numbers of the influenced nodes of a particle;
- real(8):: rpg(27,3) – The relative coordinates between a particle and its influenced nodes;
- real(8):: iJacobi – Inverse of the Jacobi determinant, which is used in the calculation of the grid nodal shape functions and their derivatives;
- real(8):: iJacobi4 – 1/(4*Dcell);
- real(8): iDcell – 1/(Dcell);
- real(8), allocatable:: SHP(:), DNDX(:), DNDY(:), DNDZ(:) – The shape functions associated with the influenced nodes of a particle, and their x-, y-, and z-derivatives;
- integer, parameter:: SNX, SNY, SNZ – The natural coordinates ($\xi_I = \pm 1, \eta_I = \pm 1, \zeta_I = \pm 1$) of the 8 vertices of a cell;
- real(8):: fricfa = 0.0 – The coefficient of friction;

- integer:: normbody – The flag to identify how to calculate the unit normal vector of a contact surface;
- integer:: contact_type – Type of the contact algorithm: 0 – Non-slip contact algorithm, and 1 – Lagrangian type contact algorithm; and
- real(8):: tot_cont_for(3) – The total contact force between two bodies at a contact node.

In addition, the module GridData also defines the following operators:

- InWhichDcell(xx) – Calculate and return the cell number in which the particle with the coordinate xx is located (In each time step, all particles are attached to a background grid so that it is necessary to know in which cell each particle is located.);
- SetGridData() – Create and initialize the background grid based on the data obtained from the input data file;
- SetContact_GridNodeData() – Create and initialize the contacted grid nodal property list;
- NShape(node1, p, ider) – Evaluate the shape functions SHP(8) and/or their derivatives DNDX(8), DNDY(8), and DNDZ(8) at a particle p (The parameter node1 denotes the number of the first grid node of the cell in which the particle p is located, ider = 0 means that only the shape functions will be calculated, ider = 1 means that only the derivatives of the shape functions will be calculated, and ider = 2 means that both the shape functions and their derivatives will be calculated.);
- NShape_GIMP (p) – Evaluate the uGIMP shape functions SHP(27) and their derivatives DNDX(27), DNDY(27), and DNDZ(27) at a particle p; and
- FindInflNode (p, icell) – Find the 27 influenced nodes of a particle p for the uGIMP, and calculate the relative coordinates between the particle p and its influenced nodes (icell is the cell number in which the particle p is located.).

4.5.3 Data Input

The module DataIn, defined in the source code file DataIn.f90, encapsulates all the operators used to input data, and to discretize and initialize the material domain. It uses the modules ParticleData, GridData, DataOut (refer to Sect. 4.5.4) and MaterialData (refer to Sect. 6.5.1), and defines the global integer variables parCounter, bodyCounter, and comCounter to count particles, bodies, and components, respectively.

The main operators defined in the module DataIn are described as follows:

- InputPara – Read data by calling the procedure GetData(), and initialize the background grid and other variables;
- GetData – Read and Parse keywords and their parameters from the input data file using the module FFI, and then respond to these keywords (A new

user-defined keyword can be implemented by adding a code segment into this operator.);

- SetMaterial – Read material data. When implementing a new material model, a code segment should be added into this operator to read its parameters;
- SetParticle – Read and discretize the material domain into particles (The material domain can be composed of particles, blocks, and spheres. When implementing a new geometric body, a code segment should be added into this operator to read the body and then discretize it into particles.);
- SetLoad – Read and initialize the external forces of all particles;
- SetVelocity – Read and initializes the initial velocities of all particles;
- SetOnOff – Read and return the on–off switch from the input data file;
- SetResOption – Read and set the flag to indicate which physical variable will be sent to the output file for postprocessing;
- SetEos – Read the type of EOS and its parameters;
- SetCurX – Find the ID of the particle whose relative coordinate to the point read from the input data file is less than 0.25*DCell, and then let the MPM3D send the physical variable of the particle to the output file for postprocessing;
- ErrorPt – Display the error message when reading more particles from the input data file than expected;
- Initial – Initialize required variables after reading all the data from the input data file;
- SetDT – Determine the time step;
- statinfor – Output statistical data, such as the total mass, total internal energy, and total kinetic energy, to the message file; and
- Setcontact – Read the contact data from the input data file (When implementing the new contact algorithm, a code segment should be added into this operator.).

4.5.4 Data Output

The module DataOut, defined in the source code file DataOut.f90, encapsulates all the operators used to output the solution results for postprocessing. It uses the modules ParticleData, GridData, and ContactGridDta, and defines the following global variables:

- integer:: nCvs – Total number of the data points in the time history data file, which can be read by TecPlot and Origin to plot time history curves;
- integer:: nAnm – Total number of the data sets in the output file, which can be read by TecPlot to plot contours, nephograms and animations;
- integer, parameter:: nVariables – Total number of the physical variables which can be written into the output file. nVariables is currently equal to 14;

- character(4), parameter:: OutputName(nVariables) – Variable list which can be outputted for postprocessing;
- logical:: WriteTecPlot – Flag to identify whether to output results to the TecoPlot data file;
- real(8):: OutTime=0.0 – Time interval for outputting simulation results;
- real(8):: ReportTime=0.0 – Time interval for reporting the simulation status during the simulation process;
- integer:: nCurves=2 – Number of the variables whose time history curves will be outputted;
- integer::nAnimate=1 – Number of the variables whose results will be outputted for postprocessing;
- integer, parameter:: MaxCurves=15 – Maximum number of the variables whose time history curves will be outputted;
- integer, parameter:: MaxAnim=10 – Maximum number of the variables whose results will be outputted for postprocessing;
- integer CurveOption(MaxCurves) – Variable ID in the variable list OutputName whose time history curves will be outputted;
- integer AnimOption(MaxAnim) – Variable ID in the variable list OutputName whose results will be outputted for postprocessing;
- integer:: CurvePoint(MaxCurves) – Particle ID whose time history curves will be outputted;
- real:: plot2d(6)=0 – Define a region in space, and only the results of those particles initially located in the region will be outputted to TecPlot for postprocessing; and
- logical:: plot2dTrue=.false. – Flag to identify whether a region has been defined by plot2d.

Furthermore, the module DataOut also provides the procedures OutCurve, OutAnim, and OutAnimPV to output the time history of a variable of a particle, to output the simulation results of the particles selected by plot2d to TecoPlot data file, and to output the simulation results of all the particles to ParaView vtu files, respectively.

4.6 MAIN SUBROUTINES

The MPM3D-F90 provides a set of keywords, as specified in Sect. 4.2. Users can use these keywords in the input data file to control the execution of the MPM3D-F90, such as choosing time integration schemes, material models, and data output methods. Users can easily add new keywords to implement their new schemes in the MPM3D-F90 without affecting the existing code.

The main program of the MPM3D-F90 is listed in this section. For the sake of clarity, the code segments used to display the program execution status have been deleted from the list. The complete code of the main program can be found in the source code file MPM3D.f90.

```
program MPM3D
  use ParticleData
  use FFI, only: iomsg, iow1, iow2
  use DataIn
  use DataOut
  implicit none

  call InputPara()    ! Input data
  call calcEnergy()   ! Calculate kinetic energy

  ! Solving
  do while(CurrentTime .le. EndTime)
     call cpu_time( t_begin )

     istep = istep+1
     CurrentTime = CurrentTime + DT
     EngInternal = 0.0

     ! Step 1: Initialize background grid nodal mass and Momentum
     call GridMomentumInitial()  ! Eq.(3.47) and Eq.(3.48)

     ! Step 2: Apply boundary conditions
     call ApplyBoundaryConditions()

     ! Step 3: Update particles stress (Only For USF)
     if(USF) then
        call ParticleStressUpdate() ! Eq.(3.50-3.51)
     end if

     ! Step 4: Calculate the grid nodal force
     call GridMomentumUpdate() ! Eq.(3.53-3.55)

     ! Step 5: Integrate momentum equations on background grids
     call IntegrateMomentum()  ! Eq.(3.56)

     ! Step 6: Detect contact grid node, calculate contact force
     ! and adjust nodal momentum
     if(Contact_type == 1) then
        call Lagr_NodContact()
     end if

     ! Step 7: Update particles position and velocity
     call ParticlePositionUpdate() ! Eq.(3.57) and Eq.(3.58)

     ! Step 8: Recalculate the grid node momentum for MUSL
     if(MUSL) then
        call GridMomentumMUSL()    ! Eq.(3.59)
        call ApplyBoundaryConditions()
     end if
```

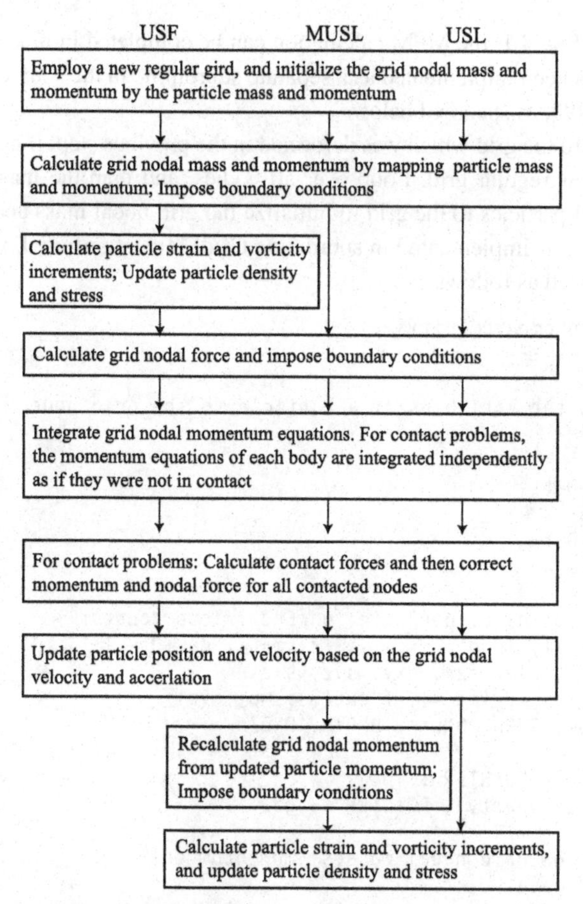
USF
MUSL
USL
Employ a new regular gird, and initialize the grid nodal mass and momentum by the particle mass and momentum
Calculate grid nodal mass and momentum by mapping particle mass and momentum; Impose boundary conditions
Calculate particle strain and vorticity increments; Update particle density and stress
Calculate grid nodal force and impose boundary conditions
Integrate grid nodal momentum equations. For contact problems, the momentum equations of each body are integrated independently as if they were not in contact
For contact problems: Calculate contact forces and then correct momentum and nodal force for all contacted nodes
Update particle position and velocity based on the grid nodal velocity and accerlation
Recalculate grid nodal momentum from updated particle momentum; Impose boundary conditions
Calculate particle strain and vorticity increments, and update particle density and stress

FIGURE 4.1 Flow chart of the MPM3D-F90.

```
    ! Step 9: Update particles stress for both USF and MUSL
    if(.NOT. USF) then
       call ParticleStressUpdate() ! Eq.(3.61-3.63)
    end if

    call calcEnergy()  ! Calculate kinetic energy

    call OutCurve()    ! out put curve and animation data

end program MPM3D
```

The MPM3D-F90 has implemented the MUSL, USL, and USF versions of the MPM with the contact algorithm, whose detailed formations can be found in Sect. 3.2.2 (if contact is off) or Sect. 3.3 (if contact is on). The flow-chart of these versions of the MPM is shown in Fig. 4.1.

As shown in Fig. 4.1, the MPM calculation can be completed in nine steps, and each step has been implemented as a separate subroutine in the source code file update_step.f90, as specified below.

Step 1. Discard the grid which was deformed in the previous step, if needed, and employ a new regular grid. Loop over all bodies, and map the mass and momentum of all particles to the grid to initialize the grid nodal mass and momentum. This step is implemented in subroutine GridMomentumInitial, whose source code is listed as follows:

```
subroutine GridMomentumInitial()
!-----------------------------------------------------------------
!-  Purpose                                                      -
!-      1. Map the variables of a particle to the grid node      -
!-----------------------------------------------------------------

  use ParticleData
  use GridData
  use MaterialData
  implicit none

  integer:: b, p, n, c, parBegin, parEnd ! loop counter
  integer:: icell, inode, ix, iy, iz, mat_, comID = 1
  real(8):: sxx, syy, szz, sxy, syz, sxz
  real(8):: fx(3), f_int(3), f_ext(3), mp_, vol_
  real(8):: shm, SHPn, DNDXn, DNDYn, DNDZn

  type(Particle), POINTER :: pt
  type(GridNodeProperty), POINTER :: gd

  ! Calculate the grid nodal masses, momentum only
  ! Reset Grid data
  grid_list%Mg = 0.0d0            ! Grid nodal mass

  grid_list%PXg(1) = 0.0d0;       ! Nodal momentum
  grid_list%PXg(2) = 0.0d0;
  grid_list%PXg(3) = 0.0d0;

  do b = 1, nb_body     ! Loop over all bodies
     parBegin = body_list(b)%par_begin
     parEnd = body_list(b)%par_End

     if(contact) comID = body_list(b)%comID ! Get comID from body

     do p = parBegin, parEnd     ! Loop over all particles (1)
        pt => particle_list(p)

        pt%icell = InWhichCell(pt%Xp)
        icell = pt%icell
        ! Particle p is out of the computational region
        if (icell < 0) cycle
```

```
        vol_ = pt%VOL
        mp_ = pt%Mass

        ! Calculate the shape functions and their derivatives
        InflNode(1:8)=CellsNode(icell,:)
        if (GIMP) then
           call FindInflNode(p,icell)
           call NShape_GIMP(p)
        else
           call NShape(CellsNode(icell,1),p,0)
        end if

        ! Loop over the grid nodes of the hexahedron
        ! in which the particle is located
        do n = 1, nb_InflNode
           ! out of the computational grid
           if (InflNode(n) .gt. nb_gridnode .or. &
               InflNode(n) .le. 0) cycle

           gd => grid_list(comID, InflNode(n))

           SHPn = SHP(n)
           shm = SHPn*mp_

           gd%Mg = gd%Mg + shm              ! the nodal mass
           gd%PXg = gd%PXg + pt%VXp*shm     ! the nodal momentum
        end do !n

     end do !p
  end do    !b
end subroutine GridMomentumInitial
```

Step 2. Impose essential boundary conditions. For a fixed boundary, set $p_{iI}^{k-1/2} = 0$. This step is implemented in subroutine ApplyBoundaryConditions, whose source code is listed as follows:

```
subroutine ApplyBoundaryConditions()
!------------------------------------------------------------------
!-  Purpose                                                       -
!-      1. Apply boundary condition                               -
!------------------------------------------------------------------
  use GridData
  use ParticleData, only: nb_component
  implicit none

  integer:: n, c ! loop counter
  type(GridNodeProperty), POINTER :: gd
  type(GridNode), POINTER :: node

  do c = 1, nb_component
     do n = 1, nb_gridnode
        gd => grid_list(c, n)
```

```
         node => node_list(n)
         ! Apply boundary conditions on computational grid
         if (node%Fix_x) then  ! Grid node n is fixed in x
                               ! direction
            gd%PXg(1) = 0.0
            gd%FXg(1) = 0.0
         end if

         if (node%Fix_y) then  ! Grid node n is fixed in y
                               ! direction
            gd%PXg(2) = 0.0
            gd%FXg(2) = 0.0
         end if

         if (node%Fix_z) then  ! Grid node n is fixed in z
                               ! direction
            gd%PXg(3) = 0.0
            gd%FXg(3) = 0.0
         end if
      end do ! n
   end do    ! c

end subroutine ApplyBoundaryConditions
```

Step 3. For the USF, calculate the particle strain increment $\Delta\varepsilon_{ijp}^{b,k-1/2}$ and vorticity increment $\Delta\Omega_{ijp}^{b,k-1/2}$, and then update the particle density ρ_p^{k+1} and stress σ_{ijp}^{k+1}. This step is implemented in subroutine ParticleStressUpdate, whose source code is listed as follows:

```
subroutine ParticleStressUpdate()
!-----------------------------------------------------------------
!-  Purpose                                                      -
!-      1. Calculate the strain rate and spin tensor             -
!-      2. Update stresses by appropriate constitution law       -
!-----------------------------------------------------------------
  use ParticleData
  use GridData
  use MaterialModel, only: Constitution
  use MaterialData
  implicit none

  integer:: b, p, n, parBegin, parEnd ! loop counter
  integer:: icell, inode, ix, iy, iz, comID = 1
  real(8):: xx(3), vx(3), ax(3), vgx(3)
  real(8):: de(6), vort(3)
  real(8):: mp_, shm, SHPn, DNDXn, DNDYn, DNDZn

  type(Particle), POINTER :: pt
  type(GridNodeProperty), POINTER :: gd

  ! Calculate the increment strain and vorticity
```

```
! de(i) comply the Voigt rule (d11, d22, d33, 2*d23, 2*d13,
! 2*d12)
do b = 1, nb_body
   parBegin = body_list(b)%par_begin
   parEnd = body_list(b)%par_End

   if(contact) comID = body_list(b)%comID  ! Get comID from body

   do p = parBegin, parEnd    ! Loop over all particles (4)
      pt => particle_list(p)

      de   = 0d0     ! Incremental strain
      vort = 0d0     ! Incremental vorticity

      icell = pt%icell    ! use old position
      ! Particle p is out of the computational region
      if (icell < 0) cycle

      ! Calculate the shape functions and their derivatives
      InflNode(1:8)=CellsNode(icell,:)
      if (GIMP) then
         call FindInflNode(p,icell)
         call NShape_GIMP(p)
      else
         call NShape(CellsNode(icell,1),p,1)
      end if

      ! Loop over all grid nodes of the hexahedron
      !  in which particle p is located
      do n = 1, nb_InflNode
         if (InflNode(n) .gt. nb_gridnode .or. InflNode(n) .le.
            0) & cycle  ! out of the computational grid
         gd => grid_list(comID, InflNode(n))
         ! If the nodal mass is not too small
         if (grid_list(comID, InflNode(n))%Mg > CutOff) then
            vgx = gd%PXg / gd%Mg     ! Grid nodal velocity

            DNDXn = DNDX(n);  DNDYn = DNDY(n);  DNDZn = DNDZ(n)
            de(1) = de(1) + DNDXn*vgx(1)             ! D11 * DT
            de(2) = de(2) + DNDYn*vgx(2)             ! D22 * DT
            de(3) = de(3) + DNDZn*vgx(3)             ! D33 * DT
            ! 2*D23 * DT
            de(4) = de(4) + (DNDYn*vgx(3) + DNDZn*vgx(2))
            ! 2*D13 * DT
            de(5) = de(5) + (DNDZn*vgx(1) + DNDXn*vgx(3))
            ! 2*D12 * DT
            de(6) = de(6) + (DNDXn*vgx(2) + DNDYn*vgx(1))

            ! W32 * DT
            vort(1) = vort(1) + (DNDYn*vgx(3) - DNDZn*vgx(2))
            ! W13 * DT
            vort(2) = vort(2) + (DNDZn*vgx(1) - DNDXn*vgx(3))
```

```
            ! W21 * DT
            vort(3) = vort(3) + (DNDXn*vgx(2) - DNDYn*vgx(1))
         end if
      end do ! n

      de = de * DT
      vort = vort * DT / 2d0

      ! Update stress by constitution law
      call Constitution(de, vort, b, p)

      if(.NOT.USF) pt%Xp = pt%XX    ! the next particle position

    end do !p
  end do    !b

end subroutine ParticleStressUpdate
```

Step 4. Calculate the grid nodal forces $f_{iI}^{b,\text{int},k}$, $f_{iI}^{b,\text{ext},k}$, and $f_{iI}^{b,k} = f_{iI}^{b,\text{int},k} + f_{iI}^{b,\text{ext},k}$. For contact problems, calculate the unit outward normal $n_{iI}^{b,k}$ of each body at the contacted grid node. This step is implemented in subroutine GridMomentumUpdate, whose source code is listed as follows:

```
subroutine GridMomentumUpdate()
!-------------------------------------------------------------------
!-  Purpose                                                        -
!-      1. Calculate the background grid nodal force               -
!-------------------------------------------------------------------
  use ParticleData
  use GridData
  use MaterialData
  implicit none

  integer:: b, p, n, c, parBegin, parEnd ! loop counter
  integer:: icell, inode, ix, iy, iz, mat_, comID = 1
  real(8):: sxx, syy, szz, sxy, syz, sxz
  real(8):: fx(3), f_int(3), f_ext(3), mp_, vol_
  real(8):: shm, SHPn, DNDXn, DNDYn, DNDZn

  type(Particle), POINTER :: pt
  type(GridNodeProperty), POINTER :: gd
  type(ContactGridNodeProperty), POINTER :: CP

  ! Calculate the grid nodal forces only

  ! Reset nodal forces
  grid_list%FXg(1) = 0.0d0;     ! Nodal forces
  grid_list%FXg(2) = 0.0d0;
  grid_list%FXg(3) = 0.0d0;

  if(contact) then
```

```
    CP_list%ndir(1) = 0.0d0
    CP_list%ndir(2) = 0.0d0
    CP_list%ndir(3) = 0.0d0
end if

do b = 1, nb_body              ! Loop over all bodies
    parBegin = body_list(b)%par_begin
    parEnd = body_list(b)%par_End

    if(contact) comID = body_list(b)%comID  ! Get comID from body
    do p = parBegin, parEnd      ! Loop over all particles
       pt => particle_list(p)

       icell = pt%icell          ! using old position

       ! Particle p is out of the computational region
       if (icell < 0) cycle

       sxx = pt%SM + pt%SDxx    ! Stresses
       syy = pt%SM + pt%SDyy
       szz = pt%SM + pt%SDzz
       sxy = pt%SDxy
       syz = pt%SDyz
       sxz = pt%SDxz
       ! External forces
       fx = pt%FXp + pt%Mass * (body_list(b)%Gravp)

       vol_ = pt%VOL
       mp_ = pt%Mass

       ! Calculate the shape functions and their derivatives
       InflNode(1:8)=CellsNode(icell,:)
       if (GIMP) then
          call FindInflNode(p,icell)
          call NShape_GIMP(p)
       else
          call NShape(CellsNode(icell,1),p,2)
       end if

       ! Loop over the grid nodes of the hexahedron
       !  in which the particle is located
       do n = 1, nb_InflNode
          if (InflNode(n) .gt. nb_gridnode .or. InflNode(n) .le.
              0) & cycle  ! out of the computational grid

          gd => grid_list(comID, InflNode(n))

          SHPn = SHP(n)
          DNDXn = DNDX(n)
          DNDYn = DNDY(n)
          DNDZn = DNDZ(n)
```

```
            f_int(1) = - (sxx*DNDXn + sxy*DNDYn + sxz*DNDZn)*vol_
            f_int(2) = - (sxy*DNDXn + syy*DNDYn + syz*DNDZn)*vol_
            f_int(3) = - (sxz*DNDXn + syz*DNDYn + szz*DNDZn)*vol_

            f_ext = fx*SHPn

            gd%FXg = gd%FXg + f_int + f_ext  !nodal force

            if(contact) then
               CP => CP_list(comID, InflNode(n))
               CP%ndir(1) = CP%ndir(1) + DNDXn*mp_
               CP%ndir(2) = CP%ndir(2) + DNDYn*mp_
               CP%ndir(3) = CP%ndir(3) + DNDZn*mp_
            end if

         end do !n

      end do !p
   end do    !b
end subroutine GridMomentumUpdate
```

Step 5. Integrate the grid nodal momentum equations. For contact problems, the momentum equations of each body are integrated independently to calculate the trail nodal momentum $\bar{p}_{iI}^{b,k+1/2}$ by Eq. (3.93) as if they were not in contact. This step is implemented in subroutine IntegrateMomentum, whose source code is listed as follows:

```
subroutine IntegrateMomentum()
!-------------------------------------------------------------------
!-  Purpose                                                        -
!-    1. Integrate the momentum equations on the                   -
!=       computational grid                                        -
!-    2. Apply boundary conditions                                 -
!-------------------------------------------------------------------
  use ParticleData
  use GridData
  implicit none

  integer:: c, n ! loop counter
  type(GridNodeProperty), POINTER :: gd
  type(GridNode), POINTER :: node

  do c = 1, nb_component
     do n = 1, nb_gridnode
        gd => grid_list(c, n)
        node => node_list(n)

        ! Integrate momentum equation
        gd%PXg = gd%PXg + gd%FXg * DT

        ! Apply boundary conditions on computational grid
```

```
        if (node%Fix_x) then  ! Grid node n is fixed in x
                              ! direction
           gd%PXg(1) = 0.0
           gd%FXg(1) = 0.0
        end if

        if (node%Fix_y) then  ! Grid node n is fixed in y
                              ! direction
           gd%PXg(2) = 0.0
           gd%FXg(2) = 0.0
        end if

        if (node%Fix_z) then  ! Grid node n is fixed in z
                              ! direction
           gd%PXg(3) = 0.0
           gd%FXg(3) = 0.0
        end if

     end do !n
  end do    !c
end subroutine IntegrateMomentum
```

Step 6. For contact problems, search for contacted nodes using the criteria Eq. (3.98). Calculate the contact forces and then correct the momentum and nodal forces for all contacted nodes. This step is implemented in subroutine Lagr_NodContact, whose source code is listed as follows:

```
subroutine Lagr_NodContact()
!---------------------------------------------------------------------
!-  Purpose                                                          -
!-      1. Establish the nodal contact criteria,                     -
!-      2. Correct  the normal vectors, and                          -
!-      3. Apply the contact force and adjust nodal velocities       -
!---------------------------------------------------------------------
  use  ParticleData
  use  GridData
  use  MaterialData

  implicit none

  integer:: p, n, c ! loop counter
  real(8):: nx, ny, nz, tt,tta,ttb,crit, crita, critb
  real(8):: nomforce, val_fslip, val_fstick, val_ffric, nodtolmg
  real(8):: fstick(3), fslip(3), cforce(3)
  integer:: abody,bbody

  type(GridNodeProperty), POINTER :: gd1
  type(GridNodeProperty), POINTER :: gd2
  type(ContactGridNodeProperty), POINTER :: CP1
  type(ContactGridNodeProperty), POINTER :: CP2

  tot_cont_for = 0.0 ! the total contact force between of bodies
```

```
! calculate contact force and adjust the nodal force and
! momentum
do n = 1, nb_gridnode
   CP1 =>  CP_list(1,n)
   CP2 =>  CP_list(2,n)
   gd1 => grid_list(1, n)
   gd2 => grid_list(2, n)

   ! recalculate the nodal normal direction
   ! if normbody 0 then using average method;
   ! if 1,using abody; if 2,using bbody
   if(normbody == 0)then
      nx = CP1%ndir(1)  - CP2%ndir(1)
      ny = CP1%ndir(2)  - CP2%ndir(2)
      nz = CP1%ndir(3)  - CP2%ndir(3)
   end if

   if(normbody == 1)then
      nx = CP1%ndir(1)
      ny = CP1%ndir(2)
      nz = CP1%ndir(3)
   end if

   if(normbody == 2)then
      nx = - CP2%ndir(1)
      ny = - CP2%ndir(2)
      nz = - CP2%ndir(3)
   end if

   ! unitize normal vector
   tt = sqrt(nx*nx + ny*ny + nz*nz)
   if(tt > epsilon(tt)) then
      nx = nx / tt
      ny = ny / tt
      nz = nz / tt
   end if

   CP1%ndir(1) = nx; ! Nodal direction for contact
   CP1%ndir(2) = ny;
   CP1%ndir(3) = nz;
   CP2%ndir = -CP1%ndir

   crit = 0.0
   ! contact criteria using the unit normal vectors
   if ( gd1%Mg > CutOff .AND. gd2%Mg > CutOff) then
      ! Eq.(3.98)
      crit = (gd1%Pxg(1)*gd2%Mg - gd2%Pxg(1)*gd1%Mg)*nx +&
             (gd1%Pxg(2)*gd2%Mg - gd2%Pxg(2)*gd1%Mg)*ny +&
             (gd1%Pxg(3)*gd2%Mg - gd2%Pxg(3)*gd1%Mg)*nz
   end if
```

```
    if(crit > epsilon(crit)) then

       tt = (gd1%Mg + gd2%Mg)*Dt

       ! calculate the normal contact force
       nomforce =crit/tt   ! Eq.(3.106)

       ! for friction contact
       if(fricfa > epsilon(fricfa)) then

          ! calculate the contact force   Eq.(3.104)
          cforce = (gd1%Pxg*gd2%Mg - gd2%Pxg*gd1%Mg)/tt

          ! calculate the tangent contact force
          fstick = cforce - nomforce*CP1%ndir
          val_fstick = sqrt( fstick(1)*fstick(1) +  &
                      fstick(2)*fstick(2) + fstick(3)*fstick(3) )
          val_fslip = fricfa*abs(nomforce)
          if(val_fslip < val_fstick) then
             cforce = nomforce + val_fslip*(fstick /val_fstick)
          end if

          ! for contact without friction
       else
          cforce = nomforce*CP1%ndir
       end if

       ! add contact force to nodal force
       gd1%Fxg = gd1%Fxg - cforce
       gd2%Fxg = gd2%Fxg + cforce

       ! adjust the nodal component by contact force
       gd1%Pxg = gd1%Pxg - cforce * Dt
       gd2%Pxg = gd2%Pxg + cforce * Dt

       tot_cont_for = tot_cont_for + cforce
    end if
  end do  !n

end subroutine Lagr_NodContact
```

Step 7. Update the particle position and velocity by mapping the grid nodal displacement and velocity increment to corresponding particles. This step is implemented in subroutine ParticlePositionUpdate, whose source code is listed as follows:

```
subroutine ParticlePositionUpdate()
!-------------------------------------------------------------------
!-  Purpose                                                        -
!-      1. Update particle position and velocity                   -
!-------------------------------------------------------------------
```

```
use ParticleData
use GridData
use MaterialModel, only: Constitution
use MaterialData
implicit none

integer:: b, p, n, parBegin, parEnd ! loop counter
integer:: icell, inode, ix, iy, iz, comID = 1
real(8):: xx(3), vx(3), ax(3), vgx(3)
real(8):: de(6), vort(3)
real(8):: mp_, shm, SHPn, DNDXn, DNDYn, DNDZn

type(Particle), POINTER :: pt
type(GridNodeProperty), POINTER :: gd

! Update particle position and velocity
do b = 1, nb_body
   parBegin = body_list(b)%par_begin
   parEnd = body_list(b)%par_End
   ! Get comID from body
   if(contact)  comID = body_list(b)%comID

   do p = parBegin, parEnd    ! Loop over all particles (2)
      pt => particle_list(p)

      icell = pt%icell
      ! Particle p is out of the computational region
      if (icell < 0) cycle

      xx = pt%Xp;  ! Particle position at time step k

      vx = 0d0
      ax = 0d0

      ! Mapping from grid to particle

      ! Calculate the shape functions and their derivatives
      InflNode(1:8)=CellsNode(icell,:)
      if (GIMP) then
         call FindInflNode(p,icell)
         call NShape_GIMP(p)
      else
         call NShape(CellsNode(icell,1),p,2)
      end if

      ! Loop over all grid nodes of the hexahedron
      !  in which particle p is located
      do n = 1, nb_InflNode
         if(InflNode(n) .gt. nb_gridnode .or. InflNode(n) .le.
            0) & cycle    ! out of the computational grid

         gd => grid_list(comID, InflNode(n))
```

```
            if (gd%Mg > CutOff) then    ! The nodal mass is not too
                                        ! small
               SHPn = SHP(n)
               vx = vx + SHPn * (gd%PXg / gd%Mg)
               ax = ax + SHPn * gd%FXg / gd%Mg
            end if

         end do ! n

         ! Time integration
         pt%XX = xx + vx * DT           ! Update particle position
         pt%VXp = pt%VXp + ax * DT      ! Update particle velocity
         if(USF)  pt%Xp = pt%XX         ! the next particle position

      end do ! p
   end do    ! b

end subroutine  ParticlePositionUpdate
```

Step 8. For the MUSL, recalculate the grid nodal momentum by mapping the updated particle momentum back to the grid nodes and imposing the essential boundary conditions. This step is implemented in subroutine GridMomentum-MUSL, whose source code is listed as follows:

```
subroutine GridMomentumMUSL()
!-------------------------------------------------------------------
!-  Purpose                                                        -
!-      1. Recalculate the grid node momentum by mapping           -
!-         the updated particle information                        -
!-      2. Apply boundary condition                                -
!-------------------------------------------------------------------
  use ParticleData
  use GridData
  use MaterialModel, only: Constitution
  use MaterialData
  implicit none

  integer:: b, c, p, n, parBegin, parEnd ! loop counter
  integer:: icell, comID = 1
  real(8):: de(6), vort(3)
  real(8):: mp_, shm, SHPn

  type(Particle), POINTER :: pt
  type(GridNodeProperty), POINTER :: gd
  type(GridNode), POINTER :: node

  grid_list%PXg(1) = 0.0d0;
  grid_list%PXg(2) = 0.0d0;
  grid_list%PXg(3) = 0.0d0;

  ! Recalculate the grid node momentum
  do b = 1, nb_body           ! Loop over all bodies
```

```
    parBegin = body_list(b)%par_begin
    parEnd = body_list(b)%par_End

    if(contact) comID = body_list(b)%comid  ! Get comID from body

    do p = parBegin, parEnd ! Loop over all particles (3)
       pt => particle_list(p)

       icell = pt%icell
       ! Particle p is out of the computational region
       if (icell < 0) cycle

       mp_ = pt%mass

       ! Calculate the shape functions
       InflNode(1:8)=CellsNode(icell,:)
       if (GIMP) then
          call FindInflNode(p,icell)
          call NShape_GIMP(p)
       else
          call NShape(CellsNode(icell,1),p,0)
       end if

       ! Loop over all grid nodes of the hexahedron
       !  in which particle p is located
       do n = 1, nb_InflNode
          if(InflNode(n) .gt. nb_gridnode .or. InflNode(n) .le.
             0) & cycle  ! out of the computational grid
          gd => grid_list(comID, InflNode(n))

          shm = SHP(n)*mp_
          gd%PXg = gd%PXg + pt%VXp*shm

       end do ! n
    end do    ! p
 end do       ! b

 ! Applying essential boundary conditions
 do c = 1, nb_component
    do n = 1, nb_gridnode
       gd => grid_list(c, n)
       node => node_list(n)
       if (node%Fix_x) then
          gd%PXg(1) = 0.0
          gd%FXg(1) = 0.0
       end if

       if (node%Fix_y) then
          gd%PXg(2) = 0.0
          gd%FXg(2) = 0.0
       end if
```

```
        if (node%Fix_z) then
           gd%PXg(3) = 0.0
           gd%FXg(3) = 0.0
        end if

     end do !n
   end do     ! c
end subroutine GridMomentumMUSL
```

Step 9. For the MUSL and USL, calculate the particle strain and vorticity increments, and update the particle density and stress by using the subroutine ParticleStressUpdate, as shown in Step 3.

4.7 NUMERICAL EXAMPLES

Several representative numerical examples of the MPM3D-F90 are presented in this section, whose input data files can be found in the folder "Data".

4.7.1 TNT Slab Detonation

A 0.1 m long TNT slab is detonated from its left end, and the detonation wave travels to its right end at the detonation speed. The left end of the slab is fixed and the right end is free. This problem is often taken as a benchmark to validate the codes simulating high explosives.

The MPM3D-F90 is used to analyze the uniaxial wave propagation problem, in which the particles are totally constrained in the y and z directions by imposing the symmetric boundary conditions on the boundary planes perpendicular to the y and z axes. The left boundary plane of the grid is also treated as the symmetric boundary.

The slab is discretized into 4000 particles, with the initial particle spacing of 0.025 mm. The grid spacing is 0.05 mm so that there are initially 2 particles in each cell. The left end is set as the detonation point. The density and detonation speed of the TNT are chosen to be 1.63×10^{-3} g/mm^3 and 6930 m/s, respectively. The JWL EOS is employed to update the pressure of the detonation products of TNT, whose parameters are listed in Table 4.2. The end-time of simulation is chosen as 0.015 ms, at which the detonation process is completed. The input data file of this example is "Deto1k.mpm".

The theoretical CJ pressure of this problem is 1.957×10^4 MPa, while the experimental pressure is 2.1×10^4 MPa. The current version of the MPM3D-F90 is unable to plot a variable along a line in space so that the time history of pressure of a few selected particles are plotted in Fig. 4.2. The peak pressure obtained by the MPM3D-F90 is about 1.90×10^4 MPa, which is very close to the theoretical CJ pressure.

TABLE 4.2 Parameters of JWL Equation of State

P_{CJ} (MPa)	A (MPa)	B (MPa)	R_1	R_2	ω	E_0 (MJ/mm^3)
2.1×10^4	3.712×10^5	3.21×10^3	4.15	0.95	0.3	6993

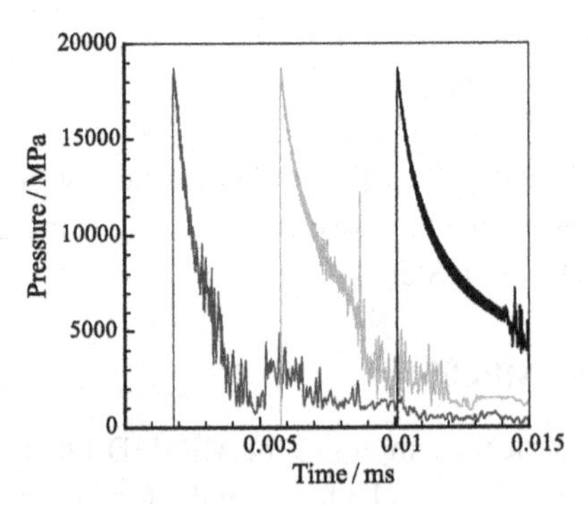

FIGURE 4.2 Time history of pressure of selected particles.

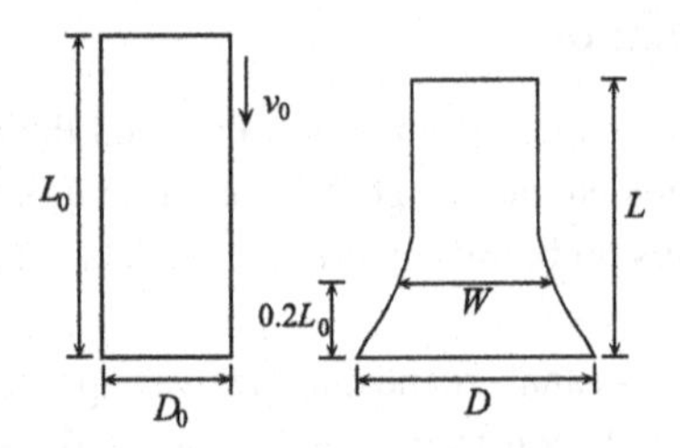

FIGURE 4.3 Taylor bar impact.

4.7.2 Taylor Bar Impact

Taylor bar impact experiments, in which a cylindrical metal bar normally impacts a rigid wall, are often used to validate the constitutive models implemented in computer codes. There are ample experimental data which can be referred to.

To quantitatively compare the final configurations of the bar as obtained by simulation and experiment, Johnson and Holmquist [118] proposed an average error as

$$\bar{\Delta} = \frac{1}{3}\left(\frac{|\Delta L|}{L} + \frac{|\Delta D|}{D} + \frac{|\Delta W|}{W}\right)$$

where L, D, and W are the final length, the diameter at the bottom, and the diameter at the section which is $0.2L_0$ away from the bottom, respectively, as shown in Fig. 4.3. ΔL, ΔD, and ΔW are the errors in L, D, and W between numerical and experimental results.

TABLE 4.3 Material Parameters

ρ (kg/m^3)	E (GPa)	ν	A (MPa)	B (MPa)	n	C
8930	117	0.35	157	425	1.0	0.0

TABLE 4.4 Comparison in Accuracy

	L (mm)	D (mm)	W (mm)	$\bar{\Delta}$
Experiment	16.2	13.5	10.1	–
MPM	16.3	13.0	9.6	0.031
FEM	16.3	13.2	10.1	0.009

TABLE 4.5 Comparison in Time Step

	Δt_{max} (μs)	Δt_{min} (μs)	Number of time steps
MPM	0.133	0.133	604
FEM	0.024	0.012	5483

In this example, a copper bar of initial length $L_0 = 25.4$ mm and diameter $D_0 = 7.6$ mm impacts a rigid wall at a velocity of 190 m/s. This problem is simulated by both the MPM and FEM. In the MPM simulation, the bar is discretized into 21,172 particles uniformly with an initial particle spacing of 0.38 mm. In the FEM simulation using LS-DYNA, the bar is discretized into 6528 elements and 7315 nodes, with the maximum element length of 0.76 mm. Johnson–Cook model is used to model the copper, whose parameters are listed in Table 4.3. The end-time of simulation is chosen as 80 μs, when the kinetic energy has reached zero. The input data file of this example is "Taylor.mpm".

Table 4.4 compares the numerical results obtained by the MPM3D-F90 and LS-DYNA with experimental results. Table 4.5 compares the maximum time step Δt_{max}, minimum time step Δt_{min}, and total number of time steps used in the MPM and FEM simulations. The time step safety factor is chosen as 0.8 in both the MPM and FEM simulations. Due to element distortion, the time step used in the FEM is reduced by 50%. This example demonstrated that the MPM is much more efficient than the FEM in solving large deformation problems.

4.7.3 Perforation of a Thick Plate

In this example, an ogive-nose high strength steel projectile impacts an aluminum target at a velocity of 400 m/s with an angle of 30° [119]. The projectile has a length of 88.9 mm and a diameter of 12.9 mm with a 3.0 caliber-radius-head. The target is an A6061-T651 plate of 26.3-mm thickness.

TABLE 4.6 Material Parameters of the Target

ρ (g/cm^3)	E (GPa)	ν	A (MPa)	B (MPa)	n	C	m
2.7	69	0.3	262	52.1	0.41	0	0.859
c_0 (m/s)	S_1	Γ_0	T_{melt} (K)	T_{room} (K)	$\varepsilon^p_{\text{fail}}$		
5350	1.34	2.0	875	293	1.6		

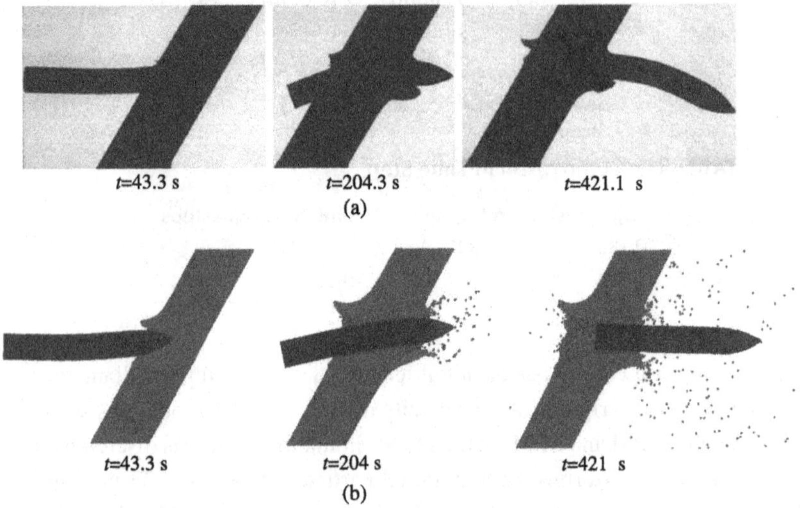

t=43.3 s t=204.3 s t=421.1 s

(a)

t=43.3 s t=204 s t=421 s

(b)

FIGURE 4.4 Configurations of the projectile and target obtained by (a) experiment and (b) MPM simulation at the striking velocity of 400 m/s [120].

The projectile is modeled by a linear elastic constitutive model with density $\rho = 7.85\,\text{g/cm}^3$, Young's modulus $E = 200$ GPa and Poisson's ratio $\nu = 0.3$. The target is modeled by the Johnson–Cook plasticity model and Mie–Grüneisen EOS with the parameters listed in Table 4.6. For the target, when the effective plastic strain of a particle reaches 1.6, the particle is labeled as a failure particle such that it is unable to sustain any deviatoric stress.

Due to symmetry, only half of the material domain is discretized in the simulation. The projectile is discretized into 13,314 particles with an initial particle spacing of 0.6 to 1.0 mm, while the target is discretized into 187,550 particles with an initial particle spacing of 1.0 mm. The grid spacing is 2.0 mm. This problem is solved with the USF version of the MPM, and the contact algorithm is turned on with the friction coefficient being zero. The input file of this example is "PeneOgive.mpm".

Fig. 4.4 compares the configurations of the projectile and target obtained by experiment and simulation at different times, which illustrates good agreement between the numerical and experimental results. The residual velocity of the

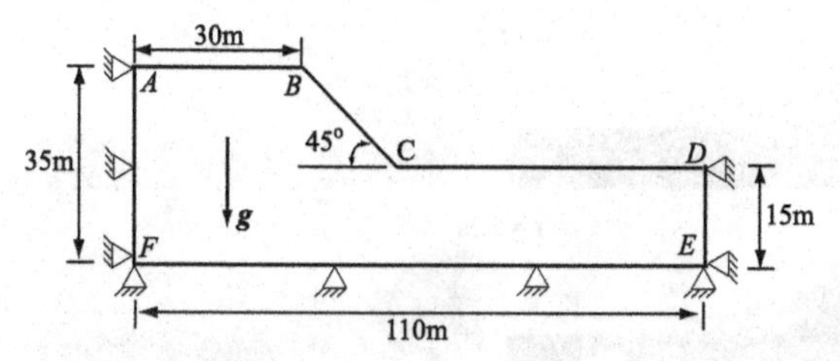

FIGURE 4.5 Soil slope under gravity.

TABLE 4.7 Material Parameters of Soil

E (MPa)	ν	ρ (g/cm^3)	$\phi°$	$\psi°$	c (kPa)
70	0.3	2.1	20	0.0	1.0
σ^t (kPa)	q_ϕ	k_ϕ (kPa)	q_ψ		
27.48	0.3545	0.974	0		

projectile obtained by the MPM simulation is 212.7 m/s which is very close to its experimental value of 217 m/s.

4.7.4 Failure of Soil Slope

The failure process of a soil slope as shown in Fig. 4.5 under gravity is simulated. The left side AF and right side DE are constrained in the horizontal direction and free in the vertical direction, while the bottom side EF is fully constrained. In this simulation, the transverse direction of the computational model is constrained to yield a plane strain state [120].

The soil is modeled by the Drucker–Prager model with the parameters listed in Table 4.7. The slope is discretized into 19,640 particles with an initial spacing of 0.5 m. The grid spacing is chosen as 1 m. For comparison, this problem is also simulated with the use of LS-DYNA with 19,640 elements and 30,273 nodes. The input file of this example is "Slopfail.mpm".

Fig. 4.6 compares the failure process obtained by the MPM3D-F90 and LS-DYNA, in which the color represents effective plastic strain. The results obtained by the MPM3D-F90 agree well with those obtained by LS-DYNA, but the computational cost of the MPM3D-F90 is much lower than that of LS-DYNA due to its element distortion, as shown in Table 4.8.

Table 4.8 compares the maximum time step size $\Delta t_{\max}$, minimum time step size $\Delta t_{\min}$, and total number of time steps used in the MPM and FEM simulations. The time step safety factor is chosen as 0.2. The time step size in the FEM simulation is reduced from its initial value of 261 to 8.21 μs due to element distortion. The same regular background grid is used in the whole simulation

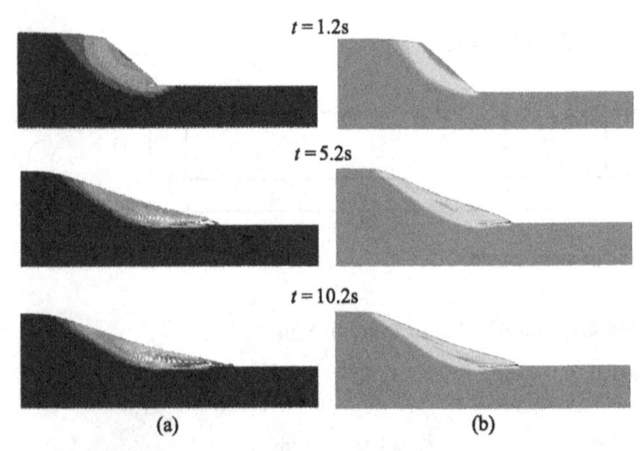

FIGURE 4.6 Slope failure process obtained by (a) MPM and (b) FEM.

TABLE 4.8 Comparison of Computational Cost

	Δt_{max} (μs)	Δt_{min} (μs)	Number of time steps	CPU time (min)
MPM	944	944	15,888	9
FEM	261	8.21	1,175,935	468

process with the MPM such that its time step size keeps constant during the simulation. The CPU time used in the MPM simulation is only 1/52 of that used in the FEM simulation.

Chapter 5

Coupling of the MPM with FEM

Contents

Both the FEM and MPM possess certain prominent advantages for different problems so that it is desirable to combine the FEM with MPM to take respective advantages of these two spatial discretization methods to advance SBES.

5.1 EXPLICIT FINITE ELEMENT METHOD

To illustrate the differences and similarities between the MPM and FEM, the basic formulation of the FEM is presented in this section.

5.1.1 Finite Element Discretization

The basic idea of the FEM is to discretize a continuous domain into a finite set of small elements of simple shapes which are interconnected at element nodes, and to approximate the global unknown function defined in the whole domain with a set of piecewisely defined approximation functions in the elements. Let $x_{iI}(t)$ and $X_{iI}(t)$ denote the coordinates of the node I at time t in the current configuration and reference configuration, respectively. The spatial coordinates $x_i(\boldsymbol{X}, t)$ of a particle $\boldsymbol{X}$ within an element at time t can be interpolated by the nodal coordinates $x_{iI}(t)$ of the element as follows:

$$x_i(\boldsymbol{X}, t) = N_I(\boldsymbol{X})x_{iI}(t) \tag{5.1}$$

where $N_I(\boldsymbol{X})$ is the value of the shape function associated with node I evaluated at the site of particle $\boldsymbol{X}$. In Eq. (5.1), the repeated nodal number index I rep-

The Material Point Method. http://dx.doi.org/10.1016/B978-0-12-407716-4.00005-3

resents a summation over its values, namely, over all the nodes of the element. The shape function $N_I(\boldsymbol{X})$ depends on the element type.

Similarly, the coordinates $X_i(t)$ of a particle $\boldsymbol{X}$ within an element at time t in the reference configuration can be interpolated by the nodal coordinates X_{iI} of the element as follows:

$$X_i(t) = N_I(\boldsymbol{X})X_{iI}. \tag{5.2}$$

The displacement of a particle $\boldsymbol{X}$ within an element can then be approximated by the nodal displacement of the element, i.e.,

$$u_i(\boldsymbol{X},t) = x_i(\boldsymbol{X},t) - X_i = N_I(\boldsymbol{X})u_{iI}(t) \tag{5.3}$$

in which

$$u_{iI}(t) = x_{iI}(x) - X_{iI} \tag{5.4}$$

is the displacement of node I.

Taking derivatives of Eq. (5.3) with respect to time t and coordinate x_j, respectively, gives

$$\ddot{u}_i(\boldsymbol{X},t) = N_I(\boldsymbol{X})\ddot{u}_{iI}(t), \tag{5.5}$$

$$u_{i,j}(\boldsymbol{X},t) = N_{I,j}(\boldsymbol{X})u_{iI}(t). \tag{5.6}$$

Similarly, the virtual displacement of a particle can also be approximated as

$$\delta u_i(\boldsymbol{X},t) = N_I(\boldsymbol{X})\delta u_{iI}(t) \tag{5.7}$$

where δu_{iI} denotes the virtual displacement of node I. Substituting Eqs. (5.5)–(5.7) into Eq. (2.65) results in

$$\delta u_{iI}\left(\int_\Omega \rho N_I N_J \ddot{u}_{iJ}\mathrm{d}V + \int_\Omega N_{I,j}\sigma_{ij}\mathrm{d}V - \int_\Omega N_I \rho b_i \mathrm{d}V - \int_{\Gamma_t} N_I \bar{t}_i \mathrm{d}\Gamma\right) = 0. \tag{5.8}$$

Since the virtual displacements equal zero on the essential boundaries and are arbitrary elsewhere, Eq. (5.8) can be reduced to

$$M_{IJ}\ddot{u}_{iJ} = f_{iI}^{\text{int}} + f_{iI}^{\text{ext}} \quad \forall I \notin \Gamma_u \tag{5.9}$$

where

$$f_{iI}^{\text{int}} = -\int_\Omega N_{I,j}\sigma_{ij}\mathrm{d}V \tag{5.10}$$

is the internal nodal force,

$$f_{iI}^{\text{ext}} = \int_{\Omega} N_I \rho b_i \mathrm{d}V + \int_{\Gamma_t} N_I \bar{t}_i \mathrm{d}\Gamma \tag{5.11}$$

is the external nodal force, and

$$M_{IJ} = \int_{\Omega} \rho N_I N_J \mathrm{d}V \tag{5.12}$$

is the consistent mass matrix. Using Eqs. (2.13) and (2.41), the integral in Eq. (5.12) can be transformed from the current configuration to the reference configuration as follows:

$$M_{IJ} = \int_{\Omega_0} \rho_0 N_I N_J \mathrm{d}V_0. \tag{5.13}$$

Thus, the mass matrix is constant so that it only needs to be calculated once in the reference configuration.

If the explicit time integration is used, a lumped mass matrix instead of the consistent matrix is employed so that Eq. (5.9) is reduced to

$$M_I \ddot{u}_{iI} = f_{iI}^{\text{int}} + f_{iI}^{\text{ext}} \quad \forall I \notin \Gamma_u \tag{5.14}$$

where M_I is the Ith diagonal element of the lumped mass matrix.

5.1.2 The FEM Formulation in Matrix Form

In computer programming, second- and fourth-order tensors are written in column and square matrices, respectively, by using the Voigt notation [107]. For example, the Cauchy stress tensor $\boldsymbol{\sigma}$ and the deformation rate tensor $\boldsymbol{D}$ are written in column matrices as

$$\boldsymbol{\sigma} = [\, \sigma_{11} \quad \sigma_{22} \quad \sigma_{33} \quad \sigma_{23} \quad \sigma_{13} \quad \sigma_{12} \,]^{\mathrm{T}}, \tag{5.15}$$

$$\boldsymbol{D} = [\, D_{11} \quad D_{22} \quad D_{33} \quad 2D_{23} \quad 2D_{13} \quad 2D_{12} \,]^{\mathrm{T}} \tag{5.16}$$

where the factor 2 in the Voigt rule for strains and strain-like tensors is used to make the energy expressions equivalent in both matrix and indicial notations, namely,

$$\rho \dot{w}^{\text{int}} = D_{ji} \sigma_{ji} = \boldsymbol{D}^{\mathrm{T}} \boldsymbol{\sigma}. \tag{5.17}$$

Similarly, the nodal vectors, such as u_{iI}, can also be converted to a column matrix $\boldsymbol{a}$ by setting $a_b = u_{iI}$ with index $b = (I-1) * n_{\text{SD}} + i$, where n_{SD} is the number of space dimensions. For example, the velocities $\dot{u}_{iI}$ of all nodes can be converted to

$$\dot{\boldsymbol{a}} = [\ \dot{\boldsymbol{a}}_1^{\mathrm{T}} \quad \dot{\boldsymbol{a}}_2^{\mathrm{T}} \quad \ldots \quad \dot{\boldsymbol{a}}_N^{\mathrm{T}}\ ]^{\mathrm{T}}, \quad \dot{\boldsymbol{a}}_I = [\ \dot{u}_{1I} \quad \dot{u}_{2I} \quad \dot{u}_{3I}\ ]^{\mathrm{T}} \tag{5.18}$$

where the column matrix $\dot{\boldsymbol{a}}$ is the global nodal velocity, and the sub-column matrix $\dot{\boldsymbol{a}}_I$ is the velocity of node I.

The deformation rate tensor

$$D_{ij} = \frac{1}{2}(\dot{u}_{i,j} + \dot{u}_{j,i}) = \frac{1}{2}(\dot{u}_{iI}N_{I,j} + \dot{u}_{jI}N_{I,i}) \tag{5.19}$$

can be converted to its column matrix form as

$$\boldsymbol{D} = \boldsymbol{B}_I\dot{\boldsymbol{a}}_I = \boldsymbol{B}\dot{\boldsymbol{a}} \tag{5.20}$$

where

$$\boldsymbol{B}_I = \begin{bmatrix} \frac{\partial N_I}{\partial x_1} & 0 & 0 & 0 & \frac{\partial N_I}{\partial x_3} & \frac{\partial N_I}{\partial x_2} \\ 0 & \frac{\partial N_I}{\partial x_2} & 0 & \frac{\partial N_I}{\partial x_3} & 0 & \frac{\partial N_I}{\partial x_1} \\ 0 & 0 & \frac{\partial N_I}{\partial x_3} & \frac{\partial N_I}{\partial x_2} & \frac{\partial N_I}{\partial x_1} & 0 \end{bmatrix}^{\mathrm{T}}, \tag{5.21}$$

$$\boldsymbol{B} = [\ \boldsymbol{B}_1 \quad \boldsymbol{B}_2 \quad \ldots \quad \boldsymbol{B}_N\ ]. \tag{5.22}$$

The column matrix form of Eq. (5.10) is given by

$$\boldsymbol{f}_I^{\mathrm{int}} = \int_V \boldsymbol{B}_I^{\mathrm{T}}\boldsymbol{\sigma}\mathrm{d}V, \tag{5.23}$$

$$\boldsymbol{f}^{\mathrm{int}} = \int_V \boldsymbol{B}^{\mathrm{T}}\boldsymbol{\sigma}\mathrm{d}V \tag{5.24}$$

where the column matrix $\boldsymbol{f}^{\mathrm{int}} = [\ (\boldsymbol{f}_1^{\mathrm{int}})^{\mathrm{T}} \quad (\boldsymbol{f}_2^{\mathrm{int}})^{\mathrm{T}} \quad \ldots \quad (\boldsymbol{f}_N^{\mathrm{int}})^{\mathrm{T}}\ ]^{\mathrm{T}}$ is the global nodal internal force, and the sub-column matrix $\boldsymbol{f}_I^{\mathrm{int}} = [\ f_{1I}^{\mathrm{int}} \quad f_{2I}^{\mathrm{int}} \quad f_{3I}^{\mathrm{int}}\ ]^{\mathrm{T}}$ is the internal force of node I.

The column matrix form of Eq. (5.11) is given by

$$\boldsymbol{f}_I^{\mathrm{ext}} = \int_V \boldsymbol{N}_I^{\mathrm{T}}\rho\boldsymbol{b}\mathrm{d}V + \int_{A_t} \boldsymbol{N}_I^{\mathrm{T}}\bar{\boldsymbol{t}}\mathrm{d}A, \tag{5.25}$$

$$\boldsymbol{f}^{\mathrm{ext}} = \int_V \boldsymbol{N}^{\mathrm{T}}\rho\boldsymbol{b}\mathrm{d}V + \int_{A_t} \boldsymbol{N}^{\mathrm{T}}\bar{\boldsymbol{t}}\mathrm{d}A \tag{5.26}$$

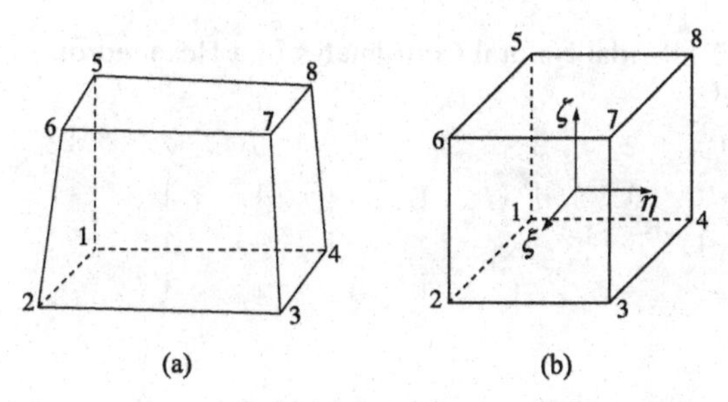

FIGURE 5.1 8-node hexahedron solid element in (a) physical coordinates, and (b) natural coordinates.

where

$$\begin{aligned}
&\boldsymbol{N}_I = N_I \boldsymbol{I}, \\
&\boldsymbol{N} = [\, \boldsymbol{N}_1 \quad \boldsymbol{N}_2 \quad \ldots \quad \boldsymbol{N}_N \,], \\
&\bar{\boldsymbol{t}} = [\, \bar{t}_1 \quad \bar{t}_2 \quad \bar{t}_3 \,]^{\mathrm{T}}, \\
&\boldsymbol{f}_I^{\mathrm{ext}} = [\, f_{1I}^{\mathrm{ext}} \quad f_{2I}^{\mathrm{ext}} \quad f_{3I}^{\mathrm{ext}} \,]^{\mathrm{T}}, \\
&\boldsymbol{f}^{\mathrm{ext}} = [\, (\boldsymbol{f}_1^{\mathrm{ext}})^{\mathrm{T}} \quad (\boldsymbol{f}_2^{\mathrm{ext}})^{\mathrm{T}} \quad \ldots \quad (\boldsymbol{f}_N^{\mathrm{ext}})^{\mathrm{T}} \,]^{\mathrm{T}}.
\end{aligned}$$

The column matrix form of Eq. (5.9) thus takes the form of

$$\boldsymbol{M}\ddot{\boldsymbol{a}} = \boldsymbol{f} \tag{5.27}$$

in which the column matrix

$$\boldsymbol{f} = \boldsymbol{f}^{\mathrm{ext}} + \boldsymbol{f}^{\mathrm{int}} \tag{5.28}$$

is the global nodal force, and the column matrix $\ddot{\boldsymbol{a}}$ is the global nodal acceleration.

5.1.3 Hexahedron Element

The shape function $N_I(\boldsymbol{X})$ depends on the element type. For example, the shape function for an 8-node hexahedron element as shown in Fig. 5.1 is given by

$$\begin{aligned}
N_I(\xi,\eta,\zeta) &= \frac{1}{8}(1+\xi_I\xi)(1+\eta_I\eta)(1+\zeta_I\zeta) \\
&= \frac{1}{8}(1+\xi_I\xi+\eta_I\eta+\zeta_I\zeta+\xi_I\eta_I\xi\eta \\
&\quad +\eta_I\zeta_I\eta\zeta+\xi_I\zeta_I\xi\zeta+\xi_I\eta_I\zeta_I\xi\eta\zeta)
\end{aligned} \tag{5.29}$$

TABLE 5.1 Nodal Natural Coordinates of a Hexahedron Element

I	1	2	3	4	5	6	7	8
ξ_I	−1	1	1	−1	−1	1	1	−1
η_I	−1	−1	1	1	−1	−1	1	1
ζ_I	−1	−1	−1	−1	1	1	1	1

where ξ, η, and ζ are the natural coordinates of a particle, and ξ_I, η_I, and ζ_I are the natural coordinates of node I supporting the particle, which take values of $(\pm 1, \pm 1)$, as shown in Table 5.1.

5.1.3.1 One-Point Gauss Quadrature

In the FEM, the integrals in Eqs. (5.23) and (5.25) are evaluated by using Gauss quadrature. The shape function (5.29) transforms a hexahedron in the xyz-coordinate system into a cube in the $\xi\eta\zeta$-coordinate system. As a result, the volume integral over a hexahedron element Ω_e can be converted into a volume integral over a cube according to Eq. (2.13) as follows:

$$\int_{\Omega_e} g(x, y, z)\mathrm{d}x\mathrm{d}y\mathrm{d}z = \int_{-1}^{1}\int_{-1}^{1}\int_{-1}^{1} g(\xi, \eta, \zeta)\,|J|\,\mathrm{d}\xi\mathrm{d}\eta\mathrm{d}\zeta. \tag{5.30}$$

Employing Gauss quadrature in Eq. (5.30) gives

$$\int_{\Omega_e} g(x, y, z)\mathrm{d}x\mathrm{d}y\mathrm{d}z = \sum_{J=1}^{n}\sum_{K=1}^{n}\sum_{L=1}^{n} g(\xi_J, \eta_K, \zeta_L)\,|J(\xi_J, \eta_K, \zeta_L)|\, w_J w_K w_L \tag{5.31}$$

where n is the number of Gauss quadrature points in each dimension, (ξ_J, η_K, ζ_L) are the natural coordinates of the quadrature point (J, K, L), and (w_J, w_K, w_L) are the corresponding weights. In one-point Gauss quadrature,

$$n = 1, \quad w_1 = 2, \quad \xi_1 = \eta_1 = \zeta_1 = 0$$

such that the volume integral over the hexahedron element Ω_e is evaluated as

$$\int_{\Omega_e} g(x, y, z)\mathrm{d}x\mathrm{d}y\mathrm{d}z \approx 8g(0, 0, 0)\,|J(0, 0, 0)|\,. \tag{5.32}$$

Letting $g = 1$ in Eq. (5.32) gives the approximation of the volume of an 8-node hexahedron element as

$$V_e \approx 8\,|J(0, 0, 0)|\,. \tag{5.33}$$

Hence, the nodal internal force Eq. (5.10) and external force Eq. (5.11) can be evaluated using the one-point Gauss quadrature as follows:

$$f_{iI}^{\text{int}} = -\sum_e N_{Ie,j}\sigma_{ije}V_e, \tag{5.34}$$

$$f_{iI}^{\text{ext}} = \sum_e N_{Ie}b_{ie}m_e + \sum_e N_{Ie}\bar{t}_i A_e. \tag{5.35}$$

5.1.3.2 Strain Matrix

The velocity $v_i(\boldsymbol{X},t)$ of a particle $\boldsymbol{X}$ within an element can be interpolated by

$$v_i(\boldsymbol{X},t) = v_i(\boldsymbol{X}(\xi,\eta,\zeta),t) = N_I(\xi,\eta,\zeta)v_{iI}(t). \tag{5.36}$$

Eq. (5.36) can be written in matrix form as

$$\boldsymbol{v} = \boldsymbol{N}^e\boldsymbol{v}^e \tag{5.37}$$

with

$$\boldsymbol{N}^e = \begin{bmatrix} N_1 & 0 & 0 & N_2 & 0 & \dots & 0 & 0 \\ 0 & N_1 & 0 & 0 & N_2 & \dots & N_8 & 0 \\ 0 & 0 & N_1 & 0 & 0 & \dots & 0 & N_8 \end{bmatrix},$$

$$\boldsymbol{v} = [\, v_1 \quad v_2 \quad v_3 \,]^{\mathrm{T}},$$

$$\boldsymbol{v}^e = [\, v_{11} \quad v_{21} \quad v_{31} \quad v_{12} \quad v_{22} \quad \dots \quad v_{24} \quad v_{24} \,]^{\mathrm{T}}.$$

The rate of deformation matrix $\boldsymbol{D} = [\, D_{11} \quad D_{22} \quad D_{33} \quad 2D_{23} \quad 2D_{13} \quad 2D_{12} \,]^{\mathrm{T}}$ of any particle within an element is given by

$$\boldsymbol{D} = \boldsymbol{L}\boldsymbol{v} \tag{5.38}$$

where the gradient operator matrix $\boldsymbol{L}$ takes the form of

$$\boldsymbol{L} = \begin{bmatrix} \frac{\partial}{\partial x} & 0 & 0 & 0 & \frac{\partial}{\partial z} & \frac{\partial}{\partial y} \\ 0 & \frac{\partial}{\partial y} & 0 & \frac{\partial}{\partial z} & 0 & \frac{\partial}{\partial x} \\ 0 & 0 & \frac{\partial}{\partial z} & \frac{\partial}{\partial y} & \frac{\partial}{\partial x} & 0 \end{bmatrix}^{\mathrm{T}}. \tag{5.39}$$

Substituting the velocity interpolation Eq. (5.37) into Eq. (5.38) leads to the rate of deformation matrix of a particle within the element e, i.e.,

$$\boldsymbol{D} = \boldsymbol{B}^e\boldsymbol{v}^e \tag{5.40}$$

where the strain-displacement matrix $\boldsymbol{B}^e$ is given by

$$\boldsymbol{B}^e = \boldsymbol{L}\boldsymbol{N}^e = [\,\boldsymbol{B}_1 \quad \boldsymbol{B}_2 \quad \ldots \quad \boldsymbol{B}_8\,] \tag{5.41}$$

with

$$\boldsymbol{B}_I = \begin{bmatrix} \frac{\partial N_I}{\partial x} & 0 & 0 & 0 & \frac{\partial N_I}{\partial z} & \frac{\partial N_I}{\partial y} \\ 0 & \frac{\partial N_I}{\partial y} & 0 & \frac{\partial N_I}{\partial z} & 0 & \frac{\partial N_I}{\partial x} \\ 0 & 0 & \frac{\partial N_I}{\partial z} & \frac{\partial N_I}{\partial y} & \frac{\partial N_I}{\partial x} & 0 \end{bmatrix}^{\mathrm{T}}. \tag{5.42}$$

The elements in the strain-displacement matrix are the derivatives of shape function N_I with respective to the spatial coordinates x_i, but the shape function N_I is a function of natural coordinate ξ, η, and ζ. The use of the chain rule yields

$$\begin{aligned} \frac{\partial N_I}{\partial \xi} &= \frac{\partial N_I}{\partial x}\frac{\partial x}{\partial \xi} + \frac{\partial N_I}{\partial y}\frac{\partial y}{\partial \xi} + \frac{\partial N_I}{\partial z}\frac{\partial z}{\partial \xi}, \\ \frac{\partial N_I}{\partial \eta} &= \frac{\partial N_I}{\partial x}\frac{\partial x}{\partial \eta} + \frac{\partial N_I}{\partial y}\frac{\partial y}{\partial \eta} + \frac{\partial N_I}{\partial z}\frac{\partial z}{\partial \eta}, \\ \frac{\partial N_I}{\partial \zeta} &= \frac{\partial N_I}{\partial x}\frac{\partial x}{\partial \zeta} + \frac{\partial N_I}{\partial y}\frac{\partial y}{\partial \zeta} + \frac{\partial N_I}{\partial z}\frac{\partial z}{\partial \zeta}, \end{aligned} \tag{5.43}$$

or in matrix form,

$$\begin{bmatrix} \frac{\partial N_I}{\partial \xi} \\ \frac{\partial N_I}{\partial \eta} \\ \frac{\partial N_I}{\partial \zeta} \end{bmatrix} = \begin{bmatrix} \frac{\partial x}{\partial \xi} & \frac{\partial y}{\partial \xi} & \frac{\partial z}{\partial \xi} \\ \frac{\partial x}{\partial \eta} & \frac{\partial y}{\partial \eta} & \frac{\partial z}{\partial \eta} \\ \frac{\partial x}{\partial \zeta} & \frac{\partial y}{\partial \zeta} & \frac{\partial z}{\partial \zeta} \end{bmatrix} \begin{bmatrix} \frac{\partial N_I}{\partial x} \\ \frac{\partial N_I}{\partial y} \\ \frac{\partial N_I}{\partial z} \end{bmatrix} = J \begin{bmatrix} \frac{\partial N_I}{\partial x} \\ \frac{\partial N_I}{\partial y} \\ \frac{\partial N_I}{\partial z} \end{bmatrix} \tag{5.44}$$

where J is the Jacobian matrix. The inverse of Eq. (5.44) is given by

$$\begin{bmatrix} \frac{\partial N_I}{\partial x} \\ \frac{\partial N_I}{\partial y} \\ \frac{\partial N_I}{\partial z} \end{bmatrix} = J^{-1} \begin{bmatrix} \frac{\partial N_I}{\partial \xi} \\ \frac{\partial N_I}{\partial \eta} \\ \frac{\partial N_I}{\partial \zeta} \end{bmatrix}. \tag{5.45}$$

It can be verified that the elements in the strain-displacement matrix at $\xi = \eta = \zeta = 0$ satisfy the following relations:

$$\begin{aligned} \frac{\partial N_7}{\partial x_i} &= -\frac{\partial N_1}{\partial x_i}, \quad \frac{\partial N_8}{\partial x_i} = -\frac{\partial N_2}{\partial x_i}, \\ \frac{\partial N_5}{\partial x_i} &= -\frac{\partial N_3}{\partial x_i}, \quad \frac{\partial N_6}{\partial x_i} = -\frac{\partial N_4}{\partial x_i} \end{aligned} \tag{5.46}$$

where the subscripts $i = 1, 2, 3$ correspond to x, y, z. The derivatives of x, y, and z with respect to ξ, η, and ζ can be obtained from Eqs. (5.1) and (5.29) as follows:

$$\begin{aligned} \frac{\partial x_i}{\partial \xi} &= \frac{\partial N_I}{\partial \xi} x_{iI} = \frac{1}{8} \sum_{I=1}^{8} \xi_I (1 + \eta \eta_I)(1 + \zeta \zeta_I) x_{iI}, \\ \frac{\partial x_i}{\partial \eta} &= \frac{\partial N_I}{\partial \eta} x_{iI} = \frac{1}{8} \sum_{I=1}^{8} \eta_I (1 + \xi \xi_I)(1 + \zeta \zeta_I) x_{iI}, \\ \frac{\partial x_i}{\partial \zeta} &= \frac{\partial N_I}{\partial \zeta} x_{iI} = \frac{1}{8} \sum_{I=1}^{8} \zeta_I (1 + \xi \xi_I)(1 + \eta \eta_I) x_{iI}. \end{aligned} \tag{5.47}$$

For the one-point Gauss quadrature, the quadrature point is $\xi_1 = \eta_1 = \zeta_1 = 0$ so that it follows that

$$\begin{aligned} \frac{\partial x_i}{\partial \xi} &= \frac{1}{8} \sum_{I=1}^{8} \xi_I x_{iI} = \frac{1}{8} [-x_{i1} + x_{i2} + x_{i3} - x_{i4} - x_{i5} + x_{i6} + x_{i7} - x_{i8}], \\ \frac{\partial x_i}{\partial \eta} &= \frac{1}{8} \sum_{I=1}^{8} \eta_I x_{iI} = \frac{1}{8} [-x_{i1} - x_{i2} + x_{i3} + x_{i4} - x_{i5} - x_{i6} + x_{i7} + x_{i8}], \\ \frac{\partial x_i}{\partial \zeta} &= \frac{1}{8} \sum_{I=1}^{8} \zeta_I x_{iI} = \frac{1}{8} [-x_{i1} - x_{i2} - x_{i3} - x_{i4} + x_{i5} + x_{i6} + x_{i7} + x_{i8}]. \end{aligned} \tag{5.48}$$

Compared with the $2 \times 2 \times 2$-point Gauss quadrature, the computational cost required to compute the strain-displacement matrix $\boldsymbol{B}$ is reduced by more than 25 times, and the number of multiplications is reduced by a factor of 16 in the calculation of strain and element nodal force. Furthermore, the cost for stress update is reduced by a factor of 8.

5.1.3.3 *Hourglass Modes*

Employing the one-point Gauss quadrature saves computational cost significantly, but it may cause spurious **zero-energy modes**, also called **hourglass**

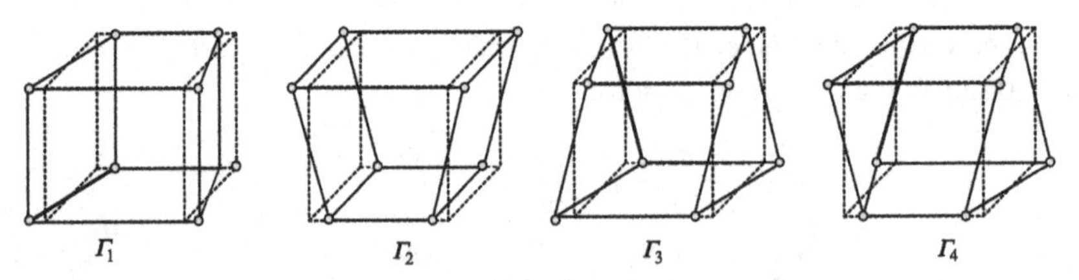

FIGURE 5.2 Hourglass modes in an 8-node hexahedron element. A total of 12 hourglass modes exist.

modes, which deteriorates the numerical solution, and even leads to the solution divergency. Hence, special measures must be taken to suppress the hourglass mode.

The shape function Eq. (5.29) of the 8-node hexahedron element can also be written in matrix form as

$$N = \frac{1}{8}(\boldsymbol{\Sigma}^{\mathrm{T}} + \boldsymbol{\Lambda}_1^{\mathrm{T}}\xi + \boldsymbol{\Lambda}_2^{\mathrm{T}}\eta + \boldsymbol{\Lambda}_3^{\mathrm{T}}\zeta + \boldsymbol{\Gamma}_1^{\mathrm{T}}\xi\eta + \boldsymbol{\Gamma}_2^{\mathrm{T}}\eta\zeta + \boldsymbol{\Gamma}_3^{\mathrm{T}}\xi\zeta + \boldsymbol{\Gamma}_4^{\mathrm{T}}\xi\eta\zeta) \quad (5.49)$$

with

$$N = [\, N_1 \quad N_2 \quad N_3 \quad N_4 \quad N_5 \quad N_6 \quad N_7 \quad N_8 \,]$$

and

$$\begin{aligned}
\boldsymbol{\Sigma} &= [\,1 \quad 1 \quad 1 \quad 1 \quad 1 \quad 1 \quad 1 \quad 1\,]^{\mathrm{T}}, \\
\boldsymbol{\Lambda}_1 &= [\,-1 \quad 1 \quad 1 \quad -1 \quad -1 \quad 1 \quad 1 \quad -1\,]^{\mathrm{T}}, \\
\boldsymbol{\Lambda}_2 &= [\,-1 \quad -1 \quad 1 \quad 1 \quad -1 \quad -1 \quad 1 \quad 1\,]^{\mathrm{T}}, \\
\boldsymbol{\Lambda}_3 &= [\,-1 \quad -1 \quad -1 \quad -1 \quad 1 \quad 1 \quad 1 \quad 1\,]^{\mathrm{T}}, \\
\boldsymbol{\Gamma}_1 &= [\,1 \quad -1 \quad 1 \quad -1 \quad 1 \quad -1 \quad 1 \quad -1\,]^{\mathrm{T}}, \\
\boldsymbol{\Gamma}_2 &= [\,1 \quad 1 \quad -1 \quad -1 \quad -1 \quad -1 \quad 1 \quad 1\,]^{\mathrm{T}}, \\
\boldsymbol{\Gamma}_3 &= [\,1 \quad -1 \quad -1 \quad 1 \quad -1 \quad 1 \quad 1 \quad -1\,]^{\mathrm{T}}, \\
\boldsymbol{\Gamma}_4 &= [\,-1 \quad 1 \quad -1 \quad 1 \quad 1 \quad -1 \quad 1 \quad -1\,]^{\mathrm{T}}.
\end{aligned}$$

The base vector $\boldsymbol{\Sigma}$ represents the rigid translation of the element, $\boldsymbol{\Lambda}_1$ represents tensile and compressive deformations, and $\boldsymbol{\Lambda}_2$ and $\boldsymbol{\Lambda}_3$ represent the shear deformations. The base vectors $\boldsymbol{\Gamma}_1$, $\boldsymbol{\Gamma}_2$, $\boldsymbol{\Gamma}_3$, and $\boldsymbol{\Gamma}_4$ are referred to as the **hourglass base vectors**. In an 8-node hexahedron element, there are 4 hourglass modes in each coordinate direction, which gives 12 hourglass modes in total. Fig. 5.2 illustrates the hourglass modes in the η-direction.

The velocity of a point within an element is given by

$$v_i(\xi, \eta, \zeta) = N_I(\xi, \eta, \zeta) v_{iI} = \boldsymbol{N} \boldsymbol{v}_i \tag{5.50}$$

where

$$\boldsymbol{v}_i = [\, v_{i1} \quad v_{i2} \quad v_{i3} \quad v_{i4} \quad v_{i5} \quad v_{i6} \quad v_{i7} \quad v_{i8} \,]^{\mathrm{T}}$$

is the element nodal velocity vector consisting of velocity components in the x_i-direction of all nodes in the element. Eqs. (5.50) and (5.49) show that the velocity $v_i(\xi, \eta, \zeta)$ of any point within the element can be expressed as a linear combination of the eight base vectors, namely, $\boldsymbol{\Sigma}$, $\boldsymbol{\Lambda}_1$ through $\boldsymbol{\Lambda}_3$, and $\boldsymbol{\Gamma}_1$ through $\boldsymbol{\Gamma}_4$.

The calculation of stress and strain requires the partial derivatives of the shape function N_I with respect to x_i, $\partial N_I/\partial x_i$, which can be calculated from $\partial N_I/\partial \xi$, $\partial N_I/\partial \eta$, and $\partial N_I/\partial \zeta$. These partial derivatives of the shape function N_I at the centroid ($\xi = \eta = \zeta = 0$) of the element can be evaluated using Eq. (5.49) as follows:

$$\frac{\partial \boldsymbol{N}}{\partial \xi} = \frac{1}{8}\boldsymbol{\Lambda}_1^{\mathrm{T}}, \quad \frac{\partial \boldsymbol{N}}{\partial \eta} = \frac{1}{8}\boldsymbol{\Lambda}_2^{\mathrm{T}}, \quad \frac{\partial \boldsymbol{N}}{\partial \zeta} = \frac{1}{8}\boldsymbol{\Lambda}_3^{\mathrm{T}} \tag{5.51}$$

in which the hourglass modes $\boldsymbol{\Gamma}_1$ through $\boldsymbol{\Gamma}_4$ vanish. Therefore, the hourglass modes result in a zero strain and thus zero stress at the centroid of an element, although they result in a nonzero strain and nonzero stress elsewhere. Since the one-point quadrature is used, the nodal internal force Eq. (5.34) is simply the product of the volume and the integrand evaluated at the centroid ($\xi = \eta = \zeta = 0$) of the element, which is zero for the hourglass modes due to the zero stress at the centroid. In other words, the hourglass modes will not generate any nodal force, i.e., they will not be resisted by the element, so that they will lead to spurious oscillations. The hourglass modes have no contribution to the strain energy of an element. That is why they are called the zero energy modes.

For example, in the Taylor bar impact simulation, the one-point Gauss quadrature results in a significant spurious oscillation due to the hourglass modes, as shown in Fig. 5.3(a). This oscillation can be effectively eliminated by introducing hourglass-resisting forces, as shown in Fig. 5.3(b).

The hourglass modes can be effectively suppressed by employing a viscous hourglass control scheme, i.e., imposing hourglass-resisting forces. It can be verified that the hourglass base vectors are orthogonal to other base vectors, i.e., $\boldsymbol{\Gamma}_k^{\mathrm{T}}\boldsymbol{\Sigma} = 0$ and $\boldsymbol{\Gamma}_k^{\mathrm{T}}\boldsymbol{\Lambda}_l = 0$, with $k = 1, 2, 3, 4$ and $l = 1, 2, 3$. If an hourglass base

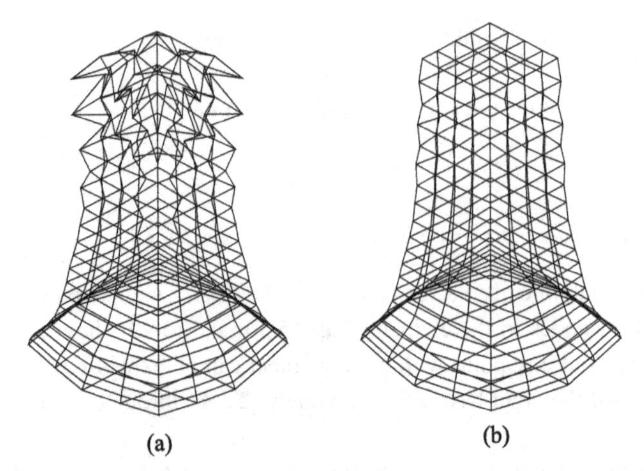

FIGURE 5.3 Oscillation caused by hourglass modes (a) without hourglass control, and (b) with hourglass control.

vector $\boldsymbol{\Gamma}_k$ is not orthogonal to the element nodal velocity vector $\boldsymbol{v}_i$, i.e.,

$$h_{ik} = \boldsymbol{\Gamma}_k^{\mathrm{T}} \boldsymbol{v}_i \neq 0, \tag{5.52}$$

the hourglass modes exist in the element velocity field so that the viscous forces should be applied at all the nodes of the element to resist the hourglass modes. The viscous forces f_{ik}^I applied at node I in the x_i-direction to resist the hourglass mode $\boldsymbol{\Gamma}_k$ should be proportional to h_{ik} and opposite to $\boldsymbol{\Gamma}_k$, namely,

$$f_{ik}^I = -\alpha_h h_{ik} \Gamma_{kI}, \quad k = 1, 2, 3, 4 \tag{5.53}$$

where Γ_{kI} is the Ith component of the hourglass base vector $\boldsymbol{\Gamma}_k$. The coefficient α_h can be determined by

$$\alpha_h = Q_h \rho V_e^{2/3} \frac{c}{4} \tag{5.54}$$

where V_e is the element volume, c is the material sound speed, and Q_h is a user-defined constant that is usually taken from the interval from 0.05 to 0.15.

The hourglass-resisting forces Eq. (5.53) are not orthogonal to rigid body rotations so that they are not well suited in the problems involving large rigid rotations. Flanagan and Belytschko proposed an hourglass-resisting force orthogonal to the rigid body rotations [121]. They defined the hourglass velocity field as

$$v_{iI}^{\mathrm{HG}} = v_i - v_{iI}^{\mathrm{LIN}} \tag{5.55}$$

with

$$v_{iI}^{\mathrm{LIN}} = \bar{v}_i + \bar{v}_{i,j}(x_{jI} - \bar{x}_j), \tag{5.56}$$

$$\bar{x}_i = \frac{1}{8}\sum_{I=1}^{8} x_{iI}, \quad \bar{v}_i = \frac{1}{8}\sum_{I=1}^{8} v_{iI}. \tag{5.57}$$

The hourglass base vectors are then defined as

$$\gamma_{kI} = \Gamma_{kI} - N_{I,i}\sum_{J=1}^{8} x_{iJ}\Gamma_{kJ}. \tag{5.58}$$

If the element nodal velocity vector v_{iI} is not orthogonal to the hourglass base vector g_{ik}, i.e.,

$$g_{ik} = \sum_{I=1}^{8} v_{iI}\gamma_{kI} \neq 0, \tag{5.59}$$

the hourglass modes exist in the element velocity field so that it is necessary to apply the hourglass-resisting forces f_{ik}^{I} which are proportional to g_{ik} and opposite to the hourglass base vectors g_{ik}, namely,

$$f_{ik}^{I} = -\alpha_h g_{ik}\gamma_{kI}, \quad k = 1, 2, 3, 4. \tag{5.60}$$

The hourglass base vectors are orthogonal to other base vectors such that the work done by the hourglass-resisting forces is negligible. The above hourglass control scheme is simple and efficient.

5.1.4 Numerical Algorithm for an Explicit FEM

Finally, the numerical algorithm for an explicit FEM within a time step can be summarized as follows, with the superscript k denoting the time step k:

1. Update the stress and density at Gauss quadrature points:
   a. Calculate the rate of deformation tensor and vorticity tensor at Gauss point p based on the element nodal velocity $v_{iI}^{k-1/2}$, i.e.,

$$D_{ijp}^{k-1/2} = (N_{Ip,j}^{k} v_{iI}^{k-1/2} + N_{Ip,i}^{k} v_{jI}^{k-1/2}), \tag{5.61}$$

$$\Omega_{ijp}^{k-1/2} = (N_{Ip,j}^{k} v_{iI}^{k-1/2} - N_{Ip,i}^{k} v_{jI}^{k-1/2}). \tag{5.62}$$

   b. Update the element density based on the volumetric strain increment with

$$\rho_p^{k+1} = \rho_p^{k}/(1 + D_{iip}^{k-1/2}\Delta t). \tag{5.63}$$

   Note that the element density can also be calculated from Eq. (2.55).

c. Update the pressure and deviatoric stress based on $D_{ijp}^{k-1/2}$ and $\Omega_{ijp}^{k-1/2}$ with the use of a constitutive model.

2. Calculate the element nodal forces and impose appropriate boundary conditions:
   Calculate the nodal internal force $f_{iI}^{k,\text{int}}$, nodal external force $f_{iI}^{k,\text{ext}}$, and nodal hourglass-resisting force $f_{iI}^{k,\text{hg}}$ using Eqs. (5.34), (5.35), and Eq. (5.53) or Eq. (5.60), respectively. The sum of $f_{iI}^{k,\text{int}}$, $f_{iI}^{k,\text{ext}}$, and $f_{iI}^{k,\text{hg}}$ gives the nodal force f_{iI}^{k}. Impose appropriate boundary conditions. For example, set $f_I^k = 0$ if node I is located at the fixed boundary.
3. Update the nodal position and velocity with

$$v_{iI}^{k+1/2} = v_{iI}^{k-1/2} + \Delta t^k f_{iI}^k / M_I, \tag{5.64}$$

$$x_{iI}^{k+1} = x_{iI}^k + \Delta t^{k+1/2} v_{iI}^{k+1/2}. \tag{5.65}$$

EFEP90 is a 3D explicit finite element code developed in the Computational Dynamics Laboratory at Tsinghua University. It serves as the companion open source code of the Chinese book entitled "Computational Dynamics" [122], and can be downloaded from http://mpm3d.comdyn.cn. The source code structure of the EFEP90 is very similar to that of the MPM3D-F90.

5.2 HYBRID FEM AND MPM

To take advantage of both the FEM and MPM, a hybrid FEM and MPM scheme is described in this section with applications to reinforced concrete (RC) structures. As a common building material, RC has been extensively used to construct civilian buildings, dams, nuclear reactor containments, and various defense structures. Therefore, it is important to investigate its responses to blast and impact loadings, where large strain, high strain rate, fracture, and crushing phenomena occur.

The MPM could be used directly to simulate the transient responses of RC structures to extreme loadings. However, the direct use of the MPM is computationally expensive if both rebars and concrete are discretized by particles with the same particle spacing due to their significant difference in size. Based on the fact that the rebars in concrete mainly sustain tensile loading, Lian et al. proposed a hybrid finite-element material-point (HFEMP) method [39] by incorporating the truss element of the FEM into the MPM model to simulate the transient responses.

As shown in Fig. 5.4, the rebars in RC are discretized into truss elements as in the FEM, while the concrete is discretized into particles as in the MPM.

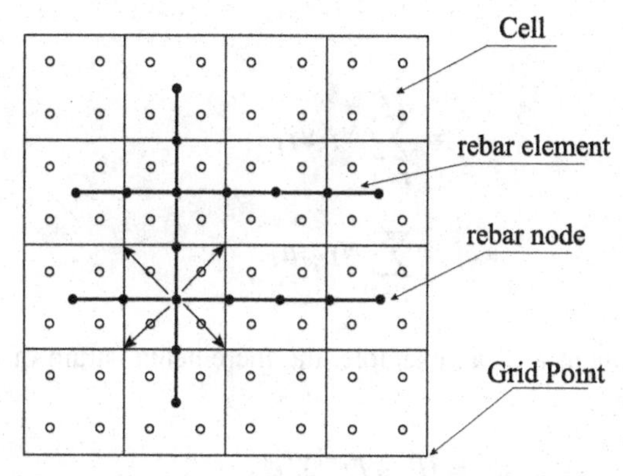

FIGURE 5.4 RC discretization in the HFEMP method. *Hollow dots* denote concrete material points, while *solid dots* denote rebar nodes, and the *solid lines* connecting *solid dots* denote rebar elements.

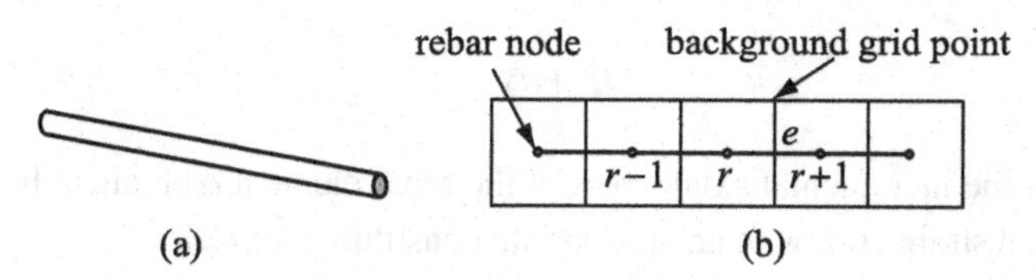

FIGURE 5.5 (a) A steel bar, and (b) discretized model with rebar elements and background grid.

All the rebar nodes and particles move in the same single-valued velocity field, approximately modeling the interaction between the rebars and concrete. Similar to the original MPM, the momentum equations in the HFEMP method are also solved on the background grid. In each time step, the momenta and forces of all particles and rebar nodes are mapped to the corresponding grid nodes to establish the nodal momentum equations. After solving the nodal momentum equations, the results are mapped from the grid nodes back to the particles and rebar nodes to update their positions and velocities. Strain increments of particles and rebar elements are calculated in different ways as discussed in detail as below. The stresses of particles and rebar elements are then updated based on respective constitutive models. The nodal force of a rebar node is obtained by accumulating the axial forces of the rebar elements connected to the node.

For the sake of demonstration, a steel bar with length L and cross-sectional area A is considered as an example. The steel bar is discretized by the rebar elements as shown in Fig. 5.5.

A variable of rebar node r, denoted by u_r, and its derivatives, $u_{r,j}$, can be obtained from its grid nodal value u_I via the standard FE shape functions as follows:

$$u_r = \sum_{I=1}^{n_g} N_{Ir} u_I, \tag{5.66}$$

$$u_{r,j} = \sum_{I=1}^{n_g} N_{Ir,j} u_I. \tag{5.67}$$

Taking rebar element e as an example, the incremental strain of the rebar element e is given by

$$\Delta\varepsilon_e^k = (l_e^k - l_e^{k-1})/l_e^{k-1} \tag{5.68}$$

where l_e^k denotes the length of the rebar element e at time t^k. The axial stress of the rebar element e is updated by

$$\sigma_e^{k+1} = \sigma_e^k + \Delta\sigma_e^k \tag{5.69}$$

where $\Delta\sigma_e^k$ is the incremental axial stress of the rebar element e obtained from the incremental strain $\Delta\varepsilon_e^k$ with an appropriate constitutive model.

The axial force of the rebar element e can be determined by

$$F_e^{k+1} = A\sigma_e^{k+1}. \tag{5.70}$$

Due to the contributions from the rebar nodes, the grid nodal mass m_I, momentum p_{iI}, internal force f_{iI}^{int}, and external force f_{iI}^{ext} in the HFEMP method take the form of

$$m_I = \sum_{p=1}^{n_p} m_p N_{Ip} + \sum_{r=1}^{n_r} m_r N_{Ir}, \tag{5.71}$$

$$p_{iI} = \sum_{p=1}^{n_p} m_p v_{ip} N_{Ip} + \sum_{r=1}^{n_r} m_r v_{ir} N_{Ir}, \tag{5.72}$$

$$f_{iI}^{\text{int}} = -\sum_{p=1}^{n_p} N_{Ip,j} \sigma_{ijp} m_p/\rho_p + \sum_{r=1}^{n_r} N_{Ir} f_{ir}^{\text{int}}, \tag{5.73}$$

$$f_{iI}^{\text{ext}} = \sum_{p=1}^{n_p} m_p N_{Ip} f_{ip} + \sum_{p=1}^{n_p} N_{Ip} \bar{t}_{ip} h^{-1} m_p/\rho_p + \sum_{r=1}^{n_r} N_{Ir} f_{ir}^{\text{ext}} \tag{5.74}$$

where n_r is the total number of rebar nodes, and m_r and v_{ir} are the lumped mass and velocity of rebar node r, respectively. f_{ir}^{ext} is the external force acting on the rebar node r, and

$$f_{ir}^{\text{int}} = \sum_{e=1}^{n_e} \Lambda_{re} F_e \cos\theta_{ie} \tag{5.75}$$

is the internal force of rebar node r. In Eq. (5.75), $\Lambda_{re} = 1$ for the left side node r of the rebar element e, $\Lambda_{re} = -1$ for other rebar nodes, and

$$\cos\theta_{ie} = (x_{i1}^k - x_{i2}^k)/l_e^k \tag{5.76}$$

is the directional cosine of the element e.

The last terms on the right side of Eqs. (5.71)–(5.74) are the contributions from the rebar nodes.

The numerical implementation of the HFEMP method can be described as follows:

1. Loop over all the particles and rebar nodes to calculate their contributions to the mass and momentum of the grid points of corresponding cells in which they are located. The mass m_I^k and momentum $p_{iI}^{k-1/2}$ of grid node I are obtained from Eqs. (5.71) and (5.72), respectively.
2. Loop over the grid points located on the boundary to reset their momentum corresponding to the essential boundary conditions.
3. This step is only used in the USF scheme. Update the stress state of particles and rebar elements as follows:
   a. Loop over all the particles to calculate their incremental strain tensor

$$\Delta\varepsilon_{ijp}^{k-1/2} = \frac{1}{2}\sum_{I=1}^{8}[N_{Ip,j}^k v_{iI}^{k-1/2} + N_{Ip,i}^k v_{jI}^{k-1/2}]\Delta t, \tag{5.77}$$

spin tensor

$$\Delta\Omega_{ijp}^{k-1/2} = \frac{1}{2}\sum_{I=1}^{8}[N_{Ip,j}^k v_{iI}^{k-1/2} - N_{Ip,i}^k v_{jI}^{k-1/2}]\Delta t^k, \tag{5.78}$$

and density

$$\rho_p^{k+1} = \rho_p^k/(1 + \Delta\varepsilon_{iip}^{k-1/2}), \tag{5.79}$$

and then to update the Cauchy stress

$$\sigma_{ijp}^{k+1} = \sigma_{ijp}^k + \dot{\sigma}_{ijp}^{k-1/2}\Delta t \tag{5.80}$$

of all the particles with an appropriate constitutive model and EOS (if needed). In the above equations, $v_{iI}^{k-1/2} = p_{iI}^{k-1/2}/m_I^k$ is the velocity of grid point I, $\dot{\sigma}_{ij}$ is the material time derivative of the Cauchy stress which is related to the Jaumann (co-rotational) stress rate σ_{ij}^{∇} by

$$\dot{\sigma}_{ij} = \sigma_{ij}^{\nabla} + \sigma_{ij}\Omega_{lj} + \sigma_{jl}\Omega_{li}. \tag{5.81}$$

The Jaumann (co-rotational) stress rate is determined from the strain rate $\dot{\varepsilon}_{ij} = (\dot{u}_{i,j} + \dot{u}_{j,i})/2$ with an appropriate constitution model.

b. Loop over all the rebar elements to calculate their incremental strain $\Delta\varepsilon_e^k$ using Eq. (5.68), and density using

$$\rho_e^{k+1} = \rho_e^k/(1 + \Delta\varepsilon_e^k), \tag{5.82}$$

and then update the axial stress σ_e^{k+1} with Eq. (5.69). Note that the stresses of rebar elements are always updated in their co-rotational coordinates.

4. Calculate the grid nodal internal force $f_{iI}^{\text{int},k}$ and external force $f_{iI}^{\text{ext},k}$ using Eqs. (5.73) and (5.74), respectively. If the USF scheme is used, $\sigma_{ijp} = \sigma_{ijp}^{k+1}$ and $\rho_p = \rho_p^{k+1}$; otherwise $\sigma_{ijp} = \sigma_{ijp}^k$ and $\rho_p = \rho_p^k$. If the node I is fixed in the ith direction, set $f_{iI}^k = f_{iI}^{\text{int},k} + f_{iI}^{\text{ext},k} = 0$ to make its corresponding acceleration $a_{iI}^k = 0$.
5. Loop over all the grid points to update their momentum using Eq. (3.56).
6. Loop over all the particles and rebar nodes to update their velocity and position by mapping the grid nodal results back to them using Eqs. (3.57) and (3.58).
7. This step is only used in the MUSL scheme. Extrapolate the new velocities of particles and rebar nodes to the grid nodes to obtain the improved nodal momentum, namely,

$$p_{iI}^{k+1/2} = \sum_{p=1}^{n_p} m_p v_{ip}^{k+1/2} N_{Ip}^k + \sum_{r=1}^{n_r} m_r v_{ir}^{k+1/2} N_{Ir}^k. \tag{5.83}$$

8. This step is only used in the MUSL and USF schemes. Update the stresses of particles and rebar elements based on the updated velocity $v_{iI}^{k+1/2}$ in a way similar to that given in step 3. For the rebar elements, the incremental strain $\Delta\varepsilon_e^{k+1}$ is calculated using Eq. (5.68).
9. Discard the deformed background grid, if needed, and define a new regular background grid. Return to step 1 for the next time step.

Lian et al. [39] simulated the perforation experiment of an ogival-nose projectile to an RC slab [123] using the HFEMP method. The geometric configuration and sizes of the projectile and the RC target with three layers of square-pattern reinforcement bars are shown in Figs. 5.6 and 5.7, respectively.

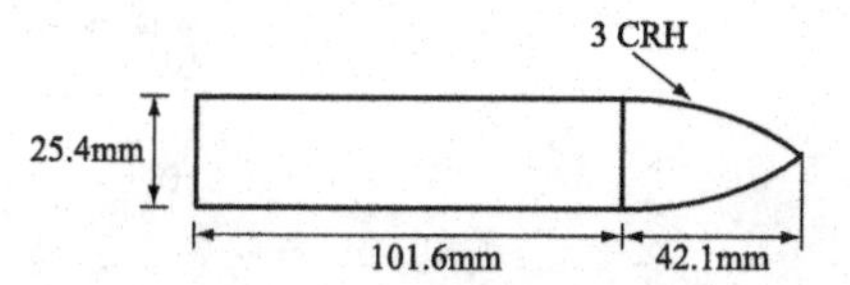

FIGURE 5.6 Projectile geometry (0.5 kg).

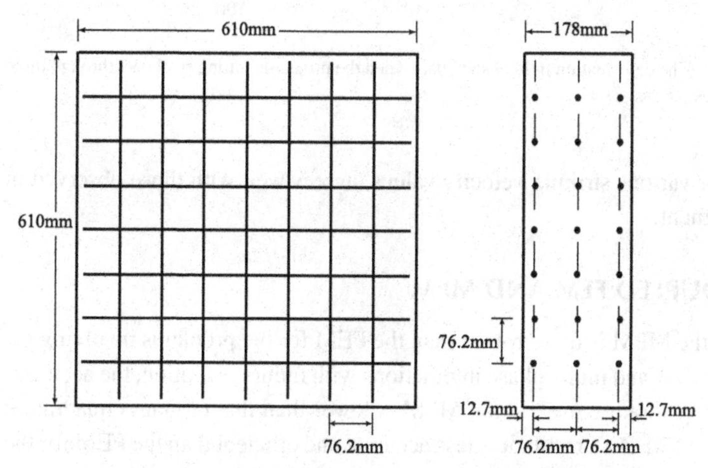

FIGURE 5.7 RC geometry with the location of steel reinforcement bars (5.59 mm diameter).

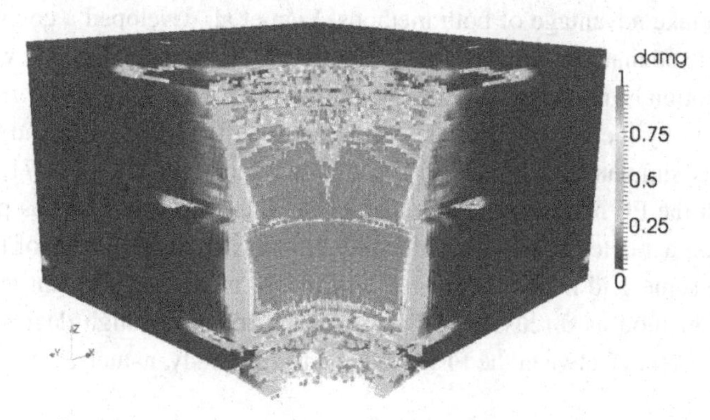

FIGURE 5.8 The damaged zone of RC after the projectile hitting steel bars through three layers.

An ideal elastoplastic model with a failure criterion was used for the rebars while the HJC model was used for concrete. The rebars were discretized by truss elements. Fig. 5.8 shows the damaged zone of the RC slab at time 0.5 ms, while Fig. 5.9 colored by the pressure value demonstrates the deformation field of the steel bars. The residual velocities of the projectile, as obtained with the HFEMP

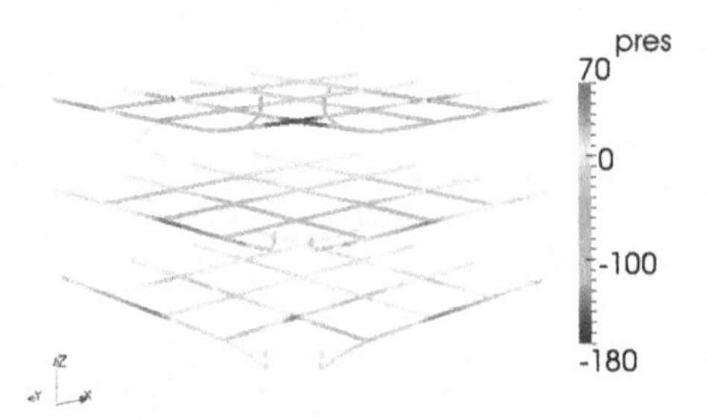

FIGURE 5.9 The deformation field of steel bars when the projectile hitting steel bars through three layers at times 0.5 ms.

method for various striking velocity values, agrees well with those observed in the experiment.

5.3 COUPLED FEM AND MPM

Although the MPM is more robust than the FEM for the problems involving severe distortions and multi-phase interactions with failure evolution, the accuracy of particle quadrature used in the MPM is lower than that of Gauss quadrature used in the FEM. As a result, it is less accurate and efficient than the FEM for the problems with small deformation. In addition, the MPM requires more computational storage because it requires both grid and particle data for the kinematic variables. To take advantage of both methods, Lian et al. developed a coupled finite-element and material-point (CFEMP) method [37], in which the body with small deformation is treated by the FEM, while the body with extreme deformation is handled by the MPM. The interaction between the FEM body and the MPM body is simulated with an MPM grid-based contact method [70,71,73, 74], in which the FE nodes located on the contact interface are treated as particles. The contact detection is determined by monitoring the velocities of two bodies at the same grid node. Therefore, the CFEMP method is different from the HFEMP method as discussed in Sect. 5.2, and requires a high degree of meshing consistency between the FEM body and MPM body, namely,

$$R = \frac{L_{\text{FEM}}}{L_{\text{MPM}}} \approx 1 \tag{5.84}$$

where L_{FEM} and L_{MPM} represent the characteristic length of FEM elements and MPM grid cells, respectively. If the meshing of the FEM body is coarser than the grid cell, the solution accuracy may be seriously deteriorated, and the MPM particles may penetrate into the FEM body. Consider the two bodies shown in Fig. 5.10 as an example in which the MPM body s is moving towards the FEM body r. Because the element size of FEM body r is much larger than the MPM

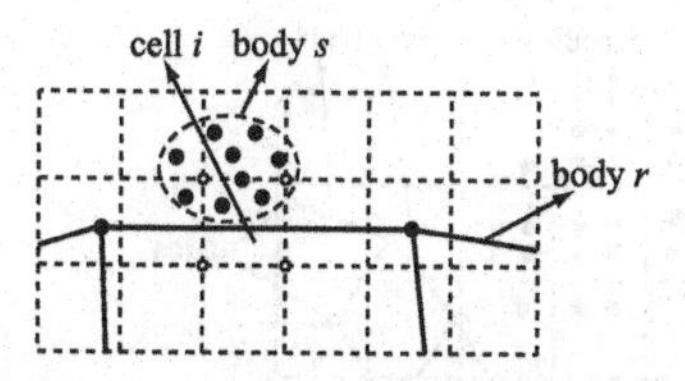

FIGURE 5.10 Illustration of the CFEMP method with inconsistent meshing.

grid cell size, only the MPM body s contributes to the grid nodal velocities in the cell i, while the FEM body r does not contribute to the gird nodal velocities. Based on the contact detection scheme used in the CFEMP method, the bodies r and s are considered as not in contact with each other at the grid nodes of cell i so that no contact force will be imposed between them in this case. Therefore, the body s will penetrate into the body r.

Consistent meshing may lead to the over-meshing in the FEM domain, which would significantly decrease the time step size and increase the computational cost as well as data storage. In order to satisfy the contact conditions exactly at the FEM element faces and to avoid the over-meshing in the FEM domain, Chen et al. [124] further improved the coupling between the FEM domain and MPM domain, based on a particle-to-surface contact method. The contact forces are calculated by using the Lagrange multiplier method, based on the penetration level between the MPM particles and FEM element faces. Moreover, a Coulomb friction model is employed to allow the relative slipping between two bodies. The improved procedure employs the master element faces rather than the grid nodes to detect whether the particles penetrate into the FEM domain so that the contact conditions are satisfied exactly between the particles and the FEM element faces, and the consistent meshing is no longer needed. Thus, the meshing of the FEM body can be much coarser than the grid cells of the MPM domain. Moreover, the computational efficiency of the improved procedure is much higher than the CFEMP method since the total number of finite elements is significantly reduced. Numerical studies have shown that the improved coupled finite-element material-point (ICFEMP) method is very robust and capable of modeling the extreme cases whose ratio of element size to grid cell size is far beyond 2.0.

For the purpose of demonstration, the element faces located on the surface of the FEM bodies are termed as segments in what follows. In the ICFEMP method, we first determine the contact pairs, namely, the MPM particles and the corresponding segments which may be penetrated by the particles, using a global search. Then the exact contact position and the gap between the contact pairs are calculated using a local search. Finally, a contact force resisting the

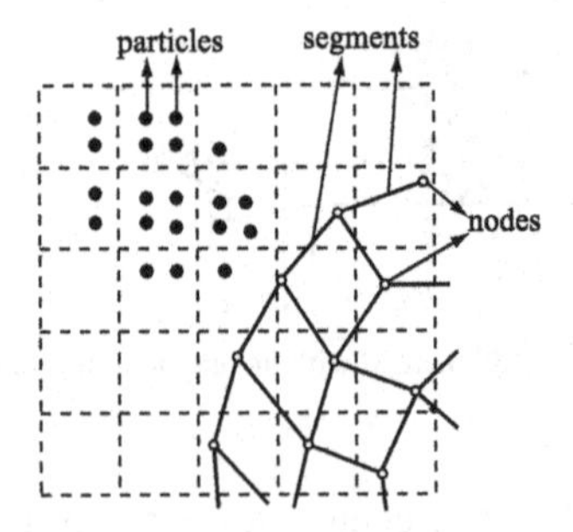

FIGURE 5.11 Bucket-sorting scheme.

penetration is imposed between the contact pair if the gap is negative. This step is neglected if the gap is equal to or greater than zero.

5.3.1 Global Search

The global search examines all segments to determine the potential contact particles. The global search costs most of the computer time in the contact algorithm so that a bucket-sorting scheme [125] as shown in Fig. 5.11 is used to minimize the cost as much as possible. The bucket-sorting scheme adopts a cell structure whose cell sizes are taken to be close to the average element size. The key part of this searching procedure is to identify all the particles in a cell, as shown below.

The cell domain can be described by the following:

$$(x_{\min}, x_{\max}, N_x), \quad (y_{\min}, y_{\max}, N_y), \quad (z_{\min}, z_{\max}, N_z) \tag{5.85}$$

where $x_{\min}$ and $x_{\max}$ denote the minimal and maximal x coordinates of the cell domain and N_x signifies the number of cells in the x-direction. Other variables are analogous to $x_{\min}$, $x_{\max}$, and N_x. The cell number in which a particle with the coordinates (x, y, z) is located can be determined by

$$I_p = I_{pz} \times N_x \times N_y + I_{py} \times N_x + I_{px} \tag{5.86}$$

with

$$\begin{aligned} I_{px} &= \text{round}(N_x(x - x_{\min})/(x_{\max} - x_{\min})), \\ I_{py} &= \text{round}(N_y(y - y_{\min})/(y_{\max} - y_{\min})), \\ I_{pz} &= \text{round}(N_z(z - z_{\min})/(z_{\max} - z_{\min})) \end{aligned} \tag{5.87}$$

where round(x) is a C++ intrinsic function which rounds down the value of x.

The MPM particles and segments located in the same cell are defined as the contact pairs.

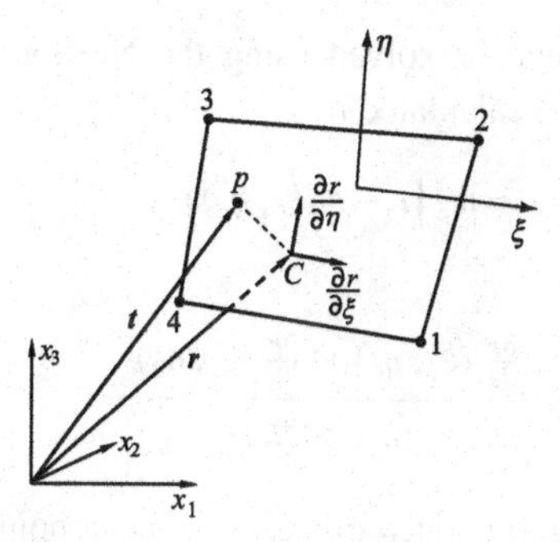

FIGURE 5.12 Local search scheme.

5.3.2 Local Search

The contact pairs detected by the global search are likely to contact with each other, but it must be further determined whether the contact occurs or not. The local search determines the exact contact position and the gap between the contact pairs.

As shown in Fig. 5.12, a segment can be described by a parametric equation [126,127] as

$$\boldsymbol{r} = f_1(\xi,\eta)\boldsymbol{i} + f_2(\xi,\eta)\boldsymbol{j} + f_3(\xi,\eta)\boldsymbol{k} \tag{5.88}$$

where $\boldsymbol{i}$, $\boldsymbol{j}$, and $\boldsymbol{k}$ denote the unit vectors in the directions of x_1, x_2, and x_3, respectively, and $f_i(\xi,\eta)$ is the corresponding global coordinate of the point (ξ,η) that can be obtained by interpolating the nodal coordinates of the segment as

$$f_i(\xi,\eta) = \sum_{J=1}^{4} \phi_J x_{iJ} \tag{5.89}$$

with $\phi_j(\xi,\eta) = \frac{1}{4}(1+\xi_i\xi)(1+\eta_i\eta)$ representing the shape function of the quadrilateral segment, and x_{iJ} the ith coordinate of the segment's jth node.

The local coordinates (ξ_c,η_c) of the contact point C on the segment can be determined from

$$\frac{\partial \boldsymbol{r}}{\partial \xi}(\xi_c,\eta_c)\cdot\left[\boldsymbol{t}-\boldsymbol{r}(\xi_c,\eta_c)\right]=0, \tag{5.90}$$

$$\frac{\partial \boldsymbol{r}}{\partial \eta}(\xi_c,\eta_c)\cdot\left[\boldsymbol{t}-\boldsymbol{r}(\xi_c,\eta_c)\right]=0 \tag{5.91}$$

where $\boldsymbol{t}$ denotes the position vector of the particle p, as shown in Fig. 5.12.

Eqs. (5.90) and (5.91) can be solved using the Newton–Rapson iterative method. Thus, the gap can be calculated by

$$g = \boldsymbol{n} \cdot \left[\boldsymbol{t} - \boldsymbol{r}(\xi_c, \eta_c) \right] \tag{5.92}$$

where

$$\boldsymbol{n} = \frac{\frac{\partial \boldsymbol{r}}{\partial \xi}(\xi_c, \eta_c) \times \frac{\partial \boldsymbol{r}}{\partial \eta}(\xi_c, \eta_c)}{\left| \frac{\partial \boldsymbol{r}}{\partial \xi}(\xi_c, \eta_c) \times \frac{\partial \boldsymbol{r}}{\partial \eta}(\xi_c, \eta_c) \right|} \tag{5.93}$$

is the unit normal vector pointing outwards at the contact point.

5.3.3 Contact Force

If $g \geqslant 0$, the particle does not penetrate into the segment so that no further treatment is needed. Otherwise, the contact force must be imposed between the particle p and the contact point C to prevent penetration. After imposing the contact force on the particle p of body s, the updated velocity $v_{ip}^{s,k+1/2}$ is given by

$$v_{ip}^{s,k+1/2} = \bar{v}_{ip}^{s,k+1/2} + \Delta t^k \frac{f_{ip}^{s,c,k}}{m_p^s} \tag{5.94}$$

where

$$\bar{v}_{ip}^{s,k+1/2} = v_{ip}^{s,k-1/2} + \Delta t^k \sum_{I=1}^{8} N_{Ip} \frac{f_{iI}^{s,k}}{m_I^s} \tag{5.95}$$

is the trial particle velocity, $v_{ip}^{s,k-1/2}$ is the particle velocity at the beginning of each time step, and $f_{ip}^{s,c,k}$ is the contact force applied on the particle p at time step t^k.

Similarly, the updated velocity $v_{ic}^{r,k+1/2}$ of the contact point C on the segment of body r can be evaluated by

$$v_{ic}^{r,k+1/2} = \bar{v}_{ic}^{r,k+1/2} + \Delta t^k \sum_{J=1}^{4} \phi_J(\xi_c, \eta_c) \frac{f_{iJ}^{r,c,k}}{m_J} \tag{5.96}$$

where

$$\bar{v}_{ic}^{r,k+1/2} = \sum_{J=1}^{4} \phi_J(\xi_c, \eta_c)\, \bar{v}_{iJ}^{k+1/2} \tag{5.97}$$

is the trial velocity of the contact point C, and

$$f_{iJ}^{r,c,k} = \phi_J(\xi_c, \eta_c) f_{ic}^{r,c,k} \tag{5.98}$$

with $f_{ic}^{r,c,k}$ being the contact force applied on the contact point C.

For a sticking contact, the updated velocities of the particle p and contact point C must satisfy the velocity continuity condition [71,72], namely

$$(v_{ic}^{r,k+1/2} - v_{ip}^{s,k+1/2}) = 0. \tag{5.99}$$

Substituting Eqs. (5.94) and (5.96) into Eq. (5.99), the contact force $f_{ic}^{r,c,k}$ for the sticking contact becomes

$$f_{ic}^{r,c,k} = \frac{m_p^s m_c^r (\bar{v}_{ic}^{s,k+1/2} - \bar{v}_{ip}^{r,k+1/2})}{(m_p^s + m_c^r)\Delta t^k} \tag{5.100}$$

where m_c^r denotes the equivalent mass of the contact point C defined by

$$\frac{1}{m_c^r} = \sum_{J=1}^{4} \frac{\phi_J^2(\xi_c, \eta_c)}{m_J}. \tag{5.101}$$

Therefore, the normal and tangential contact forces applied on the contact point C take the form of

$$f_{ic}^{\text{nor},k} = f_{jc}^{r,c,k} n_{jc}^{r,k} n_{ic}^{r,k}, \tag{5.102}$$

$$f_{ic}^{\text{tan},k} = f_{ic}^{r,c,k} - f_{ic}^{\text{nor},k} \tag{5.103}$$

where $n_{ic}^{r,k}$ denotes the unit normal vector pointing out of the segment at the contact point C.

For a slipping contact, the contact force applied at the contact point C can be written as

$$f_{ic}^{r,c,k} = f_{ic}^{\text{nor},k} + \min(\mu \| f_{ic}^{\text{nor},k} \|, \| f_{ic}^{\text{tan},k} \|) t_{ic}^{r,k} \tag{5.104}$$

where $t_{ic}^{r,k}$ denotes the unit tangential vector of the segment at the contact point C.

Finally, the contact forces applied on the four nodes of the segment can be obtained by

$$f_{iJ}^{r,c,k} = \sum_{c=1}^{n_c} \phi_J(\xi_c, \eta_c) f_{ic}^{r,c,k}, \quad J = 1, 2, 3, 4 \tag{5.105}$$

where n_c is the total number of contact points on all segments, namely, the total number of contact pairs. The contact forces applied on the grid points can be obtained by

$$f_{iI}^{s,c,k} = \sum_{p=1}^{n_p} N_{Ip} f_{ip}^{s,c,k}. \tag{5.106}$$

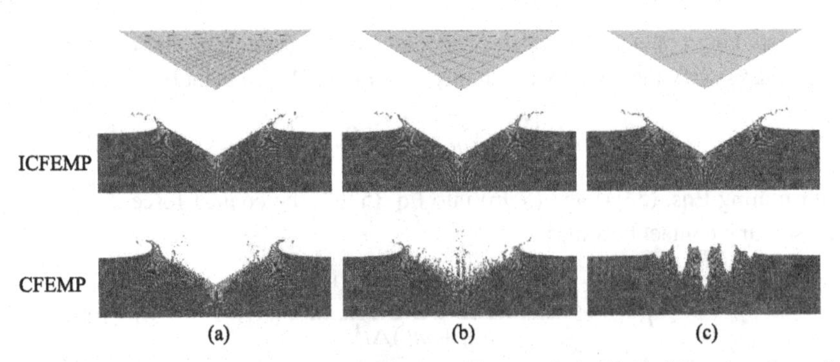

FIGURE 5.13 The water configurations at time $t = 25$ ms as obtained with different numbers of elements in the wedge: (a) 171 elements; (b) 48 elements; and (c) 3 elements.

The ICFEMP method allows a significant inconsistent meshing. For example, Chen et al. [124] simulated a wedge falling into water using the ICFEMP method, and the obtained falling velocity of the wedge agreed well with experiment data [128]. In order to examine the effects of inconsistent meshing on the solution accuracy, the wedge was meshed with 171, 48, and 3 elements, respectively. Fig. 5.13 compares the water configurations at time $t = 25$ ms as obtained by the ICFEMP method with those obtained by the CFEMP method with different numbers of elements in the wedge. The water configurations obtained with the ICFEMP method fit the wedge shape well in all the cases, but those obtained by the CFEMP method become worse as the inconsistence increases.

5.4 ADAPTIVE FEMP METHOD

To further take advantage of both the FEM and MPM, Lian et al. proposed an adaptive finite-element material-point (AFEMP) method [41]. In the AFEMP method, all bodies are initially discretized by finite elements, and then the distorted or failed elements are adaptively converted into the MPM particles when their effective plastic strain or distortion degree exceeds a user prescribed value during the simulation process. The interaction between the converted MPM particles and the remaining finite elements is implemented based on the coupling scheme proposed by Zhang et al. [36].

5.4.1 Discretization Scheme

In the AFEMP method, a material region Ω as shown in Fig. 5.14(a) is initially discretized into finite elements as shown in Fig. 5.14(b). During the simulation process, any distorted element is automatically converted into the MPM particles as shown in Fig. 5.14(c) where a regular Eulerian background grid is used to solve the nodal momentum equations. Therefore, the material region

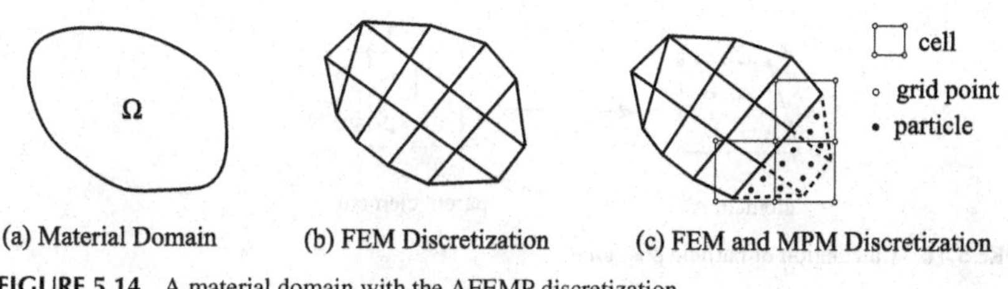

FIGURE 5.14 A material domain with the AFEMP discretization.

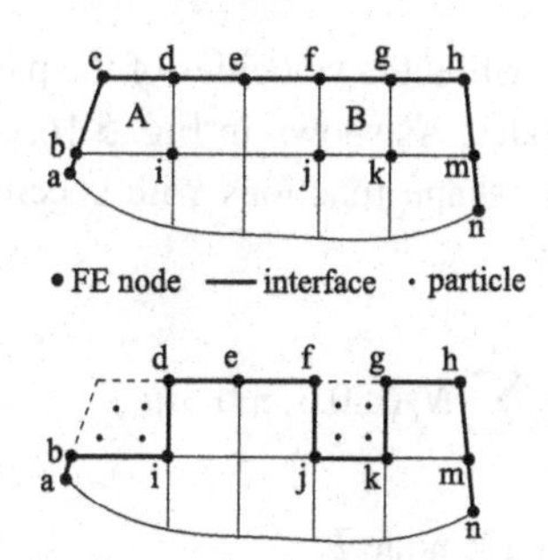

FIGURE 5.15 Conversion of finite elements to particles.

is discretized by both finite elements and particles, but the trial functions are constructed via the mesh including both the finite element mesh and MPM background grid.

5.4.2 Conversion Algorithm

Lian et al. [41] proposed an element–particle conversion algorithm to convert the distorted or failed elements into particles based on a given criterion to avoid element distortion. An element is converted into particles when either its equivalent plastic strain or its degree of element distortion exceeds a user-specified value. The degree of element distortion can be evaluated with the ratio of the minimum area over the maximum area of the element surfaces, considering the hexahedral element as an example. Of course, other criteria could also be used. Fig. 5.15 shows the quadrilateral FE mesh with a boundary defined by FE nodes, a, b, ..., n. Elements A and B are designated as the candidates for the conversion to particles. It is common that four particles are placed uniformly in a cell in the MPM for 2D problems. Therefore, elements A and B are removed from the finite element mesh and replaced by four particles, respectively.

In order to guarantee the conservation of mass, momentum, and energy, the mass, volume, and internal energy of the elements A and B are averaged equally to the four particles, respectively. The stress, strain, and other state variables of the four particles in each element are obtained from the corresponding Gauss

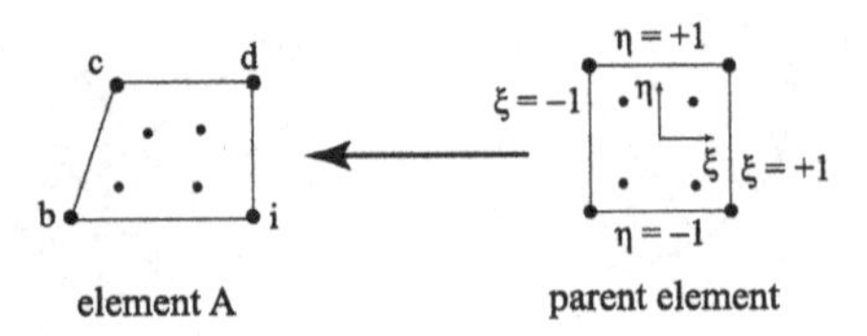

FIGURE 5.16 Calculation of particle positions.

points of the replaced element, while the velocities of the particles are the same as those of the adjacent FE nodes. As shown in Fig. 5.16, the positions of the particles are calculated via the shape functions with specified natural coordinates $(\pm 0.5, \pm 0.5)$ as follows:

$$u_{ip} = \sum_{I=1}^{4} N_I(\pm 0.5, \pm 0.5) u_{iI} \tag{5.107}$$

where u_{iI} is the position of the FE node I.

The FE node c that is not connected to any element is removed from the FE node list, while the FE nodes b, i, d, f, j, k, and g located at the interface between the MPM particles and remaining elements are labeled as the transition nodes whose nodal masses are reduced by the removal of elements.

After conversion, the one-Gauss-point quadrature is replaced by the four-particle quadrature with the conservation of mass momentum and energy. In 3D problems, one hexahedral element is replaced by eight MPM particles in a similar way.

5.4.3 Coupling Between Remaining Elements and Particles

The coupling between the remaining finite elements and particles is implemented by the transition nodes based on the background grid within the MPM framework, as proposed by Zhang et al. [36]. The momentum equations in the MPM are solved on the background grid and the incremental strains of particles are calculated from the corresponding nodal velocity field, which implies that the interaction between particles are carried out via the background grid. Therefore, the momentum equations of transition nodes are also solved on the background grid, together with those of the particles, to establish the interaction between the FEM domain and MPM one.

For the sake of clarity, consider a 2D problem shown in Fig. 5.17 as an example. The material domain is discretized with finite elements in its left part and with particles in the remaining part. The FE nodes a, b, and c located at the interface between the FEM domain and MPM one are termed as the transition nodes. In each time step, the mass, momentum, and nodal force of the transition

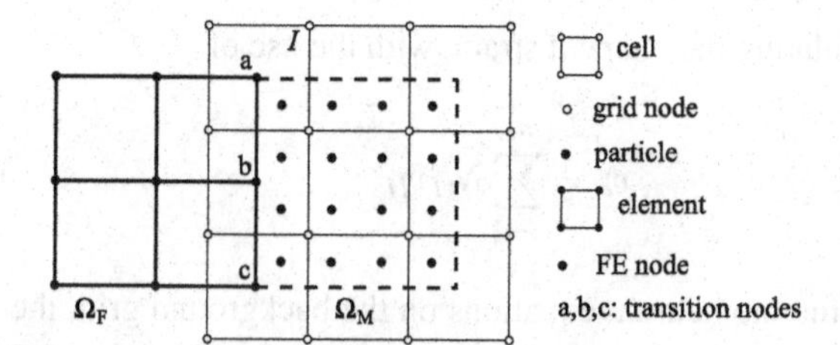

FIGURE 5.17 Coupling between the FEM and MPM.

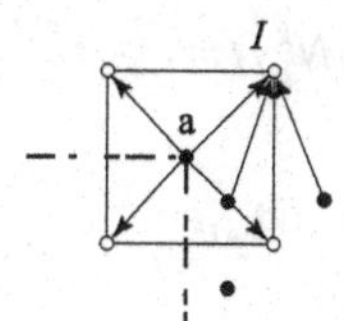

FIGURE 5.18 Grid node I.

nodes are mapped to the background grid nodes abreast with the MPM particles. Considering the grid node I shown in Fig. 5.18 as an example, the nodal mass of grid node I is given by

$$m_I = \sum_{p=1}^{n_p} N_{Ip} m_p + \sum_{t=1}^{n_t} N_{It} m_t \tag{5.108}$$

where the subscript t denotes the transition node, and n_t is the total number of transition nodes. The nodal momentum of grid node I is obtained by

$$p_{iI} = \sum_{p=1}^{n_p} N_{Ip} p_{ip} + \sum_{t=1}^{n_t} N_{It} p_{it}, \tag{5.109}$$

and the external nodal force of grid node I can be found to be

$$f_{iI}^{\text{ext}} = \sum_{p=1}^{n_p} m_p N_{Ip} b_{ip} + \sum_{t=1}^{n_t} N_{It} f_{it} \tag{5.110}$$

with f_{it} being the nodal force of the transition node without the hourglass-resisting force.

The velocity field used for the calculation of incremental strains of elements and particles must be identical. Hence, the velocities of the transition nodes must be reset by mapping the velocities of grid nodes back to the transition nodes before calculating the element strain, with the use of

$$v_{it} = \sum_{I=1}^{n_g} N_{tI} v_{iI}. \tag{5.111}$$

After solving the momentum equations on the background grid, the velocities and positions of the transition nodes are respectively updated as follows:

$$v_{it}^{k+1/2} = v_{it}^{k-1/2} + \Delta t^k \left(\sum_{I=1}^{n_g} f_{iI}^k N_{pI}^k / m_I^k + f_{it}^{\Gamma,k} / m_t \right), \tag{5.112}$$

$$x_{it}^{k+1} = x_{it}^k + \Delta t^{k+1/2} \left(\sum_{I=1}^{n_g} p_{iI}^{k+1/2} N_{pI}^k / m_I^k + \Delta t^k f_{it}^{\Gamma,k} / m_t \right) \tag{5.113}$$

where $f_{it}^{\Gamma,k}$ is the hourglass-resisting force. Therefore, the velocity field and displacement field are consistent along the interface between the FEM domain and MPM one.

In the AFEMP, a contact/friction/separation algorithm is implemented based on the background grid to handle the contact event between different bodies. Similar to the CFEMP, the AFEMP also requires a high degree of meshing consistence between the FEM body and MPM one due to the grid-based contact method used. To overcome this limitation, Chen et al. [124] proposed an improved adaptive finite-element material-point (IAFEMP) method based on a particle-to-surface contact method whose details can be found in Sect. 5.3.

In order to evaluate the accuracy and computational efficiency of the IAFEMP as compared with the AFEMP, a projectile striking an oblique thick plate with a velocity of 575 m/s at an inclined angle of 30° was investigated by Chen et al. [124]. The projectile and target are discretized with hexahedral elements with unstructured and structured arrangement, respectively. An element is automatically converted into eight MPM particles when its equivalent plastic strain exceeds the given threshold $\varepsilon_c = 0.9$. Three different cases with different projectile element sizes were considered to study the size effect on the solution accuracy of both methods. In these cases, the background grid cell size is 1 mm, while the element average size of projectile head is 2.3, 3.8, and 5.7 mm, respectively. The target is discretized by 314,600 elements in all cases, while the projectile is discretized by 12,200, 2268, and 944 elements, respectively. Numerical results show that the residual velocity of the projectile as obtained by the IAFEMP is very close to the experiment data in all the cases, but that as obtained by the AFEMP method largely depends on the meshing consistence. The

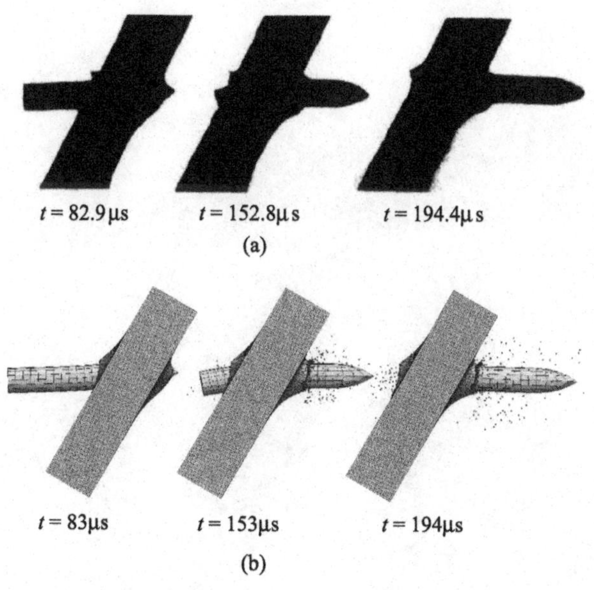

FIGURE 5.19 Perforation process at the striking velocity $v_0 = 575$ m/s: (a) experimental data, and (b) the IAFEMP results.

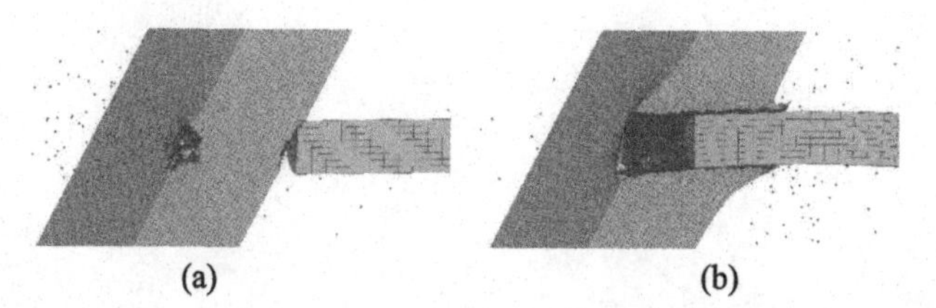

FIGURE 5.20 Sectional view of the projectile–target system as obtained by (a) the AFEMP, and (b) the IAFEMP.

perforation process as obtained by the IAFEMP agrees well with the experimental result, as shown in Fig. 5.19, where Fig. 5.19(a) shows a sequence of X-ray photographs at three different times, and Fig. 5.19(b) shows the corresponding numerical results.

For case 3, due to the significant meshing inconsistence in the AFEMP, the projectile penetrates into the plate nonphysically as shown in Fig. 5.20(a), but the IAFEMP still yields reasonable results as shown in Fig. 5.20(b).

Chapter 6

Constitutive Models

Contents

Constitutive models describe the material responses to different mechanical and/or thermal loading conditions, which provide the stress–strain relations to formulate the governing equations, together with the conservation laws and kinematic relations. Constitutive models can be divided into EOSs which relate the pressure to volume and internal energy or temperature, and strength models which relate the deviatoric stress to deviatoric strain. In addition, a failure criterion is required to identify the onset and describe the evolution of material failure. In this chapter, the essential features and numerical implementation of the constitutive models used in the open-source MPM code are discussed to facilitate the understanding and use of the code.

6.1 STRESS UPDATE

For nonlinear analyses, constitutive models are usually formulated in the rate form. The numerical algorithm for integrating the rate form of constitutive equations is called a **constitutive integration algorithm** or **stress update algorithm** [60]. The stress $\boldsymbol{\sigma}(t+\mathrm{d}t)$ at time $t+\mathrm{d}t$ can be obtained by integrating

The Material Point Method. http://dx.doi.org/10.1016/B978-0-12-407716-4.00006-5

the stress rate $\dot{\boldsymbol{\sigma}}$ as follows:

$$\boldsymbol{\sigma}(t + \mathrm{d}t) = \boldsymbol{\sigma}(t) + \dot{\boldsymbol{\sigma}}\mathrm{d}t. \tag{6.1}$$

The time derivative of the Cauchy stress tensor is not objective and depends on the frame of reference. Hence, an objective stress rate such as the Jaumann stress rate $\boldsymbol{\sigma}^{\nabla}$ (discussed in Sect. 2.5) must be used in constitutive models. The material time derivative of the Cauchy stress tensor, $\dot{\boldsymbol{\sigma}}$, is related to the Jaumann stress rate $\boldsymbol{\sigma}^{\nabla}$ via the following:

$$\dot{\boldsymbol{\sigma}} = \boldsymbol{\sigma}^{\nabla} + \boldsymbol{\sigma} \cdot \boldsymbol{\Omega}^{\mathrm{T}} + \boldsymbol{\Omega} \cdot \boldsymbol{\sigma} \tag{6.2}$$

where

$$\boldsymbol{\Omega} = \frac{1}{2}(\boldsymbol{L} - \boldsymbol{L}^{\mathrm{T}})$$

is the spin tensor, and $\boldsymbol{L}$ is the velocity gradient. The Jaumann stress rate $\boldsymbol{\sigma}^{\nabla}$ can be determined from a constitutive model based on the deformation rate $\boldsymbol{D}$.

In an explicit time integration, the stress rate at time $t^{n+1/2}$, $\dot{\boldsymbol{\sigma}}^{n+1/2}$, can be approximated by

$$\dot{\boldsymbol{\sigma}}^{n+1/2} = \boldsymbol{\sigma}^{\nabla n+1/2} + \boldsymbol{\sigma}^{n} \cdot (\boldsymbol{\Omega}^{n+1/2})^{\mathrm{T}} + \boldsymbol{\Omega}^{n+1/2} \cdot \boldsymbol{\sigma}^{n}. \tag{6.3}$$

The Cauchy stress at time t^{n+1} can then be obtained from Eqs. (6.1) and (6.3) as

$$\begin{aligned} \boldsymbol{\sigma}^{n+1} &= \boldsymbol{\sigma}^{n} + \dot{\boldsymbol{\sigma}}^{n+1/2} \Delta t^{n+1/2} \\ &= \boldsymbol{\sigma}^{R^n} + \boldsymbol{\sigma}^{\nabla n+1/2} \Delta t^{n+1/2} \end{aligned} \tag{6.4}$$

with

$$\boldsymbol{\sigma}^{R^n} = \boldsymbol{\sigma}^{n} + [\boldsymbol{\sigma}^{n} \cdot (\boldsymbol{\Omega}^{n+1/2})^{\mathrm{T}} + \boldsymbol{\Omega}^{n+1/2} \cdot \boldsymbol{\sigma}^{n})\Delta t^{n+1/2}. \tag{6.5}$$

The pressure and deviatoric stress can be updated independently due to

$$\boldsymbol{\sigma} = \boldsymbol{s} - p\mathbf{1}. \tag{6.6}$$

According to Eq. (6.4), the deviatoric stress $\boldsymbol{s}$ can be updated by

$$\boldsymbol{s}^{n+1} = \boldsymbol{s}^{R^n} + \boldsymbol{s}^{\nabla n+1/2} \Delta t^{n+1/2} \tag{6.7}$$

with

$$\boldsymbol{s}^{R^n} = \boldsymbol{s}^{n} + [\boldsymbol{s}^{n} \cdot (\boldsymbol{\Omega}^{n+1/2})^{\mathrm{T}} + \boldsymbol{\Omega}^{n+1/2} \cdot \boldsymbol{s}^{n})\Delta t^{n+1/2}. \tag{6.8}$$

The last term in Eq. (6.7) will be determined by a strength model, as discussed in Sect. 6.2.

The pressure can be updated with an EOS as

$$p = p(V, E) = p(V, T) \tag{6.9}$$

where the pressure p is positive if the material is in compression, V is the relative volume, E is internal energy per unit initial volume, and T is the temperature, as discussed in Sect. 6.3.

To find the pressure of a particle at time t^{n+1} using the EOS, the internal energy of the particle at time t^{n+1} must be calculated first by integrating the energy equation (2.87) as follows:

$$\begin{aligned} e^{n+1} &= e^n + V_0 \dot{E}^{n+1/2} \Delta t^{n+1/2} \\ &= e^n + V^{n+1/2} \boldsymbol{s}^{n+1/2} : \Delta \boldsymbol{\varepsilon}^{n+1/2} - V^{n+1/2}(p^{n+1/2} + q^{n+1/2}) \Delta \varepsilon_v^{n+1/2} \end{aligned} \tag{6.10}$$

where V_0 is the initial volume of the particle, V^n is the volume of the particle at time t^n, $\Delta \varepsilon_v^{n+1/2}$ is the volumetric strain increment, and

$$\Delta \boldsymbol{\varepsilon}^{n+1/2} = \dot{\boldsymbol{\varepsilon}}^{n+1/2} \Delta t^{n+1/2}, \tag{6.11}$$

$$V^{n+1/2} = \frac{1}{2}(V^n + V^{n+1}), \tag{6.12}$$

$$\boldsymbol{s}^{n+1/2} = \frac{1}{2}(\boldsymbol{s}^n + \boldsymbol{s}^{n+1}). \tag{6.13}$$

Noting that $V^{n+1/2} \Delta \varepsilon_v^{n+1/2} = V^{n+1} - V^n = \Delta V$, $p^{n+1/2} = (p^n + p^{n+1})/2$, Eq. (6.10) can then be rewritten as

$$e^{n+1} = e^{*n+1} - \frac{1}{2} \Delta V p^{n+1} \tag{6.14}$$

where

$$e^{*n+1} = e^n + V^{n+1/2} \boldsymbol{s}^{n+1/2} : \Delta \boldsymbol{\varepsilon}^{n+1/2} - \frac{1}{2} \Delta V p^n - \Delta V q^{n+1/2} \tag{6.15}$$

is the trial internal energy of the particle at time t^{n+1}.

If the EOS is linear in the internal energy [14], i.e.,

$$p^{n+1} = A^{n+1} + B^{n+1} E^{n+1} \tag{6.16}$$

with A^{n+1} and B^{n+1} being constants,

$$E^{n+1} = \frac{e^{n+1}}{V_0} \tag{6.17}$$

being the internal energy of the particle per unit initial volume, substitution of Eqs. (6.17) and (6.14) into Eq. (6.16) gives the pressure at time t^{n+1}, namely,

$$p^{n+1} = \frac{A^{n+1} + B^{n+1} E^{*n+1}}{1 + \frac{1}{2} B^{n+1} \frac{\Delta V}{V_0}} \tag{6.18}$$

with $E^{*n+1} = e^{*n+1}/V_0$. After solving Eq. (6.18) for pressure p^{n+1}, the internal energy of the particle at time t^{n+1} can be found from Eq. (6.14).

If the EOS is nonlinear in the internal energy, an iterative solution procedure is required [14]. The pressure can be estimated first with the EOS as

$$p^{*n+1} = p(V^{n+1}, E^{*n+1}), \tag{6.19}$$

and the estimated pressure p^{*n+1} can then be substituted into Eq. (6.14) to update the internal energy E^{n+1} of the particle at time t^{n+1}. The pressure at time t^{n+1} can be corrected by substituting the updated internal energy E^{n+1} into the EOS as

$$p^{n+1} = p(V^{n+1}, E^{n+1}) \tag{6.20}$$

until a suitable convergence criterion is satisfied. In impact and explosion simulations, the time step is very small so that only one iteration is usually required.

6.2 STRENGTH MODELS

After reviewing the previous theoretical and computational results, as well as the available experimental data, Chen and Schreyer [129] discussed the formulation and computational aspects of elastoplasticity and damage models for solids in a systematic way. A detailed discussion about fluid models with applications to extreme events could be found in the representative references [49,53,130]. In this section, the main equations and their numerical implementation are provided only for those constitutive models used in the open-source MPM code.

6.2.1 Elastic Model

For an **isotropic elastic model**, the Jaumann stress rate is related to the deformation rate via

$$\boldsymbol{\sigma}^{\nabla} = \boldsymbol{C}^{\sigma J} : \dot{\boldsymbol{\varepsilon}} \tag{6.21}$$

where

$$\boldsymbol{C}^{\sigma J} = 2G\boldsymbol{I}^{\text{dev}} + K\boldsymbol{1} \otimes \boldsymbol{1} \tag{6.22}$$

is the fourth-order elasticity tensor, $G = E/2(1+\nu)$ is the shear modulus, $K = E/3(1-2\nu)$ is the bulk modulus, and

$$I_{ijkl}^{\text{dev}} = \frac{1}{2}(\delta_{ik}\delta_{jl} + \delta_{il}\delta_{jk}) - \frac{1}{3}\delta_{ij}\delta_{kl} \tag{6.23}$$

is the fourth-order deviatoric tensor. It can be shown that for any symmetric deviatoric tensor $\boldsymbol{s}$ and symmetric tensor $\boldsymbol{\varepsilon}$, the following identities:

$$\boldsymbol{C}^{\sigma J} : \boldsymbol{s} = 2G\boldsymbol{s}, \tag{6.24}$$

$$\boldsymbol{I}^{\text{dev}} : \boldsymbol{\varepsilon} = \boldsymbol{\varepsilon}' \tag{6.25}$$

are satisfied, with $\boldsymbol{\varepsilon}'$ being the deviatoric part of the symmetric tensor $\boldsymbol{\varepsilon}$.

Eq. (6.21) can be further decomposed into

$$\boldsymbol{s}^{\nabla} = 2G\dot{\boldsymbol{\varepsilon}}', \tag{6.26}$$

$$\dot{\sigma}_m = K\dot{\varepsilon}_v \tag{6.27}$$

where $\boldsymbol{s}^{\nabla}$ is the Jaumann rate of the deviatoric stress, $\dot{\sigma}_m$ is the rate of the spherical stress, and

$$\dot{\boldsymbol{\varepsilon}}' = \dot{\boldsymbol{\varepsilon}} - \frac{1}{3}\dot{\varepsilon}_v\mathbf{1} \tag{6.28}$$

is the deviatoric strain rate, with $\dot{\varepsilon}_v = \dot{V}/V$ being the bulk strain rate.

Substituting Eqs. (6.26) and (6.27) into Eq. (6.4) yields the stress update formulation for the isotropic elastic model, namely,

$$\boldsymbol{s}^{n+1} = \boldsymbol{s}^{R^n} + 2G\dot{\boldsymbol{\varepsilon}}'^{n+1/2}\Delta t^{n+1/2}, \tag{6.29}$$

$$\sigma_m^{n+1} = \sigma_m^n + K\dot{\varepsilon}_v^{n+1/2}\Delta t^{n+1/2}. \tag{6.30}$$

The elastic model is implemented in the FORTRAN subroutine M3DM1 in Sect. 6.5.

6.2.2 Elastoplastic Models

For a metal specimen under uniaxial tension, a typical uniaxial stress–strain curve is shown in Fig. 6.1, which demonstrates the elastoplastic behavior. Above the initial yield strength, σ_y^0, plasticity occurs with a permanent (irreversible or plastic) strain upon elastic unloading, and the yield strength increases with the increase of the plastic strain, called strain hardening. There are many types of elastoplastic models, such as the J_2 flow theory for metal plasticity, the Mohr–Coulomb model and Drucker–Prager model for soil plasticity, and the Gurson

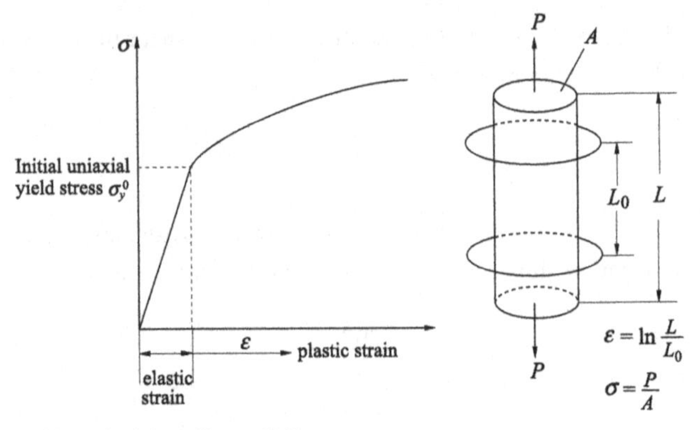

FIGURE 6.1 The uniaxial tensile test [14].

model for porous plastic solids [60]. Different models have different **yield functions** $f(\boldsymbol{\sigma}, \boldsymbol{q})$ and flow rules, where $\boldsymbol{q}$ represents a set of internal state variables, such as the effective plastic strain and void volume fraction.

For finite strain problems, the **hypoelastic–plastic model** and **hyperelastic–plastic model** are two main types of elastoplastic models. The hypoelastic–plastic model divides the deformation rate into an elastic part and a plastic part, namely,

$$\dot{\boldsymbol{\varepsilon}} = \dot{\boldsymbol{\varepsilon}}^e + \dot{\boldsymbol{\varepsilon}}^p, \tag{6.31}$$

and relates the objective stress rate, such as the Jaumann stress rate, to the elastic part of the deformation rate by

$$\boldsymbol{\sigma}^{\nabla} = \boldsymbol{C}^{\sigma J} : (\dot{\boldsymbol{\varepsilon}} - \dot{\boldsymbol{\varepsilon}}^p). \tag{6.32}$$

As a result, the hypoelastic–plastic response cannot be expressed in terms of an elastic strain energy function as the hyperelastic–plastic response.

The plastic strain rate $\dot{\boldsymbol{\varepsilon}}^p$ for non-associated plasticity is determined with a **non-associated flow law** as follows:

$$\dot{\boldsymbol{\varepsilon}}^p = \dot{\lambda}\boldsymbol{r} \tag{6.33}$$

where $\dot{\lambda} \geqslant 0$ represents the plastic loading (> 0) or neutral loading ($= 0$), with the corresponding magnitude of the plastic flow, and

$$\boldsymbol{r} = \frac{\partial \psi}{\partial \boldsymbol{\sigma}} \tag{6.34}$$

describes the plastic flow direction, with ψ being the **plastic potential function**. The plastic strain rate $\dot{\boldsymbol{\varepsilon}}^p$ is normal to the plastic potential surface, as shown in Fig. 6.2, which is called the **normality rule of plasticity**.

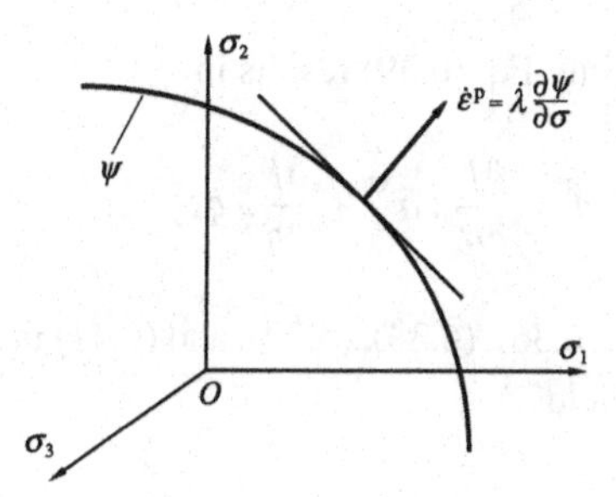

FIGURE 6.2 Plastic strain rate.

The flow rule for associated plasticity uses the yield function as the plastic potential function, i.e., $\psi \equiv f$. Thus, the plastic flow direction takes the form of

$$\boldsymbol{r} = \frac{\partial f}{\partial \boldsymbol{\sigma}}. \tag{6.35}$$

The **internal state variables** can be determined with the following equation:

$$\dot{\boldsymbol{q}} = \dot{\lambda}\boldsymbol{h}(\boldsymbol{\sigma}, \boldsymbol{q}). \tag{6.36}$$

The **Kuhn–Tucker conditions** identify plastic loading, neutral loading, and elastic unloading as follows:

$$\dot{\lambda} \geqslant 0, \quad f \leqslant 0, \quad \dot{\lambda} f = 0. \tag{6.37}$$

Eq. (6.37) indicates that the stress must be admissible (lie on or within the yield surface) and that plastic flow can take place only on the yield surface $f = 0$.

During plastic loading (i.e., $\dot{\lambda} > 0$), the stress must stay on the yield surface (i.e., $\dot{f} = 0$) so that

$$\dot{\lambda}\dot{f} = 0. \tag{6.38}$$

Eq. (6.38) is called the **plastic consistency condition** from which the loading parameter $\dot{\lambda}$ can be determined as below.

The yield function $f(\boldsymbol{\sigma}, \boldsymbol{q})$ is a function of stress and internal state variables such that

$$\dot{f} = \frac{\partial f}{\partial \boldsymbol{\sigma}} : \dot{\boldsymbol{\sigma}} + \frac{\partial f}{\partial \boldsymbol{q}} \cdot \dot{q}. \tag{6.39}$$

The following identity is satisfied if the yield function $f(\boldsymbol{\sigma}, q)$ is a function of the stress invariants [60]:

$$\frac{\partial f}{\partial \boldsymbol{\sigma}} : \dot{\boldsymbol{\sigma}} = \frac{\partial f}{\partial \boldsymbol{\sigma}} : \boldsymbol{\sigma}^{\nabla}. \tag{6.40}$$

Substituting Eq. (6.40) into Eq. (6.39) results in

$$\dot{f} = \frac{\partial f}{\partial \boldsymbol{\sigma}} : \boldsymbol{\sigma}^{\nabla} + \frac{\partial f}{\partial \boldsymbol{q}} \cdot \dot{\boldsymbol{q}}. \tag{6.41}$$

The substitution of Eqs. (6.36), (6.33), (6.32), and (6.41) into the consistency condition Eq. (6.38) then yields

$$\frac{\partial f}{\partial \boldsymbol{\sigma}} \boldsymbol{C}^{\sigma J} : (\dot{\boldsymbol{\varepsilon}} - \dot{\lambda} \boldsymbol{r}) + \dot{\lambda} \frac{\partial f}{\partial \boldsymbol{q}} \cdot \boldsymbol{h} = 0. \tag{6.42}$$

Solving Eq. (6.42) for the loading parameter $\dot{\lambda}$ gives

$$\dot{\lambda} = \frac{f_{,\sigma} : \boldsymbol{C}^{\sigma J} : \dot{\boldsymbol{\varepsilon}}}{f_{,\sigma} : \boldsymbol{C}^{\sigma J} : \boldsymbol{r} - f_{,q} \cdot \boldsymbol{h}}. \tag{6.43}$$

Finally, the substitution of Eqs. (6.33) and (6.43) into Eq. (6.32) leads to

$$\boldsymbol{\sigma}^{\nabla} = \boldsymbol{C}^{\mathrm{ep}} : \dot{\boldsymbol{\varepsilon}} \tag{6.44}$$

with the fourth-order tensor

$$\boldsymbol{C}^{\mathrm{ep}} = \begin{cases} \boldsymbol{C}^{\sigma J} & \text{if } \dot{\lambda} = 0, \\ \boldsymbol{C}^{\sigma J} - \dfrac{(\boldsymbol{C}^{\sigma J} : \boldsymbol{r}) \otimes (f_{,\sigma} : \boldsymbol{C}^{\sigma J})}{f_{,\sigma} : \boldsymbol{C}^{\sigma J} : \boldsymbol{r} - f_{,q} \cdot \boldsymbol{h}} & \text{if } \dot{\lambda} > 0 \end{cases} \tag{6.45}$$

being the continuum elastoplastic tangent moduli which can be used to formulate the acoustic tensor for discontinuous bifurcation analysis.

The hypoelastic–plastic models are simple for numerical implementation, but their responses are dissipative even though a material is supposed to be elastic. This type of models is commonly used for the materials whose elastic strains are assumed to be relatively small so that the energy dissipation is very small and can be neglected.

6.2.3 Return Mapping Algorithm

Given a deformation state $\boldsymbol{\varepsilon}^n$, the corresponding plastic deformation $\boldsymbol{\varepsilon}^{pn}$, and the internal state variable $\boldsymbol{q}^n$ at time t^n, the purpose of a constitutive integration algorithm is to find the plastic deformation $\boldsymbol{\varepsilon}^{p(n+1)}$, the internal variable $\boldsymbol{q}^{n+1}$, and $\Delta\lambda^{n+1}$ at time t^{n+1} for a prescribed strain increment $\Delta\boldsymbol{\varepsilon}^{n+1} = \dot{\boldsymbol{\varepsilon}}^{n+1/2}\Delta t^{n+1/2}$.

The simplest numerical integration scheme is the forward Euler algorithm [131–133], i.e.,

$$\begin{aligned}
\boldsymbol{\varepsilon}^{n+1} &= \boldsymbol{\varepsilon}^{n} + \Delta\boldsymbol{\varepsilon}^{n+1},\\
\boldsymbol{\varepsilon}^{p(n+1)} &= \boldsymbol{\varepsilon}^{pn} + \dot{\boldsymbol{\varepsilon}}^{pn}\Delta t = \boldsymbol{\varepsilon}^{pn} + \Delta\lambda^{n}\boldsymbol{r}^{n},\\
\boldsymbol{q}^{n+1} &= \boldsymbol{q}^{n} + \dot{\boldsymbol{q}}^{n}\Delta t = \boldsymbol{q}^{n} + \Delta\lambda^{n}\boldsymbol{h}^{n},\\
\boldsymbol{\sigma}^{n+1} &= \boldsymbol{\sigma}^{R^n} + \boldsymbol{C}^{\sigma J} : (\Delta\boldsymbol{\varepsilon}^{n+1} - \Delta\boldsymbol{\varepsilon}^{p(n+1)})
\end{aligned} \tag{6.46}$$

with $\Delta\lambda^n = \Delta t^{n+1/2}\dot{\lambda}^n$. The forward Euler algorithm is an explicit scheme, whose final stress state $\boldsymbol{\sigma}^{n+1}$ does not satisfy the yield condition at time t^{n+1}, i.e., $f^{n+1} = f(\boldsymbol{\sigma}^{n+1}, \boldsymbol{q}^{n+1}) \neq 0$. Therefore, the stress state obtained from the forward Euler algorithm gradually deviates from the yield surface and results in a significant error.

The **return mapping algorithm** [134] is the most commonly used constitutive integration algorithm which consists of two steps, namely, an elastic trial step and a plastic corrector step. In the elastic trial step, the material is assumed to be in the elastic regime so that the elastic trial stress $\boldsymbol{\sigma}^{*(n+1)}$ is calculated elastically from the total strain increment $\Delta\boldsymbol{\varepsilon}^{n+1}$. If the trial state is still in the elastic regime, the trial state is the true state. Otherwise, a plastic corrector step is performed to project the trial stress onto the yield surface at t^{n+1}. The **radial return algorithm** [135] for the J_2 flow theory is a special case of the return mapping algorithm.

6.2.3.1 *Fully Implicit Backward Euler Algorithm*

In the forward Euler algorithm, both the plastic strain rate $\dot{\boldsymbol{\varepsilon}}^p$ and the internal variable rate $\dot{\boldsymbol{q}}$ are evaluated at time t^n. On the contrary, in the backward Euler algorithm, they are evaluated at time t^{n+1}, and the final stress $\boldsymbol{\sigma}^{n+1}$ must satisfy the yield condition at time t^{n+1}, namely [60],

$$\begin{aligned}
\boldsymbol{\varepsilon}^{n+1} &= \boldsymbol{\varepsilon}^{n} + \Delta\boldsymbol{\varepsilon}^{n+1},\\
\boldsymbol{\varepsilon}^{p(n+1)} &= \boldsymbol{\varepsilon}^{pn} + \dot{\boldsymbol{\varepsilon}}^{p(n+1)}\Delta t = \boldsymbol{\varepsilon}^{pn} + \Delta\lambda^{n+1}\boldsymbol{r}^{n+1},\\
\boldsymbol{q}^{n+1} &= \boldsymbol{q}^{n} + \dot{\boldsymbol{q}}^{n+1}\Delta t = \boldsymbol{q}^{n} + \Delta\lambda^{n+1}\boldsymbol{h}^{n+1},\\
\boldsymbol{\sigma}^{n+1} &= \boldsymbol{\sigma}^{R^n} + \boldsymbol{C}^{\sigma J} : (\Delta\boldsymbol{\varepsilon}^{n+1} - \Delta\boldsymbol{\varepsilon}^{p(n+1)}),\\
f^{n+1} &= f(\boldsymbol{\sigma}^{n+1}, \boldsymbol{q}^{n+1}) = 0
\end{aligned} \tag{6.47}$$

with $\Delta\lambda^{n+1} = \dot{\lambda}^{n+1}\Delta t^{n+1/2}$.

The fourth equation in Eq. (6.47) can be rewritten as

$$\boldsymbol{\sigma}^{n+1} = \boldsymbol{\sigma}^{*(n+1)} + \Delta\boldsymbol{\sigma}^{n+1} \tag{6.48}$$

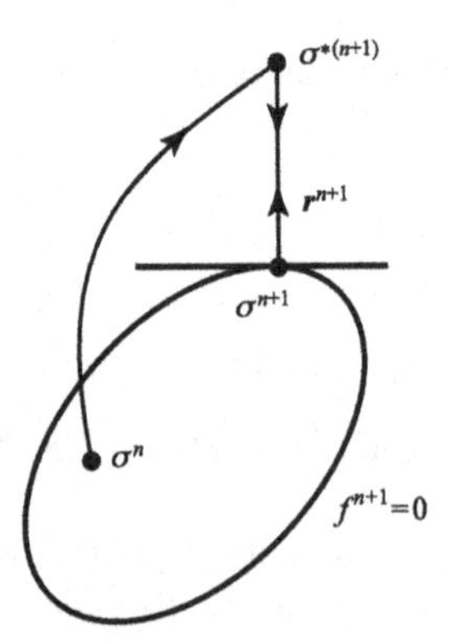

FIGURE 6.3 Return mapping algorithm for an associated plastic flow.

where

$$\boldsymbol{\sigma}^{*(n+1)} = \boldsymbol{\sigma}^{R^n} + \boldsymbol{C}^{\nabla J} : \Delta\boldsymbol{\varepsilon}^{n+1} \tag{6.49}$$

is the elastic trial stress, and

$$\Delta\boldsymbol{\sigma}^{n+1} = -\boldsymbol{C}^{\sigma J} : \Delta\boldsymbol{\varepsilon}^{p(n+1)} \tag{6.50}$$

is the plastic corrector stress that returns the elastic trial stress to the updated yield surface $f^{n+1} = 0$ along the plastic flow direction $\boldsymbol{r}^{n+1}$ at time t^{n+1}, as shown in Fig. 6.3. The final stress point $\boldsymbol{\sigma}^{n+1}$ is the closest point to the elastic trial stress $\boldsymbol{\sigma}^{*(n+1)}$ on the yield surface $f^{n+1} = 0$. Hence, this method is also called the **closest point projection method** [136,137]. The elastic trial step is driven by the strain increment $\Delta\boldsymbol{\varepsilon}^{n+1}$, while the plastic corrector step is driven by the loading parameter increment $\Delta\lambda^{n+1}$.

In the elastic trial step, the plastic strain and internal variables are fixed such that $\Delta\lambda^{n+1} = 0$. If the yield condition is convex, the plastic loading and elastic unloading can be uniquely determined by the trial state [137]. If $f^{*(n+1)} \leqslant 0$, the trial state is still elastic, i.e., $\Delta\lambda^{n+1} = 0$, so that the trial state is the true state. If $f^{*(n+1)} > 0$, the trial state is in plastic regime, i.e., $\Delta\lambda^{n+1} > 0$, such that a plastic corrector step is required to calculate $\Delta\lambda^{n+1}$ and to project the trial state onto the yield surface $f(\boldsymbol{\sigma}^{n+1}, \boldsymbol{q}^{n+1}) = 0$.

Eq. (6.47) is a nonlinear system of equations and has to be solved with an iterative scheme [60,137]. In the plastic corrector step, the strain $\boldsymbol{\varepsilon}^{n+1}$ is constant so that the iteration is performed on the loading parameter increment $\Delta\lambda$. A nonlinear equation $m(\Delta\lambda) = 0$ can be linearized using the Newton's iteration method with $\Delta\lambda^{(0)} = 0$ at the kth iteration as follows:

$$m^{(k)} + \left(\frac{\mathrm{d}m}{\mathrm{d}\Delta\lambda}\right)^{(k)} \delta\lambda^{(k)} = 0, \quad \Delta\lambda^{(k+1)} = \Delta\lambda^{(k)} + \delta\lambda^{(k)} \tag{6.51}$$

where $\delta\lambda^{(k)}$ is the increment in $\Delta\lambda^{(k)}$ at the kth iteration.

For the sake of clarity, the superscript $n+1$ will be omitted in the following. Unless otherwise indicated, all quantities are evaluated at t^{n+1}. The elastic trial state is taken as the first guess state, namely, $\Delta\lambda^{(0)} = 0$, $\boldsymbol{q}^{(0)} = \boldsymbol{q}^n$, $\boldsymbol{\varepsilon}^{p(0)} = \boldsymbol{\varepsilon}^{pn}$, $\boldsymbol{\sigma}^{(0)} = \boldsymbol{\sigma}^*$, and $f^{(0)} = f(\boldsymbol{\sigma}^*, \boldsymbol{q}^n)$.

To design the iterative solution scheme, the second, third, and fifth equations of Eq. (6.47) are rewritten as

$$\begin{aligned} \boldsymbol{a}(\Delta\lambda) &= -\boldsymbol{\varepsilon}^p + \boldsymbol{\varepsilon}^{pn} + \Delta\lambda \boldsymbol{r} = 0, \\ \boldsymbol{b}(\Delta\lambda) &= -\boldsymbol{q} + \boldsymbol{q}^n + \Delta\lambda \boldsymbol{h} = 0, \\ f(\Delta\lambda) &= f(\boldsymbol{\sigma}, \boldsymbol{q}) = 0. \end{aligned} \tag{6.52}$$

Using Eqs. (6.51) and (6.50), Eq. (6.52) can be linearized as

$$\begin{aligned} &\boldsymbol{a}^{(k)} + (\boldsymbol{C}^{\sigma J})^{-1} : \Delta\boldsymbol{\sigma}^{(k)} + \Delta\lambda^{(k)} \Delta\boldsymbol{r}^{(k)} + \delta\lambda^{(k)} \boldsymbol{r}^{(k)} = 0, \\ &\boldsymbol{b}^{(k)} - \Delta\boldsymbol{q}^{(k)} + \Delta\lambda^{(k)} \Delta\boldsymbol{h}^{(k)} + \delta\lambda^{(k)} \boldsymbol{h}^{(k)} = 0, \\ &f^{(k)} + f^{(k)}_{,\boldsymbol{\sigma}} : \Delta\boldsymbol{\sigma}^{(k)} + f^{(k)}_{,\boldsymbol{q}} \cdot \Delta\boldsymbol{q}^{(k)} = 0 \end{aligned} \tag{6.53}$$

with

$$\begin{aligned} \Delta\boldsymbol{r}^{(k)} &= \boldsymbol{r}^{(k)}_{,\boldsymbol{\sigma}} : \Delta\boldsymbol{\sigma}^{(k)} + \boldsymbol{r}^{(k)}_{,\boldsymbol{q}} \cdot \Delta\boldsymbol{q}^{(k)}, \\ \Delta\boldsymbol{h}^{(k)} &= \boldsymbol{h}^{(k)}_{,\boldsymbol{\sigma}} : \Delta\boldsymbol{\sigma}^{(k)} + \boldsymbol{h}^{(k)}_{,\boldsymbol{q}} \cdot \Delta\boldsymbol{q}^{(k)}. \end{aligned}$$

Eq. (6.53) is a system of linear equations in $\Delta\boldsymbol{\sigma}^{(k)}$, $\Delta\boldsymbol{q}^{(k)}$, and $\delta\lambda^{(k)}$. Based on the first and second equations of Eq. (6.53), $\Delta\boldsymbol{\sigma}^{(k)}$ and $\Delta\boldsymbol{q}^{(k)}$ can be expressed in terms of $\delta\lambda^{(k)}$ as

$$\left\{ \begin{matrix} \Delta\boldsymbol{\sigma}^{(k)} \\ \Delta\boldsymbol{q}^{(k)} \end{matrix} \right\} = -\boldsymbol{A}^{(k)} \tilde{\boldsymbol{a}}^{(k)} - \delta\lambda^{(k)} \boldsymbol{A}^{(k)} \tilde{\boldsymbol{r}}^{(k)} \tag{6.54}$$

with

$$\boldsymbol{A}^{(k)} = \begin{bmatrix} (\boldsymbol{C}^{\sigma J})^{-1} + \Delta\lambda^{(k)} \boldsymbol{r}^{(k)}_{,\boldsymbol{\sigma}} & \Delta\lambda^{(k)} \boldsymbol{r}^{(k)}_{,\boldsymbol{q}} \\ \Delta\lambda^{(k)} \boldsymbol{h}^{(k)}_{,\boldsymbol{\sigma}} & -\boldsymbol{I} + \Delta\lambda^{(k)} \boldsymbol{h}^{(k)}_{,\boldsymbol{q}} \end{bmatrix}^{-1},$$

$$\tilde{\boldsymbol{a}}^{(k)} = \left\{ \begin{matrix} \boldsymbol{a}^{(k)} \\ \boldsymbol{b}^{(k)} \end{matrix} \right\}, \quad \tilde{\boldsymbol{r}}^{(k)} = \left\{ \begin{matrix} \boldsymbol{r}^{(k)} \\ \boldsymbol{h}^{(k)} \end{matrix} \right\}.$$

Substituting Eq. (6.54) into the third equation of Eq. (6.53) gives

$$\delta\lambda^{(k)} = \frac{f^{(k)} - \partial \boldsymbol{f}^{(k)} \boldsymbol{A}^{(k)} \widetilde{\boldsymbol{a}}^{(k)}}{\partial \boldsymbol{f}^{(k)} \boldsymbol{A}^{(k)} \widetilde{\boldsymbol{r}}^{(k)}} \tag{6.55}$$

with

$$\partial \boldsymbol{f}^{(k)} = \begin{bmatrix} f_{\boldsymbol{\sigma}}^{(k)} & f_{\boldsymbol{q}}^{(k)} \end{bmatrix}.$$

After solving Eq. (6.55) for $\delta\lambda^{(k)}$, $\Delta\boldsymbol{\sigma}^{(k)}$ and $\Delta\boldsymbol{q}^{(k)}$ can be obtained from Eq. (6.54). The updated formulations of plastic strain, internal variables, loading parameter and stress are listed as follows:

$$\begin{aligned}
\boldsymbol{\varepsilon}^{p(k+1)} &= \boldsymbol{\varepsilon}^{p(k)} + \Delta\boldsymbol{\varepsilon}^{p(k)} = \boldsymbol{\varepsilon}^{p(k)} - (\boldsymbol{C}^{\sigma J})^{-1} : \Delta\boldsymbol{\sigma}^{(k)}, \\
\boldsymbol{q}^{(k+1)} &= \boldsymbol{q}^{(k)} + \Delta\boldsymbol{q}^{(k)}, \\
\Delta\lambda^{(k+1)} &= \Delta\lambda^{(k)} + \delta\lambda^{(k)}, \\
\boldsymbol{\sigma}^{(k+1)} &= \boldsymbol{\sigma}^{(k)} + \Delta\boldsymbol{\sigma}^{(k)}.
\end{aligned} \tag{6.56}$$

Based on the above discussions, the iterative process of the return mapping algorithm consists of the following steps:

1. Initialize $k = 0$, $\boldsymbol{\varepsilon}^{p(0)} = \boldsymbol{\varepsilon}^{pn}$, $\boldsymbol{q}^{(0)} = \boldsymbol{q}^n$, $\Delta\lambda^{(0)} = 0$, $\sigma = \boldsymbol{\sigma}^*$;
2. Compute $f^{(k)} = f(\boldsymbol{\sigma}^{(k)}, \boldsymbol{q}^{(k)})$ and $\widetilde{\boldsymbol{a}}^{(k)}$, and if $f^{(k)} < \text{TOL}_1$ and $\left\|\widetilde{\boldsymbol{a}}^{(k)}\right\| < \text{TOL}_2$, the iteration loop has converged so that the iteration is terminated;
3. Compute $\delta\lambda^{(k)}$ using Eq. (6.55);
4. Compute the stress increment $\Delta\boldsymbol{\sigma}^{(k)}$ and internal variable increment $\Delta\boldsymbol{q}^{(k)}$ using Eq. (6.54); and
5. Update $\boldsymbol{\varepsilon}^{p(k+1)}$, $\boldsymbol{q}^{(k+1)}$, $\Delta\lambda^{(k+1)}$, and $\boldsymbol{\sigma}^{(k+1)}$ according to Eq. (6.56), and letting $k = k + 1$ go to step 2 for the next iteration.

The iterative process of the return mapping algorithm is illustrated in Fig. 6.4.

The backward Euler algorithm is a fully implicit method so that an iterative loop is required in the incremental constitutive integration. The gradients of $\boldsymbol{r}$ and $\boldsymbol{h}$ are required which involve the second order derivatives of the plastic potential function such that their formulations are complicated. For some constitutive models, these gradients may not be able to be obtained analytically. For the J_2 flow theory with isotropic linear hardening, however, the analytical solution of Eq. (6.47) can be obtained so that the iteration loop is not needed, as shown in Sect. 6.2.4.

6.2.3.2 *Semi-implicit Backward Euler Algorithm*

To avoid the calculation of the gradients of $\boldsymbol{r}$ and $\boldsymbol{h}$ in each time step, $\boldsymbol{r}$ and $\boldsymbol{h}$ can be assumed to be constant and take the values at time t^n, namely

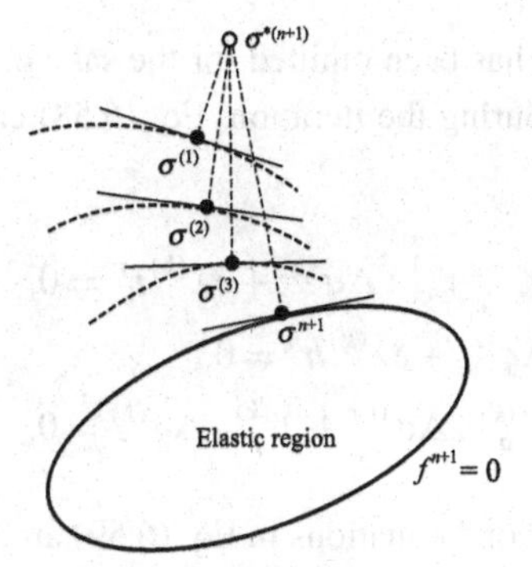

FIGURE 6.4 Iteration process of the return mapping algorithm.

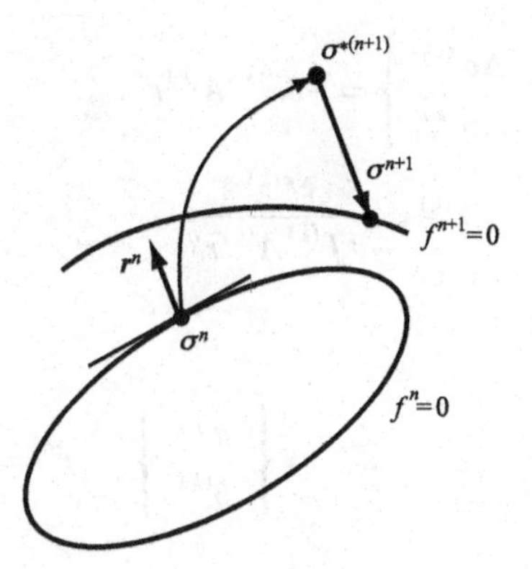

FIGURE 6.5 The semi-explicit Euler scheme.

[60,138],

$$
\begin{aligned}
\boldsymbol{\varepsilon}^{p(n+1)} &= \boldsymbol{\varepsilon}^{pn} + \Delta\lambda^{n+1}\boldsymbol{r}^n, \\
\boldsymbol{q}^{n+1} &= \boldsymbol{q}^n + \Delta\lambda^{n+1}\boldsymbol{h}^n, \\
f^{n+1} &= f(\boldsymbol{\sigma}^{n+1}, \boldsymbol{q}^{n+1}) = 0.
\end{aligned}
\tag{6.57}
$$

This scheme is a semi-implicit scheme because it is implicit in the loading parameter increment $\Delta\lambda$, but explicit in $\boldsymbol{r}$ and $\boldsymbol{h}$. During the iterative process, $\boldsymbol{r}$ and $\boldsymbol{h}$ are kept unchanged and their gradients are zero as shown in Fig. 6.5. As a result, the solution process is significantly simplified.

To establish the iterative algorithm, Eq. (6.57) is rewritten as

$$
\begin{aligned}
\boldsymbol{a}(\Delta\lambda) &= -\boldsymbol{\varepsilon}^p + \boldsymbol{\varepsilon}^{pn} + \Delta\lambda\boldsymbol{r}^n = 0, \\
\boldsymbol{b}(\Delta\lambda) &= -\boldsymbol{q} + \boldsymbol{q}^n + \Delta\lambda\boldsymbol{h}^n = 0, \\
f(\Delta\lambda) &= f(\boldsymbol{\sigma}, \boldsymbol{q}) = 0
\end{aligned}
\tag{6.58}
$$

where the superscript $n+1$ has been omitted for the sake of clarity. Because $\boldsymbol{r}$ and $\boldsymbol{h}$ are kept unchanged during the iteration, Eq. (6.58) can be linearized as follows:

$$\begin{aligned}
&\boldsymbol{a}^{(k)} + (\boldsymbol{C}^{\sigma J})^{-1} : \Delta\boldsymbol{\sigma}^{(k)} + \delta\lambda^{(k)}\boldsymbol{r}^n = 0,\\
&\boldsymbol{b}^{(k)} - \Delta\boldsymbol{q}^{(k)} + \delta\lambda^{(k)}\boldsymbol{h}^n = 0,\\
&f^{(k)} + f_{,\boldsymbol{\sigma}}^{(k)} : \Delta\boldsymbol{\sigma}^{(k)} + f_{,\boldsymbol{q}}^{(k)} \cdot \Delta\boldsymbol{q}^{(k)} = 0.
\end{aligned} \tag{6.59}$$

Note that the first and second equations in Eq. (6.59) are linear equations of $\Delta\lambda$ so that $\boldsymbol{a}^{(k)} = \boldsymbol{b}^{(k)} = 0$. Solving Eq. (6.59) yields

$$\left\{ \begin{array}{c} \Delta\boldsymbol{\sigma}^{(k)} \\ \Delta\boldsymbol{q}^{(k)} \end{array} \right\} = -\delta\lambda^{(k)}\boldsymbol{A}^{(k)}\tilde{\boldsymbol{r}}^n, \tag{6.60}$$

$$\delta\lambda^{(k)} = \frac{f^{(k)}}{\partial f^{(k)}\boldsymbol{A}^{(k)}\tilde{\boldsymbol{r}}^n} \tag{6.61}$$

with

$$\boldsymbol{A}^{(k)} = \begin{bmatrix} \boldsymbol{C}^{\sigma J} & 0 \\ 0 & -\boldsymbol{I} \end{bmatrix}^{(k)}, \quad \tilde{\boldsymbol{a}}^{(k)} = \left\{ \begin{array}{c} \boldsymbol{a}^{(k)} \\ \boldsymbol{b}^{(k)} \end{array} \right\}, \quad \tilde{\boldsymbol{r}}^n = \left\{ \begin{array}{c} \boldsymbol{r}^n \\ \boldsymbol{h}^n \end{array} \right\}.$$

6.2.4 J_2 Flow Theory

The J_2 **flow theory** as described in terms of the second invariant of deviatoric stress is widely used for metal plasticity.

6.2.4.1 *Yield Condition*

In the J_2 flow theory, the yield surface is given by

$$f(\boldsymbol{\sigma}, q_1) = \sqrt{3J_2} - \sigma_y(\bar{\varepsilon}) = 0 \tag{6.62}$$

where $q_1 \equiv \bar{\varepsilon}$ is the accumulated effective plastic strain, σ_y is the yield strength, and

$$J_2 = \frac{1}{2}\boldsymbol{s} : \boldsymbol{s} \tag{6.63}$$

is the second invariant of the deviatoric stress $\boldsymbol{s}$. Because

$$\sqrt{3J_2} = \sqrt{\frac{3}{2}\boldsymbol{s} : \boldsymbol{s}} = \bar{\sigma} \tag{6.64}$$

is the von Mises effective stress, the J_2 flow theory is also called the **von Mises yield criterion**. There is only one internal variable in the J_2 flow theory, i.e.,

$$q_1 = \bar{\varepsilon}, \quad h_1 = 1. \tag{6.65}$$

As a result, it follows that

$$\dot{\bar{\varepsilon}} = \dot{\lambda} \tag{6.66}$$

where

$$\dot{\bar{\varepsilon}} = \sqrt{\frac{2}{3}\dot{\boldsymbol{\varepsilon}}^p : \dot{\boldsymbol{\varepsilon}}^p} \tag{6.67}$$

is the von Mises effective plastic strain rate that is plastic-work-rate conjugate to the von Mises effective stress $\bar{\sigma}$, namely,

$$\bar{\sigma}\dot{\bar{\varepsilon}} = \boldsymbol{\sigma} : \dot{\boldsymbol{\varepsilon}}^p. \tag{6.68}$$

For an isotropic constitutive model as shown in Eq. (6.22), the stress update Eq. (6.48) can be decomposed into deviatoric stress update and spherical stress update, namely,

$$\boldsymbol{s}^{n+1} = \boldsymbol{s}^{*(n+1)} - 2G\Delta\boldsymbol{\varepsilon}'^{p(n+1)}, \tag{6.69}$$

$$\sigma_m^{n+1} = \sigma_m^{*(n+1)} - K\Delta\varepsilon_v^{p(n+1)} \tag{6.70}$$

where $\Delta\boldsymbol{\varepsilon}'^{p(n+1)}$ is the plastic deviatoric strain increment, $\sigma_m^{n+1} = \frac{1}{3}\sigma_{kk}^{n+1}$ is the mean stress, $\Delta\varepsilon_v^{p(n+1)} = \mathrm{tr}(\Delta\boldsymbol{\varepsilon}^{p(n+1)})$ is the plastic volumetric strain increment, and

$$\boldsymbol{s}^{*(n+1)} = \boldsymbol{s}^{R^n} + 2G\Delta\boldsymbol{\varepsilon}'^{n+1}, \tag{6.71}$$

$$\sigma_m^{*(n+1)} = \sigma_m^n + K\Delta\varepsilon_v^{n+1} \tag{6.72}$$

are the elastic trial deviatoric stress and elastic trial spherical stress, respectively. $\Delta\boldsymbol{\varepsilon}'^{n+1}$ is the deviatoric strain increment, and $\Delta\varepsilon_v^{n+1}$ is the volumetric strain increment. For the associated J_2 flow theory, $\Delta\boldsymbol{\varepsilon}'^{p(n+1)} = \Delta\boldsymbol{\varepsilon}^{p(n+1)}$ and $\Delta\varepsilon_v^{p(n+1)} = 0$.

Taking the double dot product of Eq. (6.69) with the unit normal tensor $\boldsymbol{n}$ to the yield surface and using Eqs. (6.76), (6.77), and (6.79), the von Mises effective stress can be updated with

$$\bar{\sigma}^{n+1} = \bar{\sigma}^{*(n+1)} - 3G\Delta\lambda^{n+1}. \tag{6.73}$$

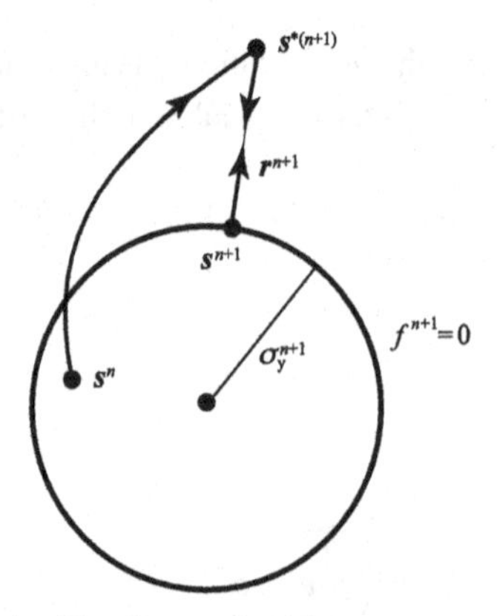

FIGURE 6.6 Radial return algorithm for associated flow.

6.2.4.2 Radial Return Algorithm

In the associated J_2 flow theory, the plastic flow is independent of hydrostatic pressure, and the yield surface in the principal stress space is a circular cylinder. With isotropic hardening, the cylinder diameter increases. With kinematic hardening, the cylinder diameter remains constant but the cylinder translates. The intersection of the yield surface with the π-plane ($\sigma_1 + \sigma_2 + \sigma_3 = 0$) is a circle whose normal vector is along the radial direction and is kept unchanged during the plastic corrector step, as shown in Fig. 6.6. Hence, the return mapping algorithm is degenerated to the **radial return algorithm**.

It can be verified that

$$\frac{\partial \boldsymbol{s}}{\partial \boldsymbol{\sigma}} = \boldsymbol{I}^{\text{dev}}, \quad \frac{\partial \bar{\sigma}}{\partial \boldsymbol{\sigma}} = \frac{3}{2\bar{\sigma}}\boldsymbol{s}. \tag{6.74}$$

For associated plastic flow, hence, the plastic flow direction can be obtained from Eq. (6.62) as

$$\boldsymbol{r} = \frac{\partial f}{\partial \boldsymbol{\sigma}} = \frac{3\boldsymbol{s}}{2\bar{\sigma}} = \sqrt{\frac{3}{2}}\boldsymbol{n} \tag{6.75}$$

where

$$\boldsymbol{n} = \sqrt{\frac{3}{2}}\frac{\boldsymbol{s}}{\bar{\sigma}} \tag{6.76}$$

is the unit normal tensor to the yield surface, which stays unchanged during the plastic corrector step so that it can also be found from the elastic trial stress, namely,

$$\boldsymbol{n} = \sqrt{\frac{3}{2}}\frac{\boldsymbol{s}^*}{\bar{\sigma}^*} \tag{6.77}$$

where

$$\bar{\sigma}^* = \sqrt{\frac{3}{2}\boldsymbol{s}^* : \boldsymbol{s}^*} \tag{6.78}$$

is the trial effective stress.

Substituting Eq. (6.75) into Eq. (6.33) results in the plastic strain increment, i.e.,

$$\Delta\boldsymbol{\varepsilon}^p = \sqrt{\frac{3}{2}}\Delta\lambda\boldsymbol{n}. \tag{6.79}$$

Eq. (6.79) indicates that the plastic strain in the J_2 flow theory is a deviatoric tensor, namely, the plastic volumetric strain $\varepsilon_v^p = 0$.

In the J_2 flow theory, $\boldsymbol{r}$ and $\boldsymbol{h}$ are constant during the plastic corrector step such that the update of plastic strain and internal variable (isotropic hardening) is linear in $\Delta\lambda$, as shown in the second and third equations in Eq. (6.47). Thus, the residuals in Eq. (6.53) become $\boldsymbol{a}^{(k)} = 0$ and $\boldsymbol{b}^{(k)} = 0$. Using Eqs. (6.75), (6.65), (6.74), and (6.76), the gradients of $\boldsymbol{r}$ and h_1 can be calculated as

$$\frac{\partial \boldsymbol{r}}{\partial \boldsymbol{\sigma}} = \frac{3}{2\bar{\sigma}}\hat{\boldsymbol{I}}, \quad \frac{\partial \boldsymbol{r}}{\partial q_1} = 0, \quad \frac{\partial h_1}{\partial \boldsymbol{\sigma}} = 0, \quad \frac{\partial h_1}{\partial q_1} = 0 \tag{6.80}$$

with

$$\hat{\boldsymbol{I}} = \boldsymbol{I}^{\text{dev}} - \boldsymbol{n} \otimes \boldsymbol{n}. \tag{6.81}$$

The gradient of the yield function (6.62) is given by

$$\frac{\partial f}{\partial \boldsymbol{\sigma}} = \boldsymbol{r}, \quad \frac{\partial f}{\partial q_1} = -\frac{\mathrm{d}\sigma_y}{\mathrm{d}\bar{\varepsilon}} = -E^p(\bar{\varepsilon}) \tag{6.82}$$

where $E^p(\bar{\varepsilon})$ is the plastic modulus.

It can be shown that

$$\boldsymbol{I}^{\text{dev}} : \hat{\boldsymbol{I}} = \hat{\boldsymbol{I}}, \quad \mathbf{1} : \hat{\boldsymbol{I}} = 0, \quad \hat{\boldsymbol{I}} : \boldsymbol{n} = 0, \tag{6.83}$$

and for the isotropic constitutive model it follows that

$$\hat{\boldsymbol{I}} : \boldsymbol{C}^{\sigma J} : \boldsymbol{n} = 0, \quad \hat{\boldsymbol{I}} : \Delta\boldsymbol{\sigma}^{(k)} = 0. \tag{6.84}$$

Substituting Eq. (6.80) into the first equation of Eq. (6.53), and then left-multiplying the $\boldsymbol{C}^{\sigma J}$ and making use of the relation $\boldsymbol{C}^{\sigma J} : \boldsymbol{r}^{(k)} = \sqrt{6}G\boldsymbol{n}^{(k)}$ gives

$$\Delta\boldsymbol{\sigma}^{(k)} = -\sqrt{6}G\delta\lambda^{(k)}\boldsymbol{n}^{(k)}. \tag{6.85}$$

Substituting Eq. (6.85) into Eq. (6.50) and noting that the plastic strain is a deviatoric tensor in the J_2 flow theory leads to

$$\Delta\boldsymbol{\varepsilon}^{p(k)} = -\frac{1}{2G}\Delta\boldsymbol{\sigma}^{(k)} = \sqrt{\frac{3}{2}}\delta\lambda^{(k)}\boldsymbol{n}^{(k)}. \tag{6.86}$$

Substituting Eqs. (6.80), (6.82), and (6.85) into the second and third equations of Eq. (6.53) results in

$$\Delta q_1^{(k)} = \Delta\bar{\varepsilon}^{(k)} = \delta\lambda^{(k)}, \tag{6.87}$$

$$f^{(k)} - 3G\delta\lambda^{(k)}\boldsymbol{n}^{(k)} : \boldsymbol{n}^{(k)} - E^p\delta\lambda^{(k)} = 0 \tag{6.88}$$

where $\boldsymbol{n}^{(k)}$ is a unit tensor such that $\boldsymbol{n}^{(k)} : \boldsymbol{n}^{(k)} = 1$. Hence, the plastic loading parameter increment can be obtained from Eq. (6.88) as

$$\delta\lambda^{(k)} = \frac{f^{(k)}}{3G + E^p}. \tag{6.89}$$

Using Eqs. (6.73) and (6.66), the yield function in the kth iteration can be written as

$$f^{(k)} = \bar{\sigma}^{(k)} - \sigma_y(\bar{\varepsilon}^{(k)}) = \bar{\sigma}^* - 3G\Delta\lambda^{(k)} - \sigma_y(\bar{\varepsilon}^{(k)}). \tag{6.90}$$

The iteration loop used in the numerical integration of the J_2 flow theory can be finally described as follows:

1. Initialize $k = 0$, $\boldsymbol{\varepsilon}^{p(0)} = \boldsymbol{\varepsilon}^{pn}$, $\bar{\varepsilon}^{(0)} = \bar{\varepsilon}^n$, $\Delta\lambda^{(0)} = 0$, $\boldsymbol{\sigma}^{(0)} = \boldsymbol{\sigma}^*$;
2. Calculate the yield function $f^{(k)}$ using Eq. (6.90), and if $|f^{(k)}| < \text{TOL}$, the iteration loop has converged so that it can be terminated;
3. Calculate the loading parameter increment $\delta\lambda^{(k)}$ using Eq. (6.89);
4. Calculate $\Delta\boldsymbol{\sigma}^{(k)}$, $\Delta\boldsymbol{\varepsilon}^{p(k)}$, and $\Delta\bar{\varepsilon}^{(k)}$ using Eqs. (6.85), (6.86), and (6.87), respectively; and
5. Update the plastic strain, stress and internal variable with

$$\begin{aligned} \boldsymbol{\varepsilon}^{p(k+1)} &= \boldsymbol{\varepsilon}^{p(k)} + \Delta\boldsymbol{\varepsilon}^{p(k)}, \\ \bar{\varepsilon}^{(k+1)} &= \bar{\varepsilon}^{(k)} + \Delta\bar{\varepsilon}^{(k)}, \\ \Delta\lambda^{(k+1)} &= \Delta\lambda^{(k)} + \delta\lambda^{(k)}, \\ \boldsymbol{\sigma}^{(k+1)} &= \boldsymbol{\sigma}^{(k)} + \Delta\boldsymbol{\sigma}^{(k)}, \end{aligned}$$

and letting $k = k + 1$ go to step 2 to continue the iteration loop.

6.2.4.3 Linear Isotropic Hardening

For linear isotropic hardening, the plastic modulus E^p is a constant. Substituting the linear hardening rule $\sigma_y^{n+1} = \sigma_y^n + E^p\Delta\bar{\varepsilon}^{(n+1)}$ and the relation $\Delta\bar{\varepsilon}^{(n+1)} =$

$\Delta\lambda^{(n+1)}$ into the yield function at time t^{n+1} gives

$$\begin{aligned} f^{n+1} &= \bar{\sigma}^{*(n+1)} - 3G\Delta\lambda^{n+1} - \sigma_y^{n+1} \\ &= \bar{\sigma}^{*(n+1)} - \sigma_y^n - (3G + E^p)\Delta\lambda^{(n+1)}. \end{aligned} \tag{6.91}$$

Thus, the yield condition $f^{n+1} = 0$ is a linear equation in $\Delta\lambda^{n+1}$, which can be solved directly for $\Delta\lambda^{n+1}$ without iteration as

$$\Delta\bar{\varepsilon}^{(n+1)} = \Delta\lambda^{n+1} = \frac{f^{*(n+1)}}{3G + E^p} \tag{6.92}$$

where

$$f^{*(n+1)} = \bar{\sigma}^{*(n+1)} - \sigma_y^n \tag{6.93}$$

is the elastic trial value of the yield function.

In the J_2 flow theory with linear isotropic hardening, the plastic volumetric strain equals to zero, and the plastic strain increment $\Delta\boldsymbol{\varepsilon}^{p(n+1/2)} = \sqrt{\frac{3}{2}}\Delta\lambda^{n+1}\boldsymbol{n}^{n+1}$ is a deviatoric tensor. Hence, the deviatoric stress can be updated according to Eqs. (6.69) and (6.75) with

$$\boldsymbol{s}^{n+1} = \boldsymbol{s}^{*(n+1)} - \sqrt{6}G\Delta\lambda^{n+1}\boldsymbol{n}^{n+1}. \tag{6.94}$$

Accordingly, the von Mises effective stress can be updated with

$$s^{n+1} = s^{*(n+1)} - 3G\Delta\lambda^{n+1}. \tag{6.95}$$

Substituting Eq. (6.77) into Eq. (6.94) and making use of Eq. (6.95) and the relation $f^{n+1} = \bar{\sigma}^{n+1} - \sigma_y^{n+1} = 0$ leads to

$$\begin{aligned} \boldsymbol{s}^{n+1} &= \frac{\boldsymbol{s}^{*(n+1)}}{\bar{\sigma}^{*(n+1)}}\left(\bar{\sigma}^{*(n+1)} - 3G\Delta\lambda^{n+1}\right) \\ &= \frac{\sigma_y^{n+1}}{\bar{\sigma}^{*(n+1)}}\boldsymbol{s}^{*(n+1)}. \end{aligned} \tag{6.96}$$

Eq. (6.96) indicates that the radial return method maps the elastic trial deviatoric stress $\boldsymbol{s}^{*(n+1)}$ onto the yield surface $f(\boldsymbol{\sigma}^{n+1}, \boldsymbol{q}^{n+1}) = 0$ at time t^{n+1}. Therefore, the numerical integration process of the J_2 flow theory with linear isotropic hardening can be summarized as follows:

1. Calculate the elastic trial stress $\boldsymbol{\sigma}^{*(n+1)}$ and the elastic trial value of yield function $f^{*(n+1)}$ using Eqs. (6.49) and (6.93), and if $|f^{*(n+1)}| \leqslant 0$, the elastic trial state is the true state so that let $\boldsymbol{\sigma}^{n+1} = \boldsymbol{\sigma}^{*(n+1)}$ and terminate the iteration loop;

2. Calculate the effective plastic strain increment $\Delta\bar{\varepsilon}^{(n+1)}$ using Eq. (6.92);

3. Update the effective plastic strain by

$$\bar{\varepsilon}^{(n+1)} = \bar{\varepsilon}^n + \Delta\bar{\varepsilon}^{(n+1)}; \quad (6.97)$$

4. Update the yield stress by

$$\sigma_y^{n+1} = \sigma_y^n + E^p \Delta\bar{\varepsilon}^{(n+1)}; \quad (6.98)$$

5. Calculate the scaling factor using the yield stress at time t^{n+1}, i.e.,

$$m = \frac{\sigma_y^{n+1}}{\bar{\sigma}^{*(n+1)}}; \quad (6.99)$$

6. Map the stress onto the yield surface with

$$\mathbf{s}^{n+1} = m\mathbf{s}^{*(n+1)}, \quad (6.100)$$

$$\bar{\sigma}^{n+1} = m\bar{\sigma}^{*(n+1)}. \quad (6.101)$$

Note that for perfect plasticity ($E^p = 0$), only the last two steps are performed.

The perfect plasticity and linear isotropic hardening plasticity are implemented into the FORTRAN subroutines M3DM2 and M3DM3 in Sect. 6.5, respectively.

The J_2 flow theory as discussed in this section could be applied in many cases. By choosing different yield surfaces, different material models could be obtained. For example,

- Letting $\sigma_y = 0$ results in a fluid model whose strength is neglected;
- Choosing σ_y as a nonzero positive constant results in the perfect plasticity;
- Letting $\sigma_y = \infty$ leads to elasticity; and
- Assuming a general yield stress function $\sigma_y = f(\bar{\varepsilon}, \dot{\bar{\varepsilon}}, T)$, strain hardening, thermal softening, and strain rate effects could be considered, such as the Johnson–Cook flow stress model.

6.2.4.4 *Johnson–Cook Flow Stress Model*

For impact and explosion problems, the strain rate effects have to be included in constitutive models. The flow stress can be expressed in the form of

$$\sigma_y = f(\bar{\varepsilon}, \dot{\bar{\varepsilon}}, T) \quad (6.102)$$

where $\bar{\varepsilon}$ is the effective plastic strain, $\dot{\bar{\varepsilon}}$ is the effective plastic strain rate, and T is the temperature.

At a low strain rate, the flow stress can be assumed to be

$$\sigma_y = \sigma_0 + k\bar{\varepsilon}^n \quad (6.103)$$

with σ_0 being the initial yield strength, n the strain hardening exponent, and k the strength coefficient. The effect of temperature on the flow stress can be expressed as

$$\sigma_y = \sigma_r(1 - T^{*m}) \tag{6.104}$$

where σ_r is the flow stress at room temperature, $T^* = (T - T_r)/(T_m - T_r) \in [0, 1]$ is the dimensionless temperature, T_r and T_m are the room temperature and melting temperature of the material, and m is a material constant.

The effect of strain rate is usually simplified as

$$\sigma_y \propto \ln \dot{\varepsilon}^p. \tag{6.105}$$

Johnson and Cook proposed a flow stress model [118,139] to account for the strain hardening, thermal softening, and strain rate effects as follows:

$$\sigma_y = (A + B\bar{\varepsilon}^n)(1 + C \ln \dot{\bar{\varepsilon}}^*)(1 - T^{*m}) \tag{6.106}$$

where A, B, n, C, and m are material constants, $\dot{\bar{\varepsilon}}^* = \dot{\bar{\varepsilon}}/\dot{\varepsilon}_0$ is the dimensionless effective plastic strain rate, $\dot{\varepsilon}_0$ is the effective plastic strain rate corresponding to the quasi-static test used to determine the yield and hardening parameters A, B and n, and $\dot{\bar{\varepsilon}} \approx \sqrt{\frac{2}{3}\dot{\varepsilon}'_{ij}\dot{\varepsilon}'_{ij}}$ is the plastic strain rate. The material constants can be determined either by torsion tests at different strain rates, Hopkinson's tests at different impact velocities and temperatures, and quasi-static tension tests or by Taylor bar impact tests.

Johnson and Cook provided the material constants in 1983 [139] for a variety of materials, such as OFHC copper, Cartridge brass, Nickel 200, Armco iron, Carpenter electrical iron, 1006 steel, 2024-T351 aluminum, 7039 aluminum, 4340 steel, S-7 tool steel, Tungsten alloy, and DU-.75Ti.

The flow stress σ_y is a nonlinear function of the effective plastic strain ε^p so that an iteration loop for finding the plastic strain increment is required to get an accurate value of the flow stress. To avoid the iteration, a Taylor series expansion with linearization about the current time can be used to solve for σ_y with sufficient accuracy. The plastic modulus is given by

$$E_p = \frac{d\sigma_y}{d\varepsilon^p} = nB(\varepsilon^p)^{n-1}(1 + C \ln \dot{\varepsilon}^*)(1 - T^{*m}).$$

The Johnson–Cook dynamic failure model can be used with the Johnson–Cook flow stress model, in which the strain at failure is expressed as [140]

$$\bar{\varepsilon}_f = (D_1 + D_2 e^{D_3\sigma^*})(1 + D_4 \ln \dot{\varepsilon}^*)(1 + D_5 T^{*m})$$

where $D_1, D_2, \ldots, D_5$ are failure parameters, $\sigma^* = \sigma_m/\bar{\sigma}$ is the triaxiality, σ_m is the mean stress, and $\bar{\sigma}$ is the von Mises effective stress. When the damage parameter

$$D = \sum \frac{\Delta\bar{\varepsilon}}{\bar{\varepsilon}_f}$$

reaches the value 1, a fracture occurs. After the fracture occurs, the deviatoric stress $\boldsymbol{s}$ is set to zero, and no hydrostatic tension is permitted. If the mean stress calculated is positive, it is reset to 0. Johnson and Cook provided failure parameters [140] in 1985 for OFHC copper, Armco iron, and 4340 steel. Note that no spatial size effect on the failure evolution is specified in the Johnson–Cook model. As a result, special caution must be taken when selecting a suitable spatial discretization scheme to calibrate model parameters against experimental data.

The Johnson–Cook plasticity model is implemented into the FORTRAN subroutine M3DM4 in Sect. 6.5.

6.2.5 Pressure-Dependent Elastoplasticity

Soils, rock, concrete, and other geomaterials are frictional materials whose plastic behaviors depend on the confining pressure. The J_2 flow models are independent of pressure such that they are not appropriate for frictional materials. Furthermore, the associated flow rule is also inappropriate for frictional materials.

6.2.5.1 Mohr–Coulomb Yield Criterion

The Mohr–Coulomb model is an extension of the Coulomb friction model for multi-axial stress–strain relations in continuum, and has been widely used in modeling granular materials. The Mohr–Coulomb model assumes that yielding in a material occurs when the shear stress τ_n and normal stress σ_n on any plane satisfy the following condition:

$$\tau_n = c - \sigma_n \tan\phi \tag{6.107}$$

where c is the cohesion, and ϕ is the angle of internal friction. In the $\tau_n\sigma_n$-plane, the line represented by Eq. (6.107) is the envelop of the Mohr circles (called the Mohr failure envelope), as shown in Fig. 6.7.

It can be obtained based on the Mohr circle as shown in Fig. 6.7 that

$$\begin{aligned}\tau_n &= \frac{1}{2}(\sigma_1 - \sigma_3)\cos\phi, \\ \sigma_n &= \frac{1}{2}(\sigma_1 + \sigma_3) + \frac{1}{2}(\sigma_1 - \sigma_3)\sin\phi\end{aligned} \tag{6.108}$$

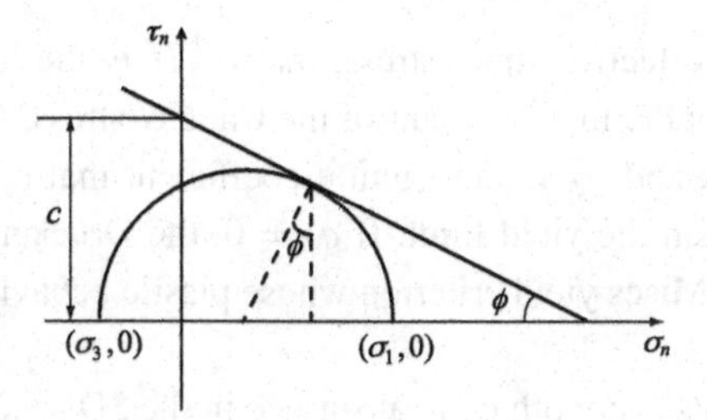

FIGURE 6.7 Yield surface of the Mohr–Coulomb model.

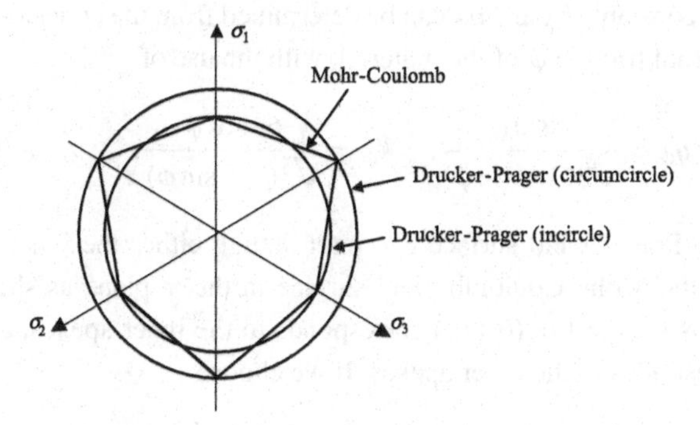

FIGURE 6.8 Mohr–Coulomb and Drucker–Prager yield surfaces in the π-plane.

where $\sigma_1 \geqslant \sigma_2 \geqslant \sigma_3$ are the principal stresses. Substituting Eq. (6.108) into Eq. (6.107), the Mohr–Coulomb yield criterion can be written as

$$f(\sigma) = \sigma_1 - \sigma_3 + (\sigma_1 + \sigma_3)\sin\phi - 2c\cos\phi = 0. \tag{6.109}$$

Eq. (6.109) represents a conical surface composed of six planes in the principal stress space. The intersection of the Mohr–Coulomb yield surface with the π-plane is an irregular hexagon, as shown in Fig. 6.8. The Mohr–Coulomb yield criterion reduces to the Tresca yield criterion when $\phi = 0°$, and reduces to the Rankine yield criterion when $\phi = 90°$.

6.2.5.2 Drucker–Prager Yield Criterion

The Mohr–Coulomb yield surface is composed of six planes so that it is convenient for analytical calculation. However, the Mohr–Coulomb yield surface is not a smooth surface whose normal is not defined at the corners, which makes it difficult for numerical calculation. The Drucker–Prager model defines a smooth yield surface with the pressure effect taken into account as follows:

$$f^s = \tau + q_\phi \sigma_m - k_\phi \tag{6.110}$$

where $\tau = \sqrt{J_2}$ is the effective shear stress, $\sigma_m = \frac{1}{3}I_1$ is the hydrostatic (or mean) stress, $I_1 = \sigma_{kk}$ is the first invariant of the Cauchy stress, k_ϕ is the yield stress under pure shear, and q_ϕ is the friction coefficient that controls the influence of the pressure on the yield limit. If $q_\phi = 0$, the Drucker–Prager yield criterion reduces to the Mises yield criterion whose plastic behavior is independent of the pressure.

Eq. (6.110) represents a smooth conical surface in the 3D space of principal stresses, whose intersection with the π-plane is a circle, as shown in Fig. 6.8.

The material constants k_ϕ and q_ϕ can be determined from the cohesion c and the angel of internal friction ϕ of the material, with the use of

$$q_\phi = \frac{6\sin\phi}{\sqrt{3}(3 \mp \sin\phi)}, \quad k_\phi = \frac{6c\cos\phi}{\sqrt{3}(3 \mp \sin\phi)}. \tag{6.111}$$

The Drucker–Prager yield surface can pass though either the inner or the outer apexes of the Mohr–Coulomb yield surface in the π-plane as shown in Fig. 6.8. The plus sign in Eq. (6.111) corresponds to the inner apexes, and the minus sign corresponds to the outer apexes. If we choose

$$q_\phi = \frac{3\tan\phi}{\sqrt{9 + 12\tan^2\phi}}, \quad k_\phi = \frac{3c}{\sqrt{9 + 12\tan^2\phi}}, \tag{6.112}$$

the Drucker–Prager yield surface inscribes the Mohr–Coulomb yield surface in the π-plane.

The associated plastic flow rule is no longer appropriate for frictional materials because it overestimates the dilatancy of frictional materials. In the Drucker–Prager model, the plastic potential ψ^s for a non-associated flow rule can be defined as

$$\psi^s = \tau + q_\psi \sigma_m \tag{6.113}$$

where q_ψ is the dilatancy coefficient related to the ratio of plastic volume change to plastic shear strain. The relationship between q_ψ and the angle of dilation ψ is the same as that between q_ϕ and ϕ in Eqs. (6.111) and (6.112). The case of $q_\psi = q_\phi$ yields the associated flow rule. If the angle of dilation q_ψ equals zero, the plastic volume does not change under shear.

Most geomaterials provide little resistance to tensile loads. With a tension cut-off, the Drucker–Prager yield surface in the 3D principal stress space is a truncated cone, as shown in Fig. 6.9(a). In the $\sigma_m\tau$-plane, the envelope of the Drucker–Prager yield surface is shown in Fig. 6.9(b).

In Fig. 6.9(b), the envelope line AB corresponds to the shear yielding whose yield function is given by Eq. (6.110), while the envelope line BC corresponds

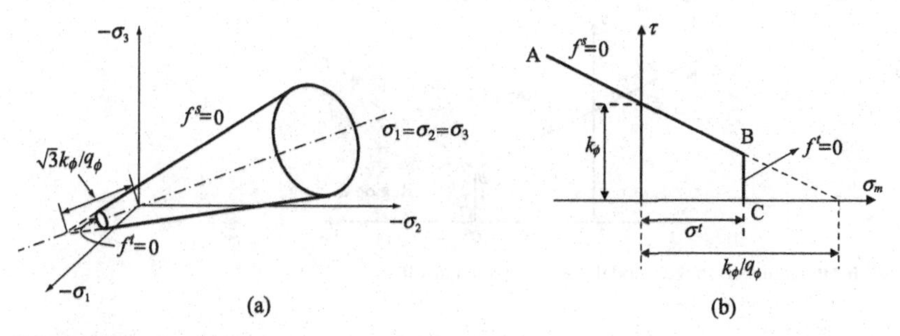

FIGURE 6.9 The Drucker–Prager yield surface: (a) in the principal stress space, and (b) in the $\sigma_m \tau$-plane.

to the tensile yielding whose yield function is given by

$$f^t = \sigma_m - \sigma^t \tag{6.114}$$

where σ^t is the tensile strength.

If the material constant q_ϕ is nonzero, the tensile strength of the material cannot exceed the maximum tensile strength $\sigma^t_{\max}$ which can be obtained from Fig. 6.9(b) as

$$\sigma^t_{\max} = \frac{k_\phi}{q_\phi}. \tag{6.115}$$

The plastic potential ψ^t for the associated flow rule corresponding to the tensile yield function Eq. (6.114) is given by

$$\psi^t = \sigma_m. \tag{6.116}$$

The Drucker–Prager plasticity model with a tension cut-off is a multi-surface model. As shown in Fig. 6.9(b), the normal of the yield surface at the corner B (singular point) in the $\sigma_m \tau$-plane is not defined. Thus, in the shadow region shown in Fig. 6.10 formed by the line BD (normal line of BC) and line BE (normal line of AB), the normal of the yield surface is not defined. Koiter assumed the plastic strain rate at the singular point to be the combination of the plastic strain rates determined by all the yield surfaces [137,141], namely,

$$\dot{\boldsymbol{\varepsilon}}^{\mathrm{p}} = \sum_{\alpha \in \mathbb{J}_{\mathrm{m}}} \dot{\lambda}^\alpha f^\alpha_{,\boldsymbol{\sigma}}(\boldsymbol{\sigma}, \boldsymbol{q}) \tag{6.117}$$

where $\mathbb{J}_{\mathrm{m}} := \{\beta \in \{1, 2, \ldots, m_{\mathrm{m}}\} | f^\beta(\boldsymbol{\sigma}, \boldsymbol{q}) = 0\}$, and m_{m} is the total number of the yield surfaces.

The FLAC code [142] employed a simplified approach which divides the region outside the elastic region into two subregions by the angle bisector BF of

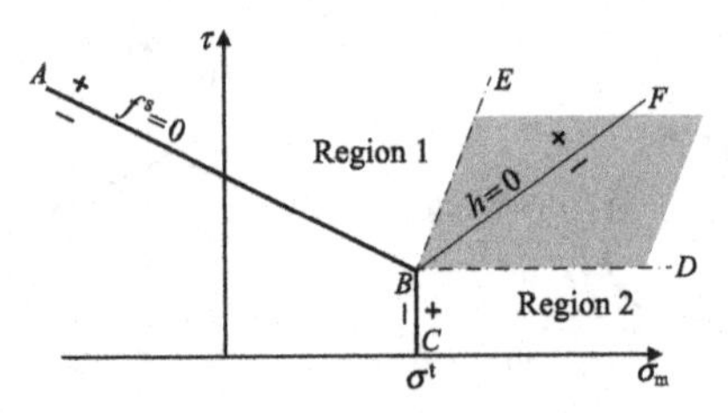

FIGURE 6.10 Drucker–Prager model with a tension cut-off.

lines BD and BE, as shown in Fig. 6.10. If the elastic trial stress point is located in region 1, the shear yielding occurs so that the plastic flow is determined by the shear plastic potential ψ^s and the trial stress point will be returned to the line $f^s = 0$. If the elastic trial stress point is located in region 2, the tensile yielding occurs such that the plastic flow is determined by the tensile plastic potential ψ^t and the trial stress point will be return to the line $f^t = 0$.

The slope of line BD is zero, while the slope of line BE is $1/q_\phi$. Based on the relationship

$$\tan\theta = \frac{2\tan\frac{\theta}{2}}{1-\tan^2\frac{\theta}{2}},$$

the slope of line BF can be found to be

$$\alpha^{\mathrm{B}} = \sqrt{1+q_\phi^2} - q_\phi. \tag{6.118}$$

Substituting $\sigma_m = \sigma^t$ into $f^s = 0$ leads to the effective shear stress at point B as

$$\tau^{\mathrm{B}} = k_\phi - q_\phi \sigma^t. \tag{6.119}$$

Thus, the equation of line BF can be written as [142]

$$h(\sigma_m, \tau) = \tau - \tau^{\mathrm{B}} - \alpha^{\mathrm{B}}(\sigma_m - \sigma^t) = 0. \tag{6.120}$$

According to Eq. (6.119), when the tensile strength $\sigma^t = k_\phi/q_\phi$, $\tau^{\mathrm{B}} = 0$ and the function $h(\sigma_m, \tau)$ becomes

$$h = \tau - \alpha^{\mathrm{B}}(\sigma_m - k_\phi/q_\phi). \tag{6.121}$$

6.2.5.3 Plastic Corrector

The elastic trial deviatoric stress $s_{ij}^{*(n+1)}$ and the elastic trial mean stress $\sigma_m^{*(n+1)}$ at time t^{n+1} can be calculated from Eqs. (6.71) and (6.72). The elastic trial effective shear stress is defined as

$$\tau^{*(n+1)} = \sqrt{\frac{1}{2} s_{ij}^{*(n+1)} s_{ij}^{*(n+1)}}. \tag{6.122}$$

According to Fig. 6.10, the elastic trial stress may satisfy one of the following four conditions:

1. If $\sigma_m^{*(n+1)} < \sigma^t$ and $f^s(\tau^{*(n+1)}, \sigma_m^{*(n+1)}) > 0$, the shear yielding occurs, and the elastic trial stress is corrected based on the plastic flow determined by the shear plastic potential ψ^s.
2. If $\sigma_m^{*(n+1)} < \sigma^t$ and $f^s(\tau^{*(n+1)}, \sigma_m^{*(n+1)}) \leqslant 0$, the elastic trial state is the true state so that no correction is required.
3. If $\sigma_m^{*(n+1)} \geqslant \sigma^t$ and $h(\tau^{*(n+1)}, \sigma_m^{*(n+1)}) > 0$, the shear yielding occurs, and the elastic trial stress is corrected based on the plastic flow determined by the shear plastic potential ψ^s.
4. If $\sigma_m^{*(n+1)} \geqslant \sigma^t$ and $h(\tau^{*(n+1)}, \sigma_m^{*(n+1)}) \leqslant 0$, the tensile yielding occurs, and the elastic trial stress is corrected based on the plastic flow determined by the tensile plastic potential ψ^t.

By checking the above four conditions, the elastic trial stress can be corrected accordingly, as detailed below.

Shear Yielding Corrector

If shear yielding occurs, the elastic trial stress should be corrected to the yield surface $f^s = 0$ based on the plastic flow determined from the shear plastic potential ψ^s. The corrected stress σ_{ij}^{n+1} must satisfy the yielding condition at time t^{n+1}, i.e.,

$$f^s(\sigma_{ij}^{n+1}) = f^s(\sigma_{ij}^{*(n+1)} - C_{ijkl}^{\sigma J} \Delta\varepsilon_{kl}^p) = 0. \tag{6.123}$$

Approximating Eq. (6.123) by the first-order Taylor series about the elastic trial stress point $\sigma_{ij}^{*(n+1)}$ gives

$$f^s(\sigma_{ij}^{*(n+1)}) - \frac{\partial f^s}{\partial \sigma_{ij}} C_{ijkl}^{\sigma J} \Delta\varepsilon_{kl}^p = 0 \tag{6.124}$$

where $\partial f^s / \partial \sigma_{ij}$ is evaluated at the trial stress point $\sigma_{ij}^{*(n+1)}$. The incremental plastic strain $\Delta\varepsilon_{kl}^p$ is then determined from the shear plastic potential ψ^s as

$$\Delta\varepsilon_{ij}^p = \Delta\lambda^s \frac{\partial \psi^s}{\partial \sigma_{ij}} \tag{6.125}$$

with $\Delta\lambda^s$ being the plastic loading parameter.

Substituting Eq. (6.125) into Eq. (6.124) leads to

$$\Delta\lambda^s = \frac{f^s(\sigma_{ij}^{*(n+1)})}{\frac{\partial f^s}{\partial \sigma_{ij}} C_{ijkl}^{\sigma J} \frac{\partial \psi^s}{\partial \sigma_{kl}}}. \tag{6.126}$$

For the isotropic elasticity tensor as shown in Eq. (6.22), we have

$$C_{ijkl}\frac{\partial \psi^s}{\partial \sigma_{kl}} = 2G\left(\frac{\partial \psi^s}{\partial \sigma_{ij}} - \frac{1}{3}\frac{\partial \psi^s}{\partial \sigma_{kl}}\delta_{kl}\delta_{ij}\right) + K\frac{\partial \psi^s}{\partial \sigma_{kl}}\delta_{kl}\delta_{ij}. \tag{6.127}$$

Substituting Eq. (6.127) into Eq. (6.126), the plastic loading parameter $\Delta\lambda^s$ can be rewritten as

$$\Delta\lambda^s = \frac{f^s(\sigma_{ij}^{*(n+1)})}{2G\frac{\partial f^s}{\partial \sigma_{ij}}\left(\frac{\partial \psi^s}{\partial \sigma_{ij}} - \frac{1}{3}\frac{\partial \psi^s}{\partial \sigma_{kl}}\delta_{kl}\delta_{ij}\right) + K\frac{\partial f^s}{\partial \sigma_{ij}}\frac{\partial \psi^s}{\partial \sigma_{kl}}\delta_{kl}\delta_{ij}}. \tag{6.128}$$

Invoking Eqs. (6.110), (6.113), and the equations

$$\frac{\partial \tau}{\partial \sigma_{ij}} = \frac{s_{ij}}{2\tau}, \quad \frac{\partial \sigma_m}{\partial \sigma_{ij}} = \frac{1}{3}\delta_{ij} \tag{6.129}$$

results in

$$\frac{\partial f^s}{\partial \sigma_{ij}} = \frac{\partial f^s}{\partial \tau}\frac{\partial \tau}{\partial \sigma_{ij}} + \frac{\partial f^s}{\partial \sigma_m}\frac{\partial \sigma_m}{\partial \sigma_{ij}} = \frac{s_{ij}}{2\tau} + \frac{q_\phi}{3}\delta_{ij}, \tag{6.130}$$

$$\frac{\partial \psi^s}{\partial \sigma_{ij}} = \frac{\partial \psi^s}{\partial \tau}\frac{\partial \tau}{\partial \sigma_{ij}} + \frac{\partial \psi^s}{\partial \sigma_m}\frac{\partial \sigma_m}{\partial \sigma_{ij}} = \frac{s_{ij}}{2\tau} + \frac{q_\psi}{3}\delta_{ij}. \tag{6.131}$$

Substituting Eqs. (6.130) and (6.131) into Eq. (6.128) and invoking the equations $s_{ij}\delta_{ij} = 0$ and $\delta_{ij}\delta_{ij} = 3$ leads to the plastic loading parameter $\Delta\lambda^s$ of the Drucker–Prager model as

$$\Delta\lambda^s = \frac{f^s(\sigma_{ij}^{*(n+1)})}{G + Kq_\phi q_\psi}. \tag{6.132}$$

Based on Eqs. (6.125) and (6.131), the deviatoric plastic strain increment $\Delta\varepsilon_{ij}^{\prime p}$, volumetric plastic strain increment $\Delta\varepsilon_{kk}^p$, and effective plastic strain increment $\Delta\varepsilon^p$ can be obtained as

$$\Delta\varepsilon_{ij}^{\prime p} = \Delta\lambda^s\frac{s_{ij}}{2\tau}, \tag{6.133}$$

$$\Delta\varepsilon_{kk}^p = \Delta\lambda^s q_\psi, \tag{6.134}$$

$$\Delta\varepsilon^p = \Delta\lambda^s\sqrt{\frac{1}{3} + \frac{2}{9}q_\psi^2}. \tag{6.135}$$

Substituting Eqs. (6.133) and (6.134) into Eqs. (6.69) and (6.70), the corrected deviatoric stress and mean stress at t^{n+1} can be found to be

$$s_{ij}^{n+1} = s_{ij}^{*(n+1)} - G\Delta\lambda^s \frac{s_{ij}^{n+1}}{\tau^{n+1}}, \tag{6.136}$$

$$\sigma_m^{n+1} = \sigma_m^{*(n+1)} - Kq_\psi \Delta\lambda^s. \tag{6.137}$$

Substituting Eqs. (6.76), (6.77), and the relation $s = \sqrt{3}\tau$ into Eq. (6.136) gives

$$\sqrt{2}\tau^{n+1} n_{ij}^{n+1} = \sqrt{2}\tau^{*(n+1)} n_{ij}^{n+1} - \sqrt{2}G\Delta\lambda^s n_{ij}^{n+1}, \tag{6.138}$$

so that

$$\tau^{n+1} = \tau^{*(n+1)} - G\Delta\lambda^s. \tag{6.139}$$

Using Eqs. (6.136) and (6.139), the deviatoric stress update formulation can be reformulated as

$$\begin{aligned} s_{ij}^{n+1} &= \frac{\tau^{n+1}}{\tau^{n+1} + G\Delta\lambda^s} s_{ij}^{*(n+1)} \\ &= \frac{\tau^{n+1}}{\tau^{*(n+1)}} s_{ij}^{*(n+1)} \end{aligned} \tag{6.140}$$

where τ^{n+1} can be obtained from the yield condition $f^{s(n+1)} = 0$ at t^{n+1} as

$$\tau^{n+1} = k_\phi - q_\phi \sigma_m^{n+1}. \tag{6.141}$$

Similar to the J_2 flow theory, the deviatoric stress is also updated by scaling the elastic trial deviatoric stress, namely, by returning the elastic trial stress along the radial direction to the yield surface.

Tensile Yielding Corrector

Taking derivatives of the tensile yielding function Eq. (6.114) and the tensile plastic potential Eq. (6.116) with respect to the stress results in

$$\frac{\partial f^t}{\partial \sigma_{ij}} = \frac{\partial f^t}{\partial \tau}\frac{\partial \tau}{\partial \sigma_{ij}} + \frac{\partial f^t}{\partial \sigma_m}\frac{\partial \sigma_m}{\partial \sigma_{ij}} = \frac{1}{3}\delta_{ij}, \tag{6.142}$$

$$\frac{\partial \psi^t}{\partial \sigma_{ij}} = \frac{\partial \psi^t}{\partial \tau}\frac{\partial \tau}{\partial \sigma_{ij}} + \frac{\partial \psi^t}{\partial \sigma_m}\frac{\partial \sigma_m}{\partial \sigma_{ij}} = \frac{1}{3}\delta_{ij}. \tag{6.143}$$

Substituting Eqs. (6.142) and (6.143) into Eq. (6.128) and replacing the superscript s with t, the plastic loading parameter can be solved as

$$\Delta\lambda^t = \frac{f^t(\sigma_{ij}^{*(n+1)})}{K} = \frac{\sigma_m^{*(n+1)} - \sigma^t}{K}. \tag{6.144}$$

The plastic strain increment can be obtained from the plastic flow rule as

$$\Delta\varepsilon_{ij}^{p} = \Delta\lambda^{t}\frac{\partial\psi^{t}}{\partial\sigma_{ij}} = \frac{1}{3}\Delta\lambda^{t}\delta_{ij}. \tag{6.145}$$

From Eq. (6.145), the plastic volumetric strain increment and effective plastic strain increment can be obtained as

$$\Delta\varepsilon_{kk}^{p} = \Delta\lambda^{t}, \tag{6.146}$$

$$\Delta\varepsilon^{p} = \frac{\sqrt{2}}{3}\Delta\lambda^{t}. \tag{6.147}$$

Substituting Eq. (6.146) into Eq. (6.70) and invoking Eq. (6.144) results in the corrected mean stress at t^{n+1} as

$$\sigma_{m}^{n+1} = \sigma_{m}^{*(n+1)} - K\Delta\lambda^{t} = \sigma^{t}. \tag{6.148}$$

For tensile yielding, only the mean stress is corrected, and the deviatoric stress stays unchanged during the plastic corrector stage (i.e., $s_{ij}^{n+1} = s_{ij}^{*n+1}$). Thus, the stress at t^{n+1} takes the form of

$$\sigma_{ij}^{n+1} = s_{ij}^{*(n+1)} + \sigma^{t}\delta_{ij} = \sigma_{ij}^{*(n+1)} + (\sigma^{t} - \sigma_{m}^{*(n+1)})\delta_{ij}. \tag{6.149}$$

6.2.5.4 Numerical Algorithm

Based on the above discussion, the numerical algorithm for the Drucker–Prager model can be summarized as follows:

1. Calculate the rotated stress $\sigma_{ij}^{R^n}$, elastic trial deviatoric stress $s_{ij}^{*(n+1)}$, and trial mean stress $\sigma_{m}^{*(n+1)}$ using Eqs. (6.5), (6.71), and (6.72), respectively, and then calculate the elastic trial shear stress $\tau^{*(n+1)}$;

2. If $\sigma_{m}^{*(n+1)} < \sigma^{t}$ and $f^{s}(\tau^{*(n+1)}, \sigma_{m}^{*(n+1)}) > 0$, calculate the loading parameter $\Delta\lambda^{s}$ and shear stress τ^{n+1} using Eqs. (6.132) and (6.141), respectively, and then update the deviatoric stress s_{ij}^{n+1} and mean stress σ_{m}^{n+1} using Eqs. (6.140) and (6.137), respectively;

3. If $\sigma_{m}^{*(n+1)} \geqslant \sigma^{t}$ and $h(\tau^{*(n+1)}, \sigma_{m}^{*(n+1)}) > 0$, calculate the loading parameter $\Delta\lambda^{s}$ and shear stress τ^{n+1} using Eqs. (6.132) and (6.141), respectively, and then update the deviatoric stress s_{ij}^{n+1} and mean stress σ_{m}^{n+1} using Eqs. (6.140) and (6.137), respectively;

4. If $\sigma_{m}^{*(n+1)} \geqslant \sigma^{t}$ and $h(\tau^{*(n+1)}, \sigma_{m}^{*(n+1)}) \leqslant 0$, update the stress σ_{ij}^{n+1} using Eq. (6.149).

The Drucker–Prager model with a tension cut-off is implemented into the FORTRAN subroutine M3DM9 in Sect. 6.5.

6.2.6 Newtonian Fluid

Ideal fluids have no shear stress and viscosity, i.e., $s_{ij} = 0$, and their pressure is determined from an EOS. Newtonian fluids are the simplest mathematical model of fluids that accounts for viscosity [130], whose viscous stress s_{ij} is related to the deviatoric strain rate $\dot{\varepsilon}'_{ij}$ by

$$s_{ij} = 2\mu\dot{\varepsilon}'_{ij} \tag{6.150}$$

where μ is the dynamic (shear) viscosity that does not depend on the velocity or stress state of the fluid. For non-Newtonian fluids, the viscosity is dependent on shear rate or shear rate history. Many common liquids and gases, such as water, alcohol, light oil, and air, can be assumed to be Newtonian. The viscosity of water at room temperature is about 1.01×10^{-3} kg/m s.

Note that when $\mu = 0$, the fluid has no resistance to shearing, and could experience extremely large shear deformation under a very small shear force, which will lead to numerical instability. Thus, $\mu = 0$ should be avoided in numerical simulation, and a critical value must be chosen with the viscosity being assumed to be very small, which is consistent with physics. Otherwise, an EOS should be used. The pressure p, density ρ, and temperature T of an ideal fluid are related by an EOS as follows:

$$p = p(\rho, T). \tag{6.151}$$

For example, the EOS of an ideal gas is given by

$$p = R\rho T \tag{6.152}$$

where R is the specific gas constant.

The model for ideal fluids is implemented into the FORTRAN subroutine M3DM7 in Sect. 6.5.

6.2.7 High Explosive

Detailed information about high explosive modeling could be found in representative references [49,53]. The model used in the open-source MPM code to simulate the detonation process of a high explosive is introduced here. Based on the Chapman–Jouguet theory, an ideal detonation consists of two processes, the steady-state detonation process and the following process including the expansion of gaseous products and their interaction with the surrounding material, as shown in Fig. 6.11.

The steady-state detonation can be seen as a shock wave moving through the explosive, whose front compresses and heats the explosive to initiate chemical

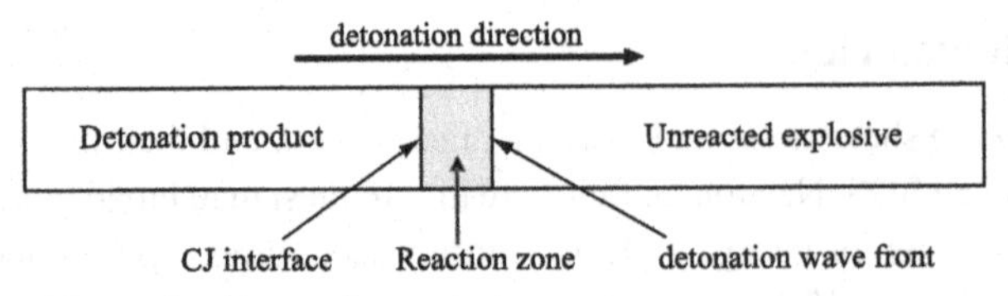

FIGURE 6.11 Detonation process of an explosive.

reaction. Because the velocity of the detonation wave is very fast, the exothermic reaction is completed within a few microseconds. The energy released by the reaction feeds the shock front and drives it forward. At the same time, the gaseous products are expanding and interact with the surrounding material. The shock front, chemical reaction, and the leading edge of the rarefaction are all in dynamic equilibrium such that they are all traveling at the same velocity which is named as the detonation velocity and is one of material constants of a specified explosive [143]. For military explosives, the detonation velocity is about 6500 to 9500 m/s, and the pressure of the detonation products could reach to tens of GPa with temperature of 3000 to 5000 K. Please refer to Sect. 2.9 for further information about the detonation wave.

In the initialization phase, a lighting time t_L is calculated for all the particles in the explosive by dividing the distance from the detonation point by the detonation velocity D. If multiple detonation points are specified, the closest point determines t_L. After the detonation, the gaseous product is controlled by an EOS. The real pressure p of the gaseous product is given by multiplying the pressure p_{EOS} obtained from the EOS for the explosive with a burn fraction F that controls the release of chemical energy for simulating detonation [14], namely,

$$p = F \cdot p_{\text{EOS}} \tag{6.153}$$

where F is the burn fraction taken as

$$F = \begin{cases} \frac{(t-t_L)D}{1.5h} & t > t_L, \\ 0 & t \leqslant t_L \end{cases} \tag{6.154}$$

where h denotes the characteristic size of a particle, and t is the current time. If F exceeds 1, it is reset to 1. It often takes several time steps for F to reach the value 1 with this calculation of the burn fraction. After reaching the value 1, F is held constant. With this method, the discontinuous detonation wave can be smoothed to be a continuous one with a rapidly changing wavefront in a narrow region to suppress the numerical oscillation induced by the discontinuities [134, 144].

Eq. (6.154) predicts the ignition due to the propagation of detonation wave. A high explosive can also be ignited due to shock compression whose burn fraction can be calculated by

$$F = \beta(1 - V) \tag{6.155}$$

with

$$\beta = \frac{\rho_0 D^2}{p_{CJ}} = \frac{1}{1 - V_{CJ}} \tag{6.156}$$

where p_{CJ} is the Chapman–Jouguet pressure provided as a material constant by users, and V_{CJ} is the Chapman–Jouguet relative volume. If both ignition mechanisms are considered, the burn fraction F in Eq. (6.153) should be chosen as the maximum value of those obtained from Eqs. (6.154) and (6.155).

Explosives can be assumed as an ideal elastoplastic material before ignition, and an ideal gas after ignition. To avoid numerical instability, the viscosity of certain value should be included in the gaseous product of explosives, as shown in Eq. (6.150).

The high explosive model is implemented into the FORTRAN subroutine M3DM6 in Sect. 6.5.

6.3 EQUATION OF STATE

An equation of state (EOS) is a relation among state variables, such as pressure, volume, temperature, or internal energy, which describes the state of matter under a given set of physical conditions. The details of the EOSs related to extreme events can be found in the representative references [49,53]. The essential features and numerical implementation of the EOSs used in the open-source MPM code are introduced in this section.

6.3.1 Polytropic Process

A polytropic process obeys the relation

$$pv^n = C \tag{6.157}$$

where $v = 1/\rho$ is the specific volume, n is the polytropic index, and C is a constant. The condition of $n = 0$ represents an isobaric process, $n = 1$ represents an isothermal process, $n = \gamma = c_p/c_v$ represents an isentropic process, and $n = \infty$ represents an isochoric process.

For an isentropic process ($n = \gamma$), the first law of thermodynamics gives

$$de = -p dv \tag{6.158}$$

where e is the specific internal energy. Integrating Eq. (6.158) leads to

$$\begin{aligned} e &= -\int_{v_0}^{v} p\mathrm{d}v = -C\int_{v_0}^{v} v^{-\gamma}\mathrm{d}v \\ &= \frac{pv - p_0 v_0}{\gamma - 1} = \frac{p}{\rho(\gamma - 1)} - \frac{p_0}{\rho_0(\gamma - 1)}. \end{aligned} \tag{6.159}$$

Eq. (6.159) can be rewritten as

$$p = \frac{\rho}{\rho_0} p_0 + (\gamma - 1)\rho e = \frac{\rho}{\rho_0}[p_0 + (\gamma - 1)E] \tag{6.160}$$

with E being the internal energy per unit initial volume.

The expansion of the gaseous products of a high explosive can be modeled as a polytropic process. In the high pressure regime, n approximately equals to 3. The value of n decreases with the expansion of the products, and finally approaches 1.4 when the products expand to the atmospheric pressure state.

6.3.2 Nearly Incompressible Fluid

If an EOS is used for a fluid with a very low Mach number, the critical time step of an explicit time integration scheme will be very small, especially when the Mach number approaches zero, which corresponds to a nearly incompressible fluid. To avoid this numerical difficulty, an artificial EOS can be employed which uses an artificial speed of sound. By choosing an appropriate value of the artificial speed of sound, the critical time step can become large, and the density fluctuation can be reduced, namely, less than 3%.

When simulating free surface flows with the SPH, Monaghan [46,99] employed the artificial EOS [145] as follows:

$$p = p_0\left[\left(\frac{\rho}{\rho_0}\right)^{\gamma} - 1\right] \tag{6.161}$$

where $\gamma = 7$, and p_0 and ρ_0 are the reference pressure and reference density, respectively. Choosing γ to be 7 can keep the density fluctuation very small to satisfy the incompressible condition. However, a small error in density may result in a significant error in pressure.

Morris et al. [146] employed the following EOS in simulating a low Reynolds number incompressible flows with the SPH, i.e.,

$$p = c^2\rho \tag{6.162}$$

where c is the artificial speed of sound, which should be much smaller than the true speed of sound of the fluid to make the critical time step large enough, but should not be too small to make the fluid nearly incompressible. Morris et al. [146] proposed to estimate the artificial speed of sound with the use of

$$c^2 \sim \max\left(\frac{V_0^2}{\delta}, \frac{\nu V_0}{L_0\delta}, \frac{F L_0}{\delta}\right) \tag{6.163}$$

where V_0 is the fluid velocity, F is the body force per unit mass, L_0 is the characteristic length, $\nu = \mu/\rho$ is the kinematic viscosity, and

$$\delta = \frac{\Delta\rho}{\rho_0}$$

is the relative density fluctuation that can be chosen as $\delta \leqslant 3\%$.

6.3.3 Linear Polynomial

The linear polynomial EOS is given by

$$p = c_0 + c_1\mu + c_2\mu^2 + c_3\mu^3 + (c_4 + c_5\mu + c_6\mu^2)E \tag{6.164}$$

where c_0, c_1, c_2, c_3, c_4, c_5, and c_6 are user-defined material constants, and

$$\mu = \frac{\rho}{\rho_0} - 1 = \frac{1}{V} - 1 \tag{6.165}$$

where $V = \rho_0/\rho$ is the relative volume, namely, the ratio of the current volume to the initial volume.

To determine the critical time step of an explicit time integration scheme, the partial derivatives of pressure p with respect to density ρ and internal energy E are required, as shown in Eq. (3.43). The derivative of μ with respect to density ρ can be obtained from Eq. (6.165) as

$$\frac{d\mu}{d\rho} = \frac{1}{\rho_0}. \tag{6.166}$$

Taking partial derivatives of Eq. (6.164) with respect to ρ and E and invoking Eq. (6.166) leads to

$$\left.\frac{\partial p}{\partial \rho}\right|_E = \left.\frac{\partial p}{\partial \mu}\right|_E \frac{d\mu}{d\rho} = \frac{1}{\rho_0}[c_1 + 2c_2\mu + 3c_3\mu^2 + (c_5 + 2c_6\mu^2)E], \tag{6.167}$$

$$\left.\frac{\partial p}{\partial E}\right|_\rho = c_4 + c_5\mu + c_6\mu^2. \tag{6.168}$$

TABLE 6.1 Material Constants in the JWL EOS for Different Explosives

Explosive	CJ parameters				Coefficients in EOS					
	ρ_0 g/cm^3	p GPa	D cm/μs	γ	E_0 kJ/cm^3	A GPa	B GPa	R_1	R_2	ω
TNT	1.63	21.0	0.693	2.727	7.0	371.2	3.23	4.15	0.95	0.30
Tetryl	1.73	28.5	0.791	2.798	8.2	586.8	10.67	4.4	1.20	0.28

Substituting Eqs. (6.167) and (6.168) into Eq. (3.43), the adiabatic sound speed of the material can be obtained as

$$c = \left\{ \frac{4G}{3\rho} + \frac{1}{\rho_0}[c_1 + 2c_2\mu + 3c_3\mu^2 + (c_5 + 2c_6\mu^2)E] + \frac{pV^2}{\rho_0}(c_4 + c_5\mu + c_6\mu^2) \right\}^{\frac{1}{2}}. \tag{6.169}$$

When the material is in the tensile regime ($\mu < 0$), the coefficients of μ^2 are set to zero, namely, $c_2 = c_6 = 0$.

When choosing $c_0 = c_1 = c_2 = c_3 = c_6 = 0$ and $c_4 = c_5 = \gamma - 1$, the linear polynomial EOS is reduced to the Gamma law EOS for ideal gas, namely,

$$p = (\gamma - 1)\rho e \tag{6.170}$$

where $\gamma = c_p/c_v$ is the adiabatic index, heat capacity ratio, or the ratio of specific heat, c_p is the heat capacity at constant pressure, and c_v is the heat capacity at constant volume.

The linear polynomial EOS is implemented into the FORTRAN subroutine eos1 in Sect. 6.5.

6.3.4 JWL

The JWL (Jones–Wilkins–Lee) EOS can be used to describe the relation among pressure, internal energy, and volume of the gaseous products of high explosives, which takes the form of

$$p = A\left(1 - \frac{\omega}{R_1 V}\right)e^{-R_1 V} + B\left(1 - \frac{\omega}{R_2 V}\right)e^{-R_2 V} + \frac{\omega E}{V} \tag{6.171}$$

where $E = \rho_0 e$ is the internal energy per unit initial volume, and ω, A, B, R_1, and R_2 are user-defined material constants. Table 6.1 lists the material constants for two kinds of explosives [8,147].

Taking partial derivatives of Eq. (6.171) with respect to ρ and E and invoking Eq. (3.41) results in

$$\begin{aligned}\left.\frac{\partial p}{\partial \rho}\right|_E &= \left.\frac{\partial p}{\partial V}\right|_E \frac{\mathrm{d}V}{\mathrm{d}\rho} \\ &= \frac{V^2}{\rho_0}\left\{\left[AR_1\left(1-\frac{\omega}{R_1 V}\right)-A\frac{\omega}{R_1 V^2}\right]e^{-R_1 V}+\right. \\ &\left.\left[BR_2\left(1-\frac{\omega}{R_2 V}\right)-B\frac{\omega}{R_2 V^2}\right]e^{-R_2 V}+\frac{\omega E}{V^2}\right\},\end{aligned} \tag{6.172}$$

$$\left.\frac{\partial p}{\partial E}\right|_\rho = \frac{\omega}{V}. \tag{6.173}$$

Substituting Eqs. (6.172) and (6.173) into Eq. (3.43), the adiabatic sound speed of the material can be obtained as

$$\begin{aligned}c = &\left\{\frac{V^2}{\rho_0}\left\{\left[AR_1\left(1-\frac{\omega}{R_1 V}\right)-A\frac{\omega}{R_1 V^2}\right]e^{-R_1 V}+\right.\right. \\ &\left.\left.\left[BR_2\left(1-\frac{\omega}{R_2 V}\right)-B\frac{\omega}{R_2 V^2}\right]e^{-R_2 V}+\frac{\omega E}{V^2}\right\}+\frac{p\omega}{\rho}\right\}^{\frac{1}{2}}.\end{aligned} \tag{6.174}$$

The JWL EOS is implemented into the FORTRAN subroutine eos3 in Sect. 6.5.

6.3.5 Mie–Grüneisen

The Mie–Grüneisen EOS used to determine the pressure in a shock-compressed solid is given as [50]

$$p = p_H + \frac{\gamma}{v}(e - e_H) \tag{6.175}$$

where p_H and e_H are the pressure and specific internal energy on the Hugoniot, respectively, which can be found from Eqs. (2.81) and (2.77), and γ is the Grüneisen parameter determined by

$$\gamma = v\left(\frac{\partial p}{\partial e}\right)_v = \frac{3\alpha v}{C_v K} \tag{6.176}$$

where $3\alpha = (1/v)(\partial v/\partial T)_p$ is the thermal expansion coefficient, $K = -(1/v)(\partial v/\partial P)_T$ is the isothermal compressibility coefficient, and $C_v = (\partial e/\partial T)_v$ is the heat capacity at constant volume. The Grüneisen parameter

γ obeys the following relation:

$$\frac{\gamma}{v} = \frac{\gamma_0}{v_0} = \text{constant} \tag{6.177}$$

where γ_0 and v_0 are the Grüneisen parameter and specific volume at the reference state, respectively.

Due to $p_H \gg p_0$, Eq. (2.81) can be rewritten as

$$p_H = \frac{c_0^2(v_0 - v)}{[v_0 - s(v_0 - v)]^2} = \frac{\rho_0 c_0^2 \mu(1+\mu)}{[1-(s-1)\mu]^2} \tag{6.178}$$

where $\mu = \rho/\rho_0 - 1 = v_0/v - 1$. Eq. (6.178) can be approximated by a Taylor series of degree 3 about $\mu = 0$ as

$$\begin{aligned} p_H &\approx p_H|_{\mu=0} + \left.\frac{\mathrm{d}p_H}{\mathrm{d}\mu}\right|_{\mu=0} \mu + \frac{1}{2}\left.\frac{\mathrm{d}^2 p_H}{\mathrm{d}\mu^2}\right|_{\mu=0} \mu^2 + \frac{1}{6}\left.\frac{\mathrm{d}p_H^3}{\mathrm{d}\mu^3}\right|_{\mu=0} \mu^3 \\ &= \rho_0 c_0^2[\mu + (2s-1)\mu^2 + (s-1)(3s-1)\mu^3]. \end{aligned} \tag{6.179}$$

Substituting the internal energy Eq. (2.77) on the Hugoniot curve into Eq. (6.175) and assuming $p_0 = 0$ and $e_0 = 0$, we have

$$p = p_H\left(1 - \frac{\gamma\mu}{2}\right) + \gamma_0 E \tag{6.180}$$

with $E = \rho_0 e$.

For expanded materials, i.e., $\mu < 0$, the pressure is defined by

$$p = \rho_0 c_0^2 \mu + \gamma_0 E. \tag{6.181}$$

The partial derivatives of Eq. (6.180) with respect to ρ and E can be obtained as

$$\left.\frac{\partial p}{\partial \rho}\right|_E = \left.\frac{\partial p}{\partial \mu}\right|_E \frac{\mathrm{d}\mu}{\mathrm{d}\rho} = \begin{cases} \left[\frac{\mathrm{d}p_H}{\mathrm{d}\mu}\left(1 - \frac{\gamma\mu}{2}\right) - p_H \frac{\gamma}{2}\right]\frac{1}{\rho_0} & \mu \geqslant 0, \\ c_0^2 & \mu < 0, \end{cases} \tag{6.182}$$

$$\left.\frac{\partial p}{\partial E}\right|_\rho = \gamma_0. \tag{6.183}$$

Thus, the adiabatic sound speed of the material can be found from Eq. (3.43) as follows:

$$c = \begin{cases} \left\{\frac{4G}{3\rho} + \frac{1}{\rho_0}\left[\frac{\mathrm{d}p_H}{\mathrm{d}\mu}\left(1 - \frac{\gamma\mu}{2}\right) - p_H\frac{\gamma}{2}\right] + \frac{pV^2}{\rho_0}\gamma_0\right\}^{\frac{1}{2}} & \mu \geqslant 0, \\ \left\{\frac{4G}{3\rho} + c_0^2 + \frac{pV^2}{\rho_0}\gamma_0\right\}^{\frac{1}{2}} & \mu < 0 \end{cases} \tag{6.184}$$

with

$$\frac{\mathrm{d}p_H}{\mathrm{d}\mu} = \rho_0 c_0^2 \frac{1+(s_1+1)\mu}{[1-(s_1-1)\mu]^3}$$
$$\approx \rho_0 c_0^2[1+2(2s-1)\mu+3(s-1)(3s-1)\mu^2]. \tag{6.185}$$

Eq. (6.178) shows that when $\mu = 1/(s-1)$, we get $p = \infty$, namely, the maximum value of μ of the material is $\mu_{\max} = 1/(s-1)$.

The Mie–Grüneisen EOS is implemented into the FORTRAN subroutine eos2 in Sect. 6.5.

6.4 FAILURE MODELS

How to predict the onset and evolution of localized failure remains a challenging task, as shown in the special issue on the recent Sandia Fracture Challenge [148]. A lack of consistency across different international research groups in addressing problems of fracture necessitates further investigations in failure modeling via integrated analytical, experimental, and numerical efforts. The focus of this section is on the introduction of the criteria used in the open-source MPM code to identify the onset of localized failure. Both continuous and discontinuous approaches are available for modeling and simulating the evolution of localized failure, depending on the level of discontinuity [149–151], etc. However, a further discussion on this topic is beyond the scope of this book. Hence, only the essential features and numerical implementation of commonly used failure criteria are given below.

In Lagrangian FEM codes, a failed element is usually removed from the computational process (element erosion) to overcome the numerical difficulties associated with the element distortions. However, the erosion will cause a loss of mass, strength, and internal energy, due to the nonlocal nature of failure, so that it is not a physics-based modeling procedure. In the MPM, there is no element distortion so that erosion is no longer required. In the open-source MPM code, a failed particle is unable to sustain shear or tensile stress. In each time step, hence, the deviatoric stresses of all failed particles are set to zero. In addition, the mean stress of a failed particle is also set to zero if its hydrostatic pressure is negative. Note that this treatment does not consider the nonlocal nature of failure evolution for computational convenience so that special caution must be taken when calibrating the model parameters against experimental data based on a specific arrangement of background grid and particle spacing. A recent study on the improved decohesion modeling with the MPM has circumvented the mesh-dependency of numerical solutions [151].

6.4.1 Effective Plastic Strain Failure Model

A particle is assumed to be failed if its effective plastic strain ε^p exceeds a user-defined critical value $\varepsilon^p_{\max}$, i.e., $\varepsilon^p > \varepsilon^p_{\max}$. This criterion is usually used to model ductile materials. Refer to Sect. 6.2.2 for the calculation of the effective plastic strain ε^p.

6.4.2 Hydrostatic Tensile Failure Model

A particle is assumed to be failed if it is in tension and its hydrostatic pressure p is less than a user-defined critical value (negative) $p_{\min}$, which is usually adopted to model brittle failure.

The effective plastic strain and hydrostatic tensile failure models can be used together, namely, if a particle satisfies the condition of $\varepsilon^p > \varepsilon^p_{\max}$ or $p < p_{\min}$, the particle is assumed to be failed.

6.4.3 Maximum Principal/Shear Stress Failure Model

When the maximum principal stress or maximum shear stress of a particle exceeds a user-defined value, the particle is assumed to be failed.

The principal stress can be found from the following characteristic equation:

$$\lambda^3 - I_1\lambda^2 - I_2\lambda - I_3 = 0 \tag{6.186}$$

where

$$I_1 = \sigma_{kk}, \tag{6.187}$$

$$I_2 = -\frac{1}{2}(\sigma_{ii}\sigma_{kk} - \sigma_{ik}\sigma_{ik}), \tag{6.188}$$

$$I_3 = \det(\boldsymbol{\sigma}) \tag{6.189}$$

are the three invariants of the Cauchy stress. It can be shown that Eq. (6.186) possesses three real roots, σ_1, σ_2, and σ_3. Assuming $\sigma_1 \geqslant \sigma_2 \geqslant \sigma_3$, the maximum shear stress can be written as

$$\tau_{\max} = \frac{\sigma_1 - \sigma_3}{2}. \tag{6.190}$$

6.4.4 Maximum Principal/Shear Strain Failure Model

When the maximum principal strain or maximum shear strain of a particle exceeds a user-defined value, the particle is assumed to be failed.

6.4.5 Effective Strain Failure Model

When the effective strain ε_{eff} of a particle exceeds a user-defined value, the particle is assumed to be failed. The effective strain ε_{eff} is defined as

$$\varepsilon_{\text{eff}} = \left(\frac{2}{3}\boldsymbol{\varepsilon} : \boldsymbol{\varepsilon}\right)^{\frac{1}{2}}. \tag{6.191}$$

The above two criteria are suitable for strain-based constitutive modeling, such as strain-based damage models, in failure analyses.

6.5 COMPUTER IMPLEMENTATION OF MATERIAL MODELS

The main purpose of constitutive modeling is to update the stress state $\boldsymbol{\sigma}^{n+1}$ for the given strain rate, vorticity, and internal state variables based on the previous state, as discussed in Sect. 6.1. The deviatoric stress s^{n+1} is updated using Eq. (6.7), and the pressure p^{n+1} is updated using Eq. (6.18). In the MPM3D-F90, the module MaterialData encapsulates the variables of the material models initialized by the user input data file, while the module MaterialModel encapsulates the inherent variables and operators of the material models.

6.5.1 Module MaterialData

The module MaterialData encapsulates the variables of material models which are initialized by the user input data file, such as the material parameters, the total number of detonation points, and their positions. When implementing new constitutive models and EOSs, the new required variables could be added into the module.

The module MaterialData defines a derived data type Material with the following elements:

- integer::MatType – Type of strength model (1-Elasticity, 2-Perfect elastoplasticity, 3-Linear isotropic hardening, 4-Johnson–Cook plasticity, 5-Simplified Johnson–Cook plasticity, 6-Simplified Johnson–Cook plasticity with failure, 7-Johnson–Cook plasticity with failure, 8-high explosive, 9-Null material, and 10-Drucker–Prager plasticity);
- integer::EosType – Type of EOSs (1-Polynomial, 2-Mie–Grüneisen, and 3-JWL);
- real(8)::Density – Initial density of a particle;
- real(8)::Young – Young's modulus;
- real(8)::Poisson – Poisson's ratio;
- real(8)::Mp – Particle mass;

- real(8)::Yield0 – Initial yield strength;
- real(8)::TangMod – Tangential modulus;
- real(8)::roomt – Room temperature;
- real(8)::Melt: Melting temperature;
- real(8)::SpecHeat – Specified heat;
- real(8)::B_jc, n_jc, C_jc, m_jc – Parameters for Johnson–Cook plasticity;
- real(8)::epso – The effective plastic strain-rate of the quasi-static test, $\dot{\varepsilon}_0$, used to determine the yield and hardening parameters in the Johnson–Cook plasticity model;
- real(8)::prd – The tensile pressure value to initiate damage;
- real(8)::epf – The effective plastic strain at failure;
- real(8)::D – Detonation velocity;
- real(8)::cEos(10) = 0 – The constants in the EOSs, which are C0, C1, C2, C3, C4, C5, C6 for the linear polynomial EOS, C1, C2, C3, C4 for the Grüneisen EOS, and A, B, R1, R2, W, E0, V0 for the JWL EOS;
- real(8)::Wavespd – The sound speed of a NULL material. In MPM3D-F90, the sound speed of linearly elastic material, Eq. (3.45), is used to determine the critical time step. Thus, for a NULL material, an artificial sound speed has to be specified by users; and
- real(8)::q_fai, k_fai, q_psi, ten_f – The parameters for the Drucker–Prager model.

The module MaterialData also defines the following global variables:

- integer::nb_mat = 0 – Total number of material sets defined in the user input data file;
- type(Material), allocatable::mat_list(:) – The list of materials, which stores the nb_mat materials;
- logical::Jaum = .true. – Flag to identify whether to use Jaumann stress rate in the stress update;
- integer, parameter::maxDeto = 10 – The maximum number of detonation points;
- integer::nDeto = 0 – Total number of detonation points defined in the user input data file, which must be less than or equal to maxDeto;
- real(8)::DetoX(maxDeto) = 0.0, DetoY(maxDeto) = 0.0, DetoZ(maxDeto) = 0.0 – The x, y, and z coordinates of the detonation points; and
- real(8)::bq1 = 0.0, bq2 = 0.0 – The artificial bulk coefficients.

The source code of the module MaterialData can be found in the file Material.f90.

6.5.2 Module MaterialModel

The module MaterialModel encapsulates the variables and corresponding operators for each material model, which consists of a strength model and an EOS for extreme events. The strength model is used to update the deviatoric stress, while the EOS is used to update the pressure. The strength models implemented in the module MaterialModel include elasticity, elastoplasticity, Johnson–Cook plasticity, Drucker–Prager plasticity, null material, and high explosive material. The EOSs implemented in the module MaterialModel include the polynomial, Mie–Grüneisen, and JWL EOSs. New variables and operators could be added in the module MaterialModel to implement new strength models and/or EOSs.

The module MaterialModel defines the following global variables:

- integer::mid – The material ID;
- integer::etype_ – Type of EOS;
- integer::mtype_ – Type of strength model;
- real(8)::young_, poisson, yield0_, tangmod_ – The elastic modulus, Poisson's ratio, initial yielding strength, and elastoplastic tangent modulus, respectively;
- real(8)::den0_ – The initial density of material;
- real(8)::den_ – The current density of material;
- real(8)::vold – The volume of a particle at the nth time step;
- real(8)::vol_ – The current volume of a particle;
- real(8)::vol0_ – The initial volume of a particle;
- real(8)::dvol – Half of the volume increment of a particle;
- real(8)::dinc(6) – The strain increment of a particle;
- real(8)::sm – The mean stress of a particle;
- real(8)::sd(6) – The deviatoric stress of a particle;
- real(8)::sig(6) – The stress of a particle;
- real(8)::dsm – The pressure increment of a particle;
- real(8)::bqf – The bulk viscosity force;
- real(8)::seqv – The effective stress;
- real(8)::epeff_ – The effective plastic strain;
- real(8)::sig_y – The current yield stress;
- real(8)::depeff – The effective plastic strain increment;
- real(8)::ratio – For hardening calculation;
- real(8)::G2, K3, PlaMod – 2*G, 3*K, and plastic hardening modulus;
- real(8)::specheat_ – Specific heat capacity;
- real(8)::tmprt – Temperature;
- real(8)::iener – Internal energy
- real(8)::specen – Internal energy per initial volume;

- real(8)::mu – mu = den_/den0_ −1;
- real(8)::rv – Relative volume
- real(8)::bfac – Burn fraction; and
- real(8)::cp – Sound speed.

The module MaterialModel defines the following operators:

- Constitution(de, vort, b, p) – Update the stress of particle p by invoking corresponding strength model and EOS. The parameters de, vort, b, and p are the strain increment, vortex increment, body index, and particle ID, respectively;
- sigrot (vort, sig, sm, sd) – Rotate the stress state when the Jaumann rate is used in the stress update, as shown in Sect. 6.1;
- elastic_devi – Update the deviatoric stress of the particle using the elasticity model Eq. (6.29);
- elastic_p – Update the mean stress of the particle using the elasticity model Eq. (6.30);
- lieupd – Update the internal energy of the particle whose strength model is not used with an EOS, using Eq. (6.10);
- hieupd – Update the internal energy of the particle whose strength model is used with an EOS, using Eq. (6.15);
- EquivalentStress – Calculate the effective stress of the particle using Eq. (6.64);
- M3DM1 – The elasticity model, as shown in Sect. 6.2.1;
- M3DM2 – The perfect elastoplasticity model, as shown in Sect. 6.2.4;
- M3D3 – The linear isotropic hardening model, as shown in Sect. 6.2.4;
- M3DM4 (mat, DT, tmprt) – The Johnson–Cook plasticity model in which the parameters mat, DT, and tmprt are the material ID, time step size, and current temperature, respectively, as shown in Sect. 6.2.4.4;
- M3DM5(mat, DT) – The simplified Johnson–Cook plasticity model that must be used with an EOS, as shown in Sect. 6.2.4.4;
- M3DM6(mat, DT) – The high explosive model that must be used with an EOS, as shown in Sect. 6.2.7;
- M3DM7 – The null material model that must be used with an EOS for fluid, as shown in Sect. 6.2.6;
- M3DM8(mat, DT) – The Johnson–Cook plasticity model that must be used with an EOS, as shown in Sect. 6.2.4.4;
- M3DM9(mat, DT) – The perfect Drucker–Prager plasticity model, as shown in Sect. 6.2.5;
- seleos(failure) – Update the pressure of a particle by invoking an EOS. The parameter failure indicates whether the particle is failed;
- eos1(failure) – The polynomial EOS, as shown in Sect. 6.3.3;

- eos2(failure) – The Mie–Grüneisen EOS, as shown in Sect. 6.3.5;
- eos3(failure) – The JWL EOS, as shown in Sect. 6.3.4;
- bulkq – Calculate the artificial bulk viscosity force, as shown in Sect. 2.8.2.

The source code of the module MaterialModel can be found in the file Constitution.f90.

Chapter 7

Multiscale MPM

Contents

Research efforts have been made over the recent years to advance the MPM for multiscale modeling and simulation. A multi-level refinement scheme has been designed for the GIMP Method [83], a hierarchical approach has been proposed in which material points at the fine level in the MPM are allowed to directly couple with the atoms in molecular dynamics (MD) simulations [152], and a sequential procedure has been developed to formulate the EOS based on the MD results for the macroscopic MPM simulation [153]. However, the need for a transition region between different spatial scales in the hierarchical/sequential or multi-level refinement approaches limits its application to the cases where discrete nano/microstructures of certain sizes interact with each other in composite systems. To circumvent this limitation, a particle-based multiscale scheme has been proposed, in which MD is linked with Cluster Dynamics (CD) via a hierarchical way for sub-micron scale simulations while CD is coupled with the MPM concurrently for microscale simulations [154]. With this approach, the longitudinal impact response between two microrods with different nanostructures has been explored based on the size effect on the impact responses of discrete nanostructures [155,156]. However, the CD potential is formulated from the L-J MD potential, and much work remains to be performed to formulate the CD potential for better modeling the interactions among metallic nanostructures.

To extend the recent work [154] to more applicable cases, the particle-based multiscale procedure has been improved with a concurrent link between the MPM and Dissipative Particle Dynamics (DPD) for microscale simulations, and a hierarchical bridge from MD to DPD for nanoscale simulations [157]. In particular, an effective interfacial scheme between the DPD and MPM is designed for concurrent simulations, as described below.

The Material Point Method. http://dx.doi.org/10.1016/B978-0-12-407716-4.00007-7

7.1 GOVERNING EQUATIONS AT DIFFERENT SCALES

To better understand the particle-based multiscale simulation procedure, the governing equations at different scales are first summarized in this section.

At molecular scale, the dynamics of a system with N molecules is governed by the following equation of motion for each molecule i (assumed to be a rigid particle):

$$m_i \frac{\mathrm{d}^2 \boldsymbol{r}_i}{\mathrm{d}t^2} = m_i \boldsymbol{a}_i = -\frac{\partial}{\partial \boldsymbol{r}_i} U_{\text{tot}} (\boldsymbol{r}_1, \boldsymbol{r}_2, \ldots, \boldsymbol{r}_N) \tag{7.1}$$

in which m_i, $\boldsymbol{r}_i$, and $\boldsymbol{a}_i$ are the mass, position, and acceleration vectors of molecule i, respectively, and U_{tot} represents the total potential energy that depends on all the molecular positions and could be divided into two parts, namely, nonbonded molecular interaction and intramolecular interaction. To demonstrate the essential feature of the multiscale simulation procedure, only the nonbonded molecular interaction is considered here.

With the molecular details being coarse-grained at the sub-micron scale, the DPD governing equations for each rigid particle i can be summarized as follows:

$$m_i \frac{\mathrm{d}^2 \boldsymbol{r}_i}{\mathrm{d}t^2} = m_i \boldsymbol{a}_i = \boldsymbol{f}_i^C + \boldsymbol{f}_i^D + \boldsymbol{f}_i^R, \tag{7.2}$$

$$\boldsymbol{f}_i^C = \sum_{i \neq j} -\nabla U (\boldsymbol{r}_{ij}) \boldsymbol{e}_{ij}, \tag{7.3}$$

$$\boldsymbol{f}_i^D = \sum_{i \neq j} -\gamma_{ij} w^D (\boldsymbol{r}_{ij}) \boldsymbol{v}_{ij}, \tag{7.4}$$

$$\boldsymbol{f}_i^R = \sum_{i \neq j} \sigma_{ij} w^R (\boldsymbol{r}_{ij}) \, dW_{ij} \boldsymbol{e}_i \tag{7.5}$$

where $\boldsymbol{f}_i^C$, $\boldsymbol{f}_i^D$, and $\boldsymbol{f}_i^R$ represent respectively the conservative force (in a form similar to MD), dissipative force, and random force vector acting on particle i, with m_i, $\boldsymbol{r}_i$, and $\boldsymbol{a}_i$ being respectively the mass, position, and acceleration vector of particle i, U the inter-particle potential, γ_{ij} and σ_{ij} the force magnitudes, w^D and w^R the weight functions of $\boldsymbol{r}_{ij} = \boldsymbol{r}_i - \boldsymbol{r}_j$, $\boldsymbol{v}_{ij} = \boldsymbol{v}_i - \boldsymbol{v}_j$ the velocity difference vector between particle i and j, $\boldsymbol{e}_i = \boldsymbol{r}_i / |\boldsymbol{r}_i|$ the normalized position vector of particle i, $\boldsymbol{e}_{ij} = \boldsymbol{r}_{ij} / |\boldsymbol{r}_{ij}|$ the normalized connection vector between particle i and j, and W_{ij} the independent d-dimensional Wiener processes. The above three kinds of force vectors have their respective cutoff radius which is several orders larger than that in MD, depending on the coarse-graining level in different cases. In fact, all the discrete approaches such as MD and DPD adopt discrete forcing functions with different values of cutoff radius, while the continuous approaches such as the FEM use the internal force vectors obtained from the constitutive models defined over specific material elements. As can be found from the MPM formulation, the MPM is a continuum-based particle method such that it possess useful features of both discrete and continuous approaches, as shown next.

At microscale (continuum level), the governing differential equations discretized with the MPM take the form of

$$m_i^t a_i^t = \left(f_i^t\right)^{\text{int}} + \left(f_i^t\right)^{\text{ext}} \tag{7.6}$$

for a lumped mass matrix, where the internal force vector is given by

$$\left(f_i^t\right)^{\text{int}} = -\sum_{p=1}^{N_p} M_p s_p^{s,t} \cdot G_i\left(x_p^t\right) \tag{7.7}$$

with the particle stress $s_p^{s,t} = s^s\left(x_p^t, t\right)$ and the gradient of shape function $G_i\left(x_p^t\right) = \nabla N_i|_{x_p^t}$, and the external force vector is defined to be

$$\left(f_i^t\right)^{\text{ext}} = c_i^t + b_i^t \tag{7.8}$$

with c_i^t and b_i^t denoting the specific traction and body force vectors evaluated at the grid nodes, respectively.

As can be seen from the above, Eqs. (7.1), (7.2), and (7.6) have similar form although they are formulated at different scales with different domains of influence. The right-hand side (forcing function) of these equations includes the internal interactions among discrete particles (molecules, DPD particles or material points), as well as the external forces. The difference between the MPM and DPD/MD is that Eq. (7.6) is evaluated at the background grid nodes, instead of the material points. As a result, the strain and stress fields in the MPM could be easily determined with the use of the gradient of nodal basis function and constitutive model, respectively, instead of finding a representative domain of certain cutoff radius to determine the strain and stress as required in the DPD and MD. Thus, the particle-based multiscale simulation procedure consists of a concurrent link between the MPM and DPD particles to simulate mesoscale/microscale responses, and a hierarchical bridge from MD to DPD to characterize the DPD forcing function for nano and sub-micron scale simulations based on the MD solutions. In the concurrent MPM and DPD computational domain, a particle is a DPD one if its forcing function is defined as shown in Eq. (7.2) for which there is a cutoff distance. Within the MPM framework, the DPD cutoff distance should be larger than the cubic cell size for 3D problems. It will be demonstrated that the DPD details could be effectively coarse-grained with the use of the mapping and re-mapping procedure via a coarse MPM background grid. The reason is because the lower order shape functions could smooth out the higher order responses. A particle is a material point if the constitutive model at continuum level is used to calculate the internal force vector as shown in Eq. (7.7). It is possible for a single MPM cell to include both DPD and MPM particles at a given time so that the mapping and re-mapping procedure in the MPM yields a computational homogenization scheme over the cell domain. The concurrent link between the MPM and DPD enables the seamless integration of constitutive modeling at continuum level with discrete DPD forcing functions. The simplicity of the particle-based simulation procedure provides a robust way for zoom-in to molecular details and zoom-out to microscale responses.

7.2 SOLUTION SCHEME FOR CONCURRENT SIMULATIONS

The specific solution steps for a concurrent MPM and DPD simulation are described below.

7.2.1 Preprocessor

1. Discretize a continuum body into a finite set of N_p material points with respect to the original configuration of the body. Each material point carries its original material properties, which is a DPD particle if the DPD forcing function is active, or an MPM particle if the constitutive model formulated at the continuum level is used for calculating the internal force vector. The material points are followed throughout the deformation process of the body. An arbitrary background grid of certain spatial resolution is employed to find the natural coordinates of any material point, and the grid cell that contains the material point.

2. Initialize all the state variables at the material points, input the control parameters for the computer code, and equilibrate the system of material points as required for the concurrent MPM and DPD simulation due to the use of the DPD forcing functions.

7.2.2 Central Processing Unit

The detailed steps in each temporal increment are listed as follows:

1. For each material point (include both DPD particles and MPM particles), perform the mapping operation from the point to the corresponding cell nodes.

Map the mass from the material points to the nodes of the cell containing these points

$$m_i^t = \sum_{p=1}^{N_p} M_p N_i\left(\boldsymbol{x}_p^t\right) \tag{7.9}$$

where m_i^t is the mass at node i at time t, M_p the material point mass, N_i the shape function associated with node i, and $\boldsymbol{x}_p^t$ the location of the material point at t. Map the momentum from the material points to the nodes of the cell containing these points

$$(m\boldsymbol{v})_i^t = \sum_{p=1}^{N_p} (M\boldsymbol{v})_p^t N_i(\boldsymbol{x}_p^t) \tag{7.10}$$

where $(m\boldsymbol{v})_i^t$ denotes the nodal momentum at node i at t, and $(M\boldsymbol{v})_p^t$ the material point momentum at t. Find the internal force vector at the cell nodes for the MPM particles associated with that cell

$$(\boldsymbol{f}_i^t)^{\text{int}} = -\sum_{p=1}^{N_p} \boldsymbol{G}_i(\boldsymbol{x}_p^t) \cdot \boldsymbol{s}_p^t \frac{M_p}{\rho_p^t} \tag{7.11}$$

in which $\boldsymbol{G}_i(\boldsymbol{x}_p^t)$ is the gradient of the shape function associated with node i evaluated at $\boldsymbol{x}_p^t$, $\boldsymbol{s}_p^t$ the particle stress tensor at t, and ρ_p^t the particle mass density at t. For the DPD particles related to that cell, Eq. (7.11) is replaced with the following one:

$$(\boldsymbol{f}_i^t)^{\text{int}} = \sum_{p=1}^{N_p} (\boldsymbol{f}_{pj}^C + \boldsymbol{f}_{pj}^D + \boldsymbol{f}_{pj}^R) N_i(\boldsymbol{x}_p^t) \tag{7.12}$$

where $\boldsymbol{f}_{pj}^C$, $\boldsymbol{f}_{pj}^D$, and $\boldsymbol{f}_{pj}^R$ represent respectively the conservative, dissipative, and random force vectors acting on particle p by particle j ($j = 1, 2, \ldots, N_p^j$) with N_p^j being the number of the DPD particles within the cutoff radius of particle p.

2. Apply essential and natural boundary conditions to the cell nodes, and compute the nodal force vector

$$\boldsymbol{f}_i^t = (\boldsymbol{f}_i^t)^{\text{int}} + (\boldsymbol{f}_i^t)^{\text{ext}}. \tag{7.13}$$

3. Update the momenta at the cell nodes

$$(m\boldsymbol{v})_i^{t+\Delta t} = (m\boldsymbol{v})_i^t + \boldsymbol{f}_i^t \Delta t. \tag{7.14}$$

4. For each material point, perform the mapping operation from the nodes of the cell containing the material point to that point.

Map the nodal accelerations back to the material point

$$\boldsymbol{a}_p^t = \sum_{i=1}^{N_n} \frac{\boldsymbol{f}_i^t}{m_i^t} N_i(\boldsymbol{x}_p^t). \tag{7.15}$$

Map the current nodal velocities back to the material point

$$\bar{\boldsymbol{v}}_p^{t+\Delta t} = \sum_{i=1}^{N_n} \frac{(m\boldsymbol{v})_i^{t+\Delta t}}{m_i^t} N_i\left(\boldsymbol{x}_p^t\right). \tag{7.16}$$

Compute the current material point position

$$\boldsymbol{x}_p^{t+\Delta t} = \boldsymbol{x}_p^t + \bar{\boldsymbol{v}}_p^{t+\Delta t}\Delta t \tag{7.17}$$

that represents a backward integration.

Compute the material point displacement

$$\boldsymbol{u}_p^{t+\Delta t} = \boldsymbol{x}_p^{t+\Delta t} - \boldsymbol{x}_p^0. \tag{7.18}$$

As can be seen from Eqs. (7.16) and (7.17), nodal shape functions are used to map the nodal velocity continuously to the interior of the grid cell so that the positions of the material points are updated by moving them in a single-valued, continuous velocity field.

5. Map the updated material point momenta back to the nodes of the cell containing these material points

$$(m\boldsymbol{v})_i^{t+\Delta t} = \sum_{p=1}^{N_p} (M\boldsymbol{v})_p^{t+\Delta t} N_i\left(\boldsymbol{x}_p^t\right). \tag{7.19}$$

6. Find the updated nodal velocities

$$\boldsymbol{v}_i^{t+\Delta t} = \frac{(m\boldsymbol{v})_i^{t+\Delta t}}{m_i^t}. \tag{7.20}$$

7. Apply the essential boundary conditions to the nodes of the cells containing the boundary material points.

8. Find the current gradient of particle velocity

$$\boldsymbol{L}_p^{t+\Delta t} = \sum_{i=1}^{N_n} \boldsymbol{v}_i^{t+\Delta t} \boldsymbol{G}_i\left(\boldsymbol{x}_p^t\right) \tag{7.21}$$

and the particle strain increment

$$\Delta\boldsymbol{e}_p = \left(\text{sym}\boldsymbol{L}_p^{t+\Delta t}\right)\Delta t \tag{7.22}$$

so that the stress increment could be obtained from the constitutive model for the given strain increment to update the stress tensor of the MPM particle

$$\boldsymbol{s}_p^{t+\Delta t} = \boldsymbol{s}_p^t + \Delta\boldsymbol{s}. \tag{7.23}$$

9. Identify which grid cell each material point belongs to, and update the natural coordinates of the material point. This is the convective phase for the next time increment.

10. Go to Step 1 for the next time increment, if the required termination time has not been reached. Otherwise, go to Postprocessor for processing the output files.

7.3 INTERFACIAL TREATMENT

For the DPD and MPM particles in the interfacial region, a simple interfacial treatment can be employed to capture the essential physics by smoothing the mismatch between DPD and MPM particles. The internal force due to the DPD and MPM particles in the interfacial region is calculated with the following equation:

$$\left(f_i^t\right)^{\text{int}} = \sum_{p=1}^{N_p^d}\left(f_{pj}^C + f_{pj}^D + f_{pj}^R\right)N_i\left(x_p^t\right) - \sum_{p=N_p^d+1}^{N_p^d+N_p^m} G_i\left(x_p^t\right)\cdot s_p^t \frac{M_p}{\rho_p^t} \tag{7.24}$$

with N_p^d and N_p^m being the number of the DPD and MPM particles within the interfacial region, respectively. For each DPD particle in the interfacial region, f_{pj}^C, f_{pj}^D, and f_{pj}^R in the first term of Eq. (7.24) are determined as follows:

$$f_{pj}^C = \sum_{k=1,k\neq j}^{N_p^j} -\nabla U\left(r_{jk}\right)e_{jk}, \tag{7.25}$$

$$f_{pj}^D = \sum_{k=1,k\neq j}^{N_p^j} -\gamma_{jk}w^D\left(r_{jk}\right)v_{jk}, \tag{7.26}$$

$$f_{pj}^R = \sum_{k=1,k\neq j}^{N_p^j} \sigma_{jk}w^R\left(r_{jk}\right)dW_{jk}e_j \tag{7.27}$$

where N_p^j is the total number of the particles within the cutoff radius of DPD particle p. In combination with Eq. (7.13), Eqs. (7.11) and (7.12) are used to find the internal forces due to the MPM and DPD particles outside the interfacial region, respectively. It can be found from Eqs. (7.25)–(7.27) that each DPD particle inside the interfacial region can feel the interactions from the MPM particles (treated as the DPD particles with corresponding forcing functions) within its cutoff radius. On the other hand, the use of Eq. (7.24) includes the internal force contributions from both DPD and MPM particles located within the interfacial region. Hence, each interfacial MPM particle is also connected

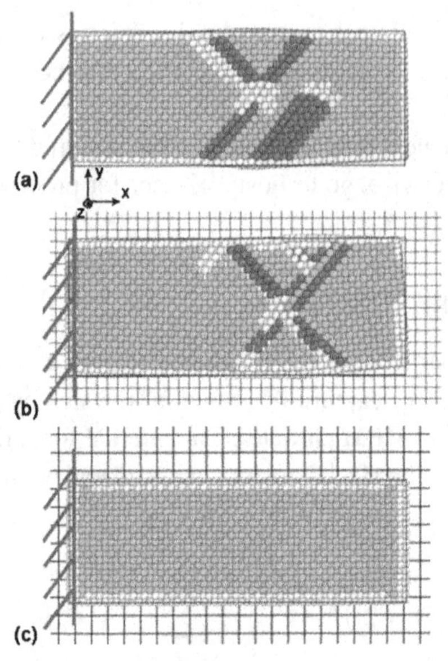

FIGURE 7.1 Deformed configurations of the Cu target at 10 ps after impact occurs, simulated with (a) a DPD only model, and the model coupling DPD particles with the MPM background grid of (b) cell size 2 Å and (c) cell size 8 Å, respectively [157].

with its neighbor DPD particles via the mapping and re-mapping process within the MPM framework.

7.4 DEMONSTRATION

Consider first a copper target under impact loading. As shown in Fig. 7.1, the MPM mapping and re-mapping operation would coarsen the DPD deformation details if the background grid resolution is decreased (i.e., the cell size is increased). As can be observed from Fig. 7.1(c), a large cell size leads to the loss of the detailed deformation pattern as shown in Fig. 7.1(a) that is produced by DPD-only simulation. In contrast, the use of a small cell size could produce more details about the evolution of localization and the deformation pattern close to that predicted by the DPD-only model, as shown in Fig. 7.1(b). Thus, the spatial resolution of the MPM background grid is related to the coarse-graining level of DPD simulations without affecting the essential feature of wave propagation in the impacted target.

To further confirm the above findings, consider a copper rod under dynamic tensile loading with the use of the DPD-only model and the model coupling the DPD particles with the MPM background grid, as illustrated in Fig. 7.2. The rod has the dimension of 216 Å × 72.3 Å × 72.3 Å along the x, y, and z coordinates. The two ends of thickness 15 Å each are treated as a rigid body with

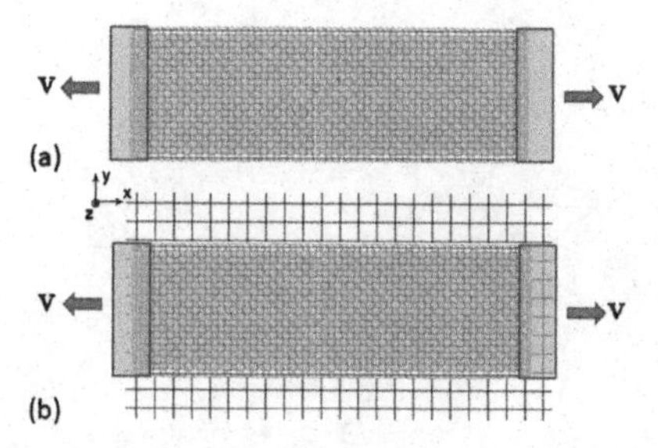

FIGURE 7.2 The geometric sketch of a Cu rod subjected to tensile loading at both ends, with (a) a DPD only model and (b) the model coupling DPD particles with the MPM background grid [157].

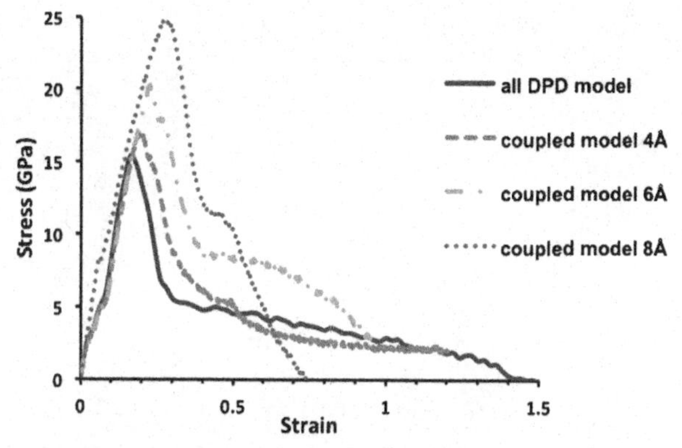

FIGURE 7.3 Stress–strain relations of the Cu rod under tension at a strain rate of 0.02/ps, simulated with a DPD only model and the models coupling DPD particles with the MPM background grid of different cell sizes [157].

a constant velocity applied in the opposite directions. As shown in Fig. 7.3 for the stress–strain relations of the copper rod, the elastic responses are consistent for both the DPD-only model and the model coupling DPD particles with the MPM background grid of different cell sizes, which implies that the elastic modulus is independent of the spatial resolution of the MPM background grid. Hence, it could be confirmed that the elastic wave speed is not affected by the coupling between the DPD and MPM background grid. Moreover, it can be found that the peak stress approaches the value predicted by the DPD-only model with the decrease of the MPM cell size.

More examples and related references could be found from the recent papers for concurrent MPM/DPD simulations of failure evolution, solid–fluid interaction, and solid-state sintering [157,158] in order to demonstrate the robustness and potential of the proposed procedure for multiscale simulations which combines discrete and continuous approaches in a single computational domain.

Chapter 8

Applications of the MPM

Contents

Based on the theoretical framework and numerical implementation as discussed in the previous chapters, this chapter further demonstrates the features of the MPM with practical applications. Since the MPM possesses the advantages of both Lagragian and Eulerian descriptions, no element distortion exists in the MPM so that the MPM is robust in solving problems involving extremely large deformations and moving discontinuities. Applications of the MPM in extreme loading cases such as transient fracture, hyper-velocity impact, penetration, explosion, and multiphase interaction involving failure evolution have been attracting much attention in recent years, as described below. In addition, the on-going research efforts on multiscale analyses and special computational schemes are also introduced here.

8.1 FRACTURE EVOLUTION

Simulating the evolution of fracture more accurately and effectively can be considered as one of the original driving forces for developing novel spatial discretization methods such as meshfree methods and the MPM. Cracks are strong discontinuities, which is in contradiction with the continuity requirement of finite element approximation. Embedding cracks in finite elements and allowing arbitrary propagation directions are therefore a challenging task for the *traditional* FEM. Remeshing is usually required for simulating crack propagation, which is not only a computationally heavy burden but also involves the mesh generation for complex geometry, not to mention the possible accuracy loss due to the mapping process between the old mesh and new mesh.

The MPM, though employing the finite element shape function, does not suffer from the above difficulties. Discontinuities can be described in two ways. One is to abandon the single-valued velocity field property near the crack by

The Material Point Method. http://dx.doi.org/10.1016/B978-0-12-407716-4.00008-9

using two or more background meshes, and the other is to use failed material points to approximately describe the crack.

In the first approach, Nairn and his collaborators [66,159] developed the formulations containing multiple velocity fields for two- and three-dimensional fracture problems. They named their method as "CRAcks with Material Points (CRAMP)" [66]. For the nodes far way from the crack path, the nodal variables are assembled in the classical way. For each node adjacent to the crack, the material points influenced by the node are divided into two groups, depending on whether the material point and the node are at the same side. Only the material points on the same side can contribute to nodal equations. A function $v(p, I)$ identifying whether the node I and the material point p are at the same side is employed in CRAMP [66,159]. Taking a 2D case, for example, $v(p, I) = 1$ if the material point is at the "upper" side of the crack path or both the node and the material point are far away from the crack path; $v(p, I) = 2$ if the material point is at the "lower" side of the crack path. The value of $v(p, I)$ and the position of the node will determine whether point p will contribute to the nodal equations of node I. The crack surface in the CRAMP method is described with line segments in 2D and triangle patches in 3D cases [66,159]. The essential idea of the CRAMP method can be extended to other problems with discontinuities or interfaces [160]. However, the use of the function $v(p, I)$ may be more suitable for a single crack and the elements which the crack cuts through, while a more precise description for multiple cracks and more complex fracture problems is still under development.

Typical fracture parameters, such as the energy release rate, the stress intensity factor, or the J-integral, are critical to the fracture analysis. The criteria for crack propagation are also based on these parameters. Tan and Nairn [82] first discussed the details of calculating the energy release rate with the MPM. High numerical accuracy is always desired for the calculation due to the singularity at the crack tip. Tan and Nairn [82] proposed a scheme to increase the numerical accuracy, which contains a multiple-sized background mesh and the refinement of material points. Guo and Nairn [161] discussed the calculation of J-integral and then dynamic stress intensity factor. The stresses at the material points are mapped (extrapolated) to corresponding nodes. The boundary part in J-integral is calculated with nodal physical variables, and the domain part is calculated with the physical variables at material points. The CRAMP results of stress intensity factors of typical problems were validated with the theoretical results and the solutions from other numerical methods. The influences of integration path and background mesh size were also investigated [161].

The MPM for crack propagation under transient loading conditions was developed [162,163] based on the above schemes. The propagation criterion based on the J-integral was adopted, and new material points were inserted in front of the crack tip when the criterion was satisfied. The cohesive law was used to describe the mechanical behavior of the material points in the processing zone. Bardenhagen et al. [162] investigated the influences of various computational parameters and physical parameters in a mode-I problem. The parameters included the time step size, the computational domain for the J-integral, the parameter of the cohesive law, and the fracture strength. Daphalapurkar [163] studied the dynamic propagation of mode-II crack with the GIMP and cohesive modeling.

Another approach for fracture simulation uses failed material points to approximately represent cracks instead of an explicit description of discontinuities. Such a scheme is referred to as the implicit fracture simulation scheme in the context. The formation of failed material points represents the initiation, propagation, and branching of cracks. The position of the crack and the interaction between two sides of the crack, therefore, do not need to be explicitly stated. After the discontinuous bifurcation analysis identifies the transition from continuous to discontinuous failure, a decohesion model is used to predict the failure evolution. A representative decohesion model consists of the following set of equations [164,165]:

$$\text{(Stress-strain relationship)} \quad \boldsymbol{\sigma}^{\nabla} = \boldsymbol{C} : (\dot{\boldsymbol{\varepsilon}} - \dot{\boldsymbol{\varepsilon}}^{\mathrm{d}}), \tag{8.1}$$

$$\text{(Traction equilibrium)} \quad \dot{\boldsymbol{\tau}} = \dot{\boldsymbol{\sigma}} \cdot \boldsymbol{n}, \tag{8.2}$$

$$\text{(Decohesion evolution)} \quad \dot{\boldsymbol{u}}^{\mathrm{d}} = \dot{\lambda}^{\mathrm{d}} \boldsymbol{m}, \tag{8.3}$$

$$\text{(Strain–displacement relationship)} \quad \dot{\boldsymbol{\varepsilon}}^{\mathrm{d}} = \frac{\dot{\lambda}^{\mathrm{d}}}{2L_e} \left(\boldsymbol{n} \otimes \boldsymbol{m} + \boldsymbol{m} \otimes \boldsymbol{n}\right), \tag{8.4}$$

$$\text{(Consistency condition)} \quad F^{\mathrm{d}} = \tau^e - U_0[1 - (\lambda^{\mathrm{d}})^q] = 0 \tag{8.5}$$

where the decohesion displacement $\boldsymbol{u}^{\mathrm{d}}$ is the displacement jump between the two sides of a failure surface, $\dot{\boldsymbol{\varepsilon}}^{\mathrm{d}}$ is the strain rate tensor associated with the decohesion displacement, $\boldsymbol{n}$ is the unit normal vector to the failure surface, $\boldsymbol{\tau}$ is the traction vector, and $\dot{\lambda}^{\mathrm{d}}$ is a dimensionless variable characterizing the evolution of decohesion. L_e is an effective length, which can be defined as the ratio of a material volume to the area of the failure surface inside the volume. The vector $\boldsymbol{m}$ has the displacement dimension and represents the direction in which the two sides of the failure surfaces detach. The effective traction takes the form of $\tau^e = \boldsymbol{\tau} \cdot \boldsymbol{m}$. U_0 is the reference surface energy. The parameter q determines the convexity of the function relating λ^d with τ^e. The meaning of the other variables can be found in the preceding chapters.

Once the detachment direction $\boldsymbol{m}$ is determined, Eqs. (8.1)–(8.5) can be solved. A simple but effective way is to use the associated evolution equation [165], i.e.,

$$\boldsymbol{m} = \overline{u}_0 \frac{\boldsymbol{A}_\mathrm{d} \cdot \boldsymbol{\tau}}{(\boldsymbol{\tau} \cdot \boldsymbol{A}_\mathrm{d} \cdot \boldsymbol{\tau})^{1/2}} \tag{8.6}$$

where $\overline{u}_0 = U_0/\overline{\tau}_0$, and $\boldsymbol{A}_\mathrm{d}$ is a positive second-order tensor which determines the failure mode. For two-dimensional problems, $\boldsymbol{A}_\mathrm{d}$ can be written in the following form:

$$\boldsymbol{A}_\mathrm{d} = \overline{\tau}_0^2 \begin{bmatrix} \frac{1}{\tau_\mathrm{np}^2} & \\ & \frac{1}{\tau_\mathrm{tp}^2} \end{bmatrix}. \tag{8.7}$$

The non-associated evolution equation [165] can also be used to determine $\boldsymbol{m}$.

Chen et al. [166,167] adopted the MPM to simulate failure evolution of brittle solids under impact loading. Sulsky and Schreyer [168] investigated the spalling failure of brittle material with the above decohesion model. Similar to the role of the failure criteria in modeling crack initiation and propagation, identifying the transition from continuous (microcracking) to discontinuous (macrocracking) processes is also very important in describing the complete failure evolution process. To make the governing equations well-posed with the least computational expenses, Chen et al. [164,169] proposed the discontinuous bifurcation analysis in combination with decohesion modeling to determine when the failure mode of a material point becomes discontinuous and to which direction the evolution of failure will go. The bifurcation analysis is based on the continuum tangent stiffness tensor obtained from a suitable local elastoplastic constitutive model. Chen et al. [170] combined a decohesion model and heat conduction analysis to study the material failure under local heating, where the finite difference method was used to find the temperature field while the MPM was employed to complete the thermomechanical solution process. Shen et al. [171] investigated the fragmentation of the glass under impact loading. The Drucker–Prager model was used to describe the compression stage, the rate-dependent damage model was used to describe the tension stage, and the decohesion model was used to describe the material behavior as discontinuous bifurcation occurred.

Sulsky et al. [172] studied the movement and deformation of sea ice in Beaufort Sea based on the decohesion model and heat conduction analysis. Both the land and sea were discretized with material points in the simulation. Velocity boundary conditions were enforced for the boundary of the sea region according to the remote measurement from satellite data, while the material points for the land are fully fixed. Satellite and meteorological observed data were used to specify the initial conditions and the initial notches inside the sea ice.

The position and opening displacement of the discontinuities obtained by the MPM simulation qualitatively agreed with the observed data [172] except for one small region where the difference might be caused by the perturbation of the wind and ocean current.

To perform large-scale simulation of failure evolution, Yang et al. [151] improved the decohesion model for metals in combination with J_2 plasticity model. They replaced the bifurcation analysis with the following limit strength criterion to identify the onset of decohesion:

1. Mode I failure will occur if $\sigma_1 \geqslant \sigma_{\max}$ and $\tau_t < \tau_{\max}$;
2. Mode II failure will occur if $\sigma_1 < \sigma_{\max}$ and $\tau_t \geqslant \tau_{\max}$;
3. Mixed mode failure will occur if $\sigma_1 \geqslant \sigma_{\max}$ and $\tau_t \geqslant \tau_{\max}$.

Here σ_1 is the maximum principal stress, τ_t is the maximum shear stress, and $\sigma_{\max}$ and $\tau_{\max}$ are the tensile limit strength and the shear limit strength, respectively. As result, the failure mode would change during the evolution of failure due to its nonlocal nature, instead of being fixed as assumed in the previous investigations. The vector $\boldsymbol{n}$ corresponding to each failure model is then calculated by the following:

$$
\begin{aligned}
&\boldsymbol{n} = \boldsymbol{n}_1 && \text{(for mode I)},\\
&\boldsymbol{n} = \frac{\boldsymbol{n}_2 \times (\boldsymbol{n}_1 + \boldsymbol{n}_3)}{\|\boldsymbol{n}_2 \times (\boldsymbol{n}_1 + \boldsymbol{n}_3)\|} && \text{(for mode II)},\\
&\boldsymbol{n} = \bar{\boldsymbol{n}}, \quad \bar{\boldsymbol{n}} = \boldsymbol{n}_1 + \frac{\boldsymbol{n}_2 \times (\boldsymbol{n}_1 + \boldsymbol{n}_3)}{\|\boldsymbol{n}_2 \times (\boldsymbol{n}_1 + \boldsymbol{n}_3)\|} && \text{(for mixed mode)}
\end{aligned}
$$

where $\boldsymbol{n}_i$ $(i = 1, 2, 3)$ is the unit vectors along the ith principal direction. Yang et al. [151] simulated the Sandia fracture challenge problem [148], which is a compact tension specimen with initial holes, using the MPM with the above improved decohesion model. The final configurations of the MPM results and the experimental results are consistent as shown in Fig. 8.1.

Gupta et al. [173] compared the performances of different meshfree particle methods, including the element free Galerkin method, the SPH method, and the MPM, for the dynamic fracture problems. Various criteria were tested for the problems of uniform tension, four-point bending, and crack branching. Results indicated that the proper fracture criterion is more important in correctly predicting the fracture behavior, and the difference between the results from different meshfree particle methods is not crucial. Wang et al. [174] calculated the stress intensity factor of the crack of mixed modes with irregular point discretization, with the results comparing well with finite element results. Gilabert et al. [175] simulated the initiation and propagation of cracks near the quartz inclusion in brittle materials such as ceramics and glasses. Their work adopted

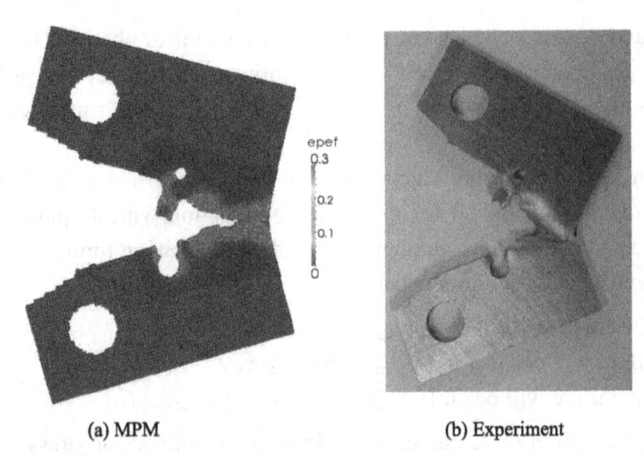

FIGURE 8.1 Comparison of the results from (a) the MPM simulation with improved decohesion model and (b) the experiment.

a local Rankine criterion and the energy release rate as the criteria for predicting crack initiation and propagation. Their simulation results indicated that the crack was first initiated in the inclusion and then propagated into the matrix, which was consistent with the experimental observation. They further investigated the influences of different inclusion shapes on the fracture mode [176]. They found that cracks surrounded the inclusion for the circular inclusion, several cracks cut through the inclusion for the ellipse inclusion, only one crack occurred and ran through the inclusion for the rectangular case, and the cracks were close to the vertices if the inclusion had triangular shape. All the modes were observed in the experiments.

8.2 IMPACT

Impact problems often involve strong nonlinearities including large deformation, fracture, fragmentation, or even phase transformation. The applications of the MPM in different impact problems have been receiving much attention in the recent years. The existence of extreme deformation will lead to a sharp decrease of the characteristic discretization length, i.e., the element size or the distance between nodes, during the impact. The critical time step size of the explicit FEM and typical explicit meshfree methods (EFGM or SPH) would therefore decrease sharply due to the decrease of the characteristic discretization length. Since it is controlled by the background mesh size in the MPM, the critical time step size of the MPM will not decrease much even when undergoing extremely large compression. Therefore, the MPM has advantages in efficiency over the explicit FEM and other meshfree particle methods. Ma et al. [47,177] showed that the cost of the MPM could be about one fourth of that of the SPH method when solving impact problems. Numerical fractures would occur in the SPH simulation of the Taylor bar impact problem, but the MPM results do not show any numerical fracture.

There are different responses for different impact velocities. The elastic and plastic deformations dominate the responses when the impact velocity is low. Local failure will happen when the structure is under an impact of intermediate velocity, such as a bird impact on an aircraft and the projectile penetrating the target plate. The mechanism and the configuration of failure and the influences on the loading capacity of the structure are major issues of the medium-velocity impact problems. When the impact velocity further increases, the influence of material strength is sharply decreased, and serious local failure such as the cratering, spalling, phase transformation, and debris cloud will occur. When the impact pressure is much higher than the material strength, which is usually called the hyper-velocity impact, the material behavior is close to fluid flow. Material melting and even material vaporization may happen when the pressure and temperature are very high.

Sulsky et al. [178] derived the axisymmetric formulation for the MPM and applied it in the analysis of the upsetting process and Taylor bar impact problem. Wang et al. [179] investigated the responses of the steel-porous aluminum–steel sandwich composite structure to the flyer impact. Zhang et al. [79,180] simulated the perforation of the steel target by the tungsten projectile with the MPM of modified shape functions. Proper contact algorithms are essential in simulating impact problems of medium- and low-velocities. The automatic single-valued velocity field in the original MPM may not be accurate enough because only non-slip conditions could be enforced. Ma et al. [74] simulated perforation problems with the MPM equipped with the contact algorithms based on local multiple background grids. The indicators such as the residual velocity were identical to experimental results. Huang et al. [73] further compared the results of the original MPM and the MPM with a contact algorithm for perforation simulation, and they found that the results with the contact algorithm were much better. Li et al. [181] studied the influences of the impact velocity on the failure pattern of brittle disc, and showed that the simulation results agreed well with the experiment and that circular cracks appeared when the impact velocity was relatively low. A single crack penetrated into the disc at a medium impact velocity. Many cracks appeared and cut the disc into several fragments when the impact velocity was further increased.

For the cases with high-velocity and hyper-velocity, Ma et al. [177,182] investigated the hyper-velocity impact on the thin plate and thick plate with the MPM. The results agreed well with the empirical formula and those from the

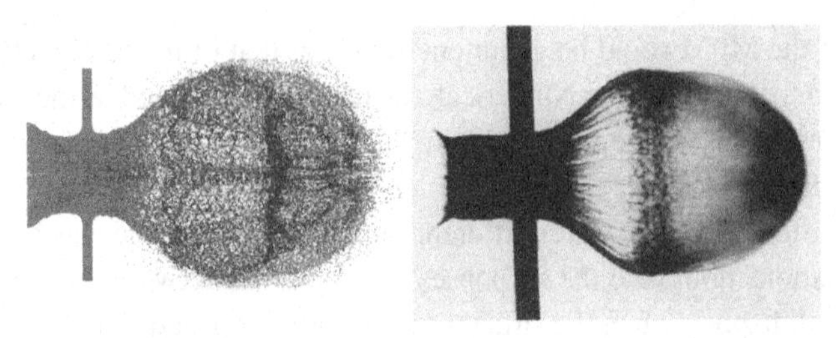

FIGURE 8.2 The configuration of debris cloud [183]: (left) MPM results; (right) experimental results.

other numerical simulations. Huang et al. [183] studied the effects of the background mesh size and the number of material points on the debris cloud occurred in the hyper-velocity impact. They obtained the debris cloud with the MPM simulation of 13 million material points. The sizes of the debris cloud are identical to the experiment observation, as compared in Fig. 8.2. Pan et al. [184] investigated the deformation and the collapse of the hole in high explosive under impact loading. Simulation demonstrated that the hole collapsed due to the microjet of the material on the top of the hole. Xu et al. [185,186] investigated the responses of porous material under impact, and the effects of the impact velocity, the porous ratio, the impact intensity, and the hole size were thoroughly studied. Chen et al. [187] simulated the spalling of the Armco steel during the impact.

The impact response is usually highly localized. The deformation in the impact region will be extremely large, while the far-field region may only undergo small plastic deformation or even elastic deformation. Such a feature makes the finite element material point method quite appropriate for the impact simulation. Zhang et al. [36] simulated the hyper-velocity impact problem with the finite element material point method. Lian et al. simulated the perforation of metal targets with the coupled finite element material point (CFEMP) method [37] and the adaptive finite element material point (AFEMP) method [41]. The objects or the regions undergoing small deformation were simulated with the FEM, while those undergoing large deformation or fracture were simulated with the MPM. Fig. 8.3 illustrates an example in which a rod made of tungsten alloy penetrates a steel plate [41]. The initial length of the rod is $L_0 = 75$ mm, the diameter is $D_0 = 5$ mm, and the in-plane dimensions of the target are 150 mm × 150 mm. Two cases were studied: the thickness of the plate and the impact velocity were chosen as 5 mm and 1500 m/s, respectively, in case 1, while they were 10 mm and 2500 m/s, respectively, in case 2. Only a half of the model needs to be simulated owing to the symmetry, as shown in Fig. 8.3. Both the projectile and the target were initially discretized with eight-node cuboid elements, and the finite elements were automatically converted into material points when the effective

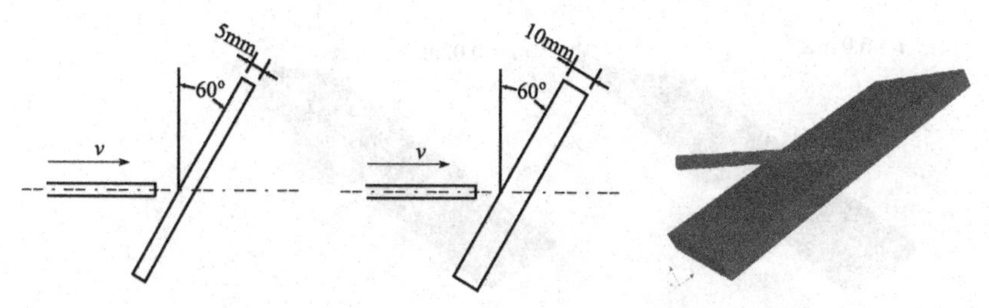

FIGURE 8.3 Schematic diagram of the long rod perforation example (left) and the discretization model of material points (right) [41].

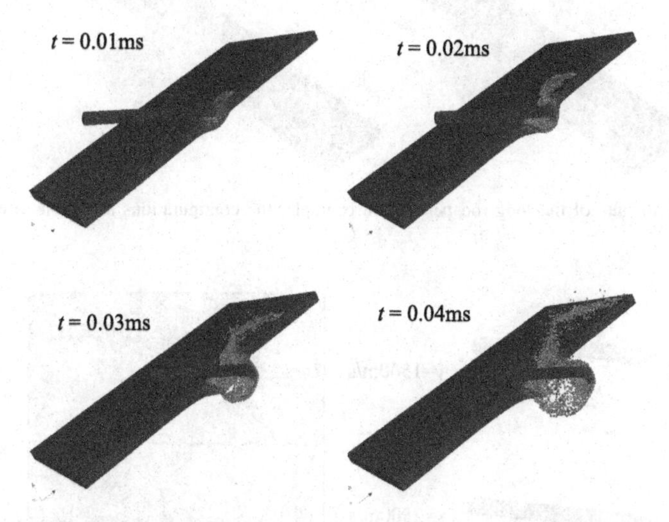

FIGURE 8.4 Results of the long rod perforation example, the configurations at various times (case 1) [41].

plastic strain reached the threshold value. The Johnson–Cook strength model and the Mie–Grüneisen equation of state were employed.

Figs. 8.4 and 8.5 demonstrate the configurations of cases 1 and 2 at different times [41]. The elements in the impact region were continuously converted into finite elements as the projectile penetrated into the target. Hence, the MPM was automatically adopted for the stage of large deformation, and possible element distortion in the impact region was fully avoided. Comparisons of the final configurations of numerical and experimental results are shown in Fig. 8.6, and the non-dimensional residual rod length and velocity are compared in Table 8.1. The results of the original MPM are also listed in Table 8.1. Both the original MPM and AFEMP methods obtained identical to experimental results.

Table 8.2 compares the maximum time step size (Δt_{max}), the minimum time step size (Δt_{min}), the number of time steps, and the CPU time of the AFEMP

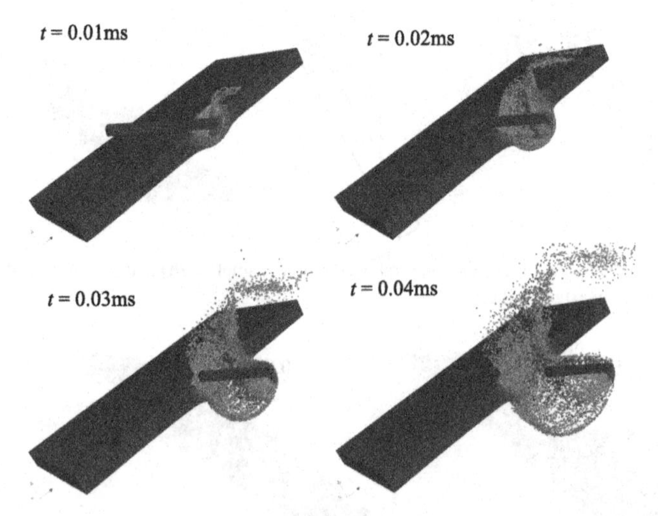

FIGURE 8.5 Results of the long rod perforation example, the configurations at various times (case 2) [41].

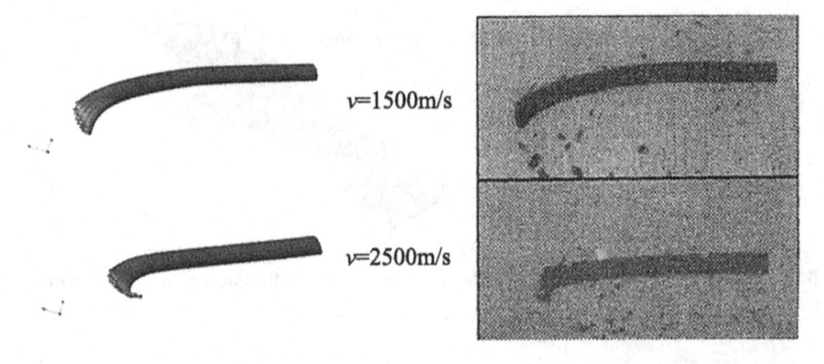

FIGURE 8.6 Comparison of the final configurations of the long rods after perforation [41]: (left) numerical results; (right) experimental results. The top two figures are results of case 1, and the bottom ones are results of case 2.

TABLE 8.1 Result Comparison of the Long Rod Penetration Example [41]

Case	Method	Dimensionless residual length	Dimensionless residual velocity
1	Experiment	0.85	0.97
	AFEMP	0.82	0.96
	MPM	0.85	0.96
2	Experiment	0.76	0.99
	AFEMP	0.72	0.97
	MPM	0.74	0.97

TABLE 8.2 Efficiency Comparison for Simulating the Long Rod Penetration [41]

Method	$\Delta t_{max}/\mu s$	$\Delta t_{min}/\mu s$	The number of time steps	CPU time/s
AFEMP	3.67×10^{-2}	1.52×10^{-2}	2208	838
MPM	6.45×10^{-2}	3.71×10^{-2}	1180	2241

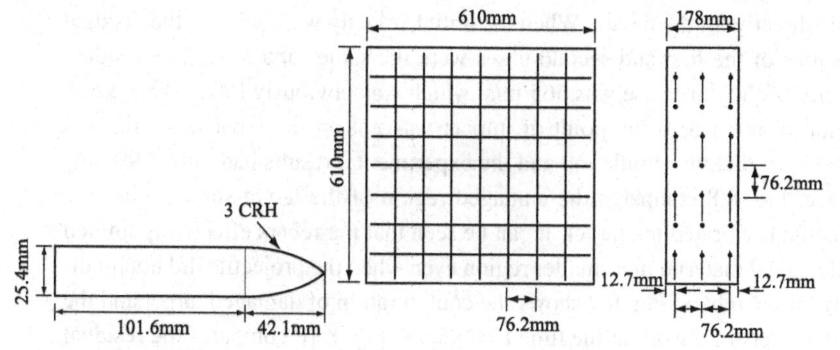

FIGURE 8.7 The geometry of the projectile and the rebar distribution [39].

method with those in the original MPM. The minimum time step size of the AFEMP method is only decreased by 58.6% during the penetration, which is because the failure elements were converted into material points as the penetration evolved. The maximum time step size of the AFEMP method is less than that of the MPM because the initial characteristic length of finite element is less than the nodal distance of MPM background mesh. Although the total number of time steps in the AFEMP computation is larger than that in the MPM computation, the efficiency is greatly improved as the computational cost of one single FEM time step is less than the cost of one MPM time step in the phase of small deformation. The total CPU time of AFEMP simulation is only 37% of MPM CPU time.

The perforation of the reinforced concrete can be well simulated with the hybrid finite-element material-point (HFEMP) method [39]. Discretizing the rebars with material points will result in an extremely large number of degrees of freedom. Simulating the rebars with bar element is a much better choice, and the HFEMP method provides a seamless way to link the bar element and the material points through the background mesh. A projectile of mass 0.5 kg and diameter 25.4 mm perforated a reinforced concrete target plate of dimensions 610 mm × 610 mm × 178 mm. Fig. 8.7 shows the geometry of the projectile and the rebar distribution [39]. The experimental results did not indicate obvious plastic deformation in the projectile except for a very small region near the projectile tip where the material was a little abrased. Hence, the elastic material model was used for the projectile. The HJC elastic–plastic model was used for the concrete, and the ideal elastic–plastic model was used for the rebars.

Three kinds of simulation were performed [39] to study the functionality of the rebars in resisting perforation. The first used the plain concrete target; the second used the reinforced concrete target, but the projectile did not directly strike the rebars; and the third used the reinforced concrete target, and the projectile directly hit the rebars. When the initial velocity was 749 m/s, the residual velocities of the first and second cases were the same (585 m/s). The residual velocity of the third case was 565 m/s, which was obviously lower. The experimental result, where the point of impact was not on the rebar network, was 615 m/s, so that the simulation and the experimental results had only 4.9% difference. Fig. 8.8 compares the damaged region of the target surface after the projectile perforated the target. It can be seen that the rebars effectively limited the damaged material in a smaller region even when the projectile did not hit directly on the rebars. Fig. 8.9 shows the configuration of damaged target and the fractured rebar network at the time $t = 0.5$ ms. Fig. 8.10 compares the residual velocities from the simulation and the experiment. A small deviation from the experimental residual velocity was observed when the initial velocity is high, but the results for low impact velocities were very close.

8.3 EXPLOSION

Explosion problems always contain extremely large deformation and fragmentation of materials, which is one of the major application fields of the MPM. The modeling and simulation of explosion problems usually have two aspects. The first is how the material responds and fails under explosion loading. The explosively driven flyer, metal jetting, and explosion welding are typical examples. The second aspect is how to describe the evolution of denotation field. The MPM has been widely applied in both areas, and has demonstrated its robustness.

Hu and Chen [93] validated the MPM with typical explosion problems including the shock tube and the Sedov–Taylor detonation wave problems. They then evaluated the fracture of the concrete wall under explosion loading. All the materials, including the air, explosive, and wall were discretized with material points in the simulation which was straightforward because there was no need to employ different methods for different regions even though the material properties in different regions were quite different. Requirement of high accuracy in simulating the explosion field usually demands discretization refinement, while the volume of the explosive usually varies greatly during the explosion. It would be much better if the discretization could be adapted automatically in the explo-

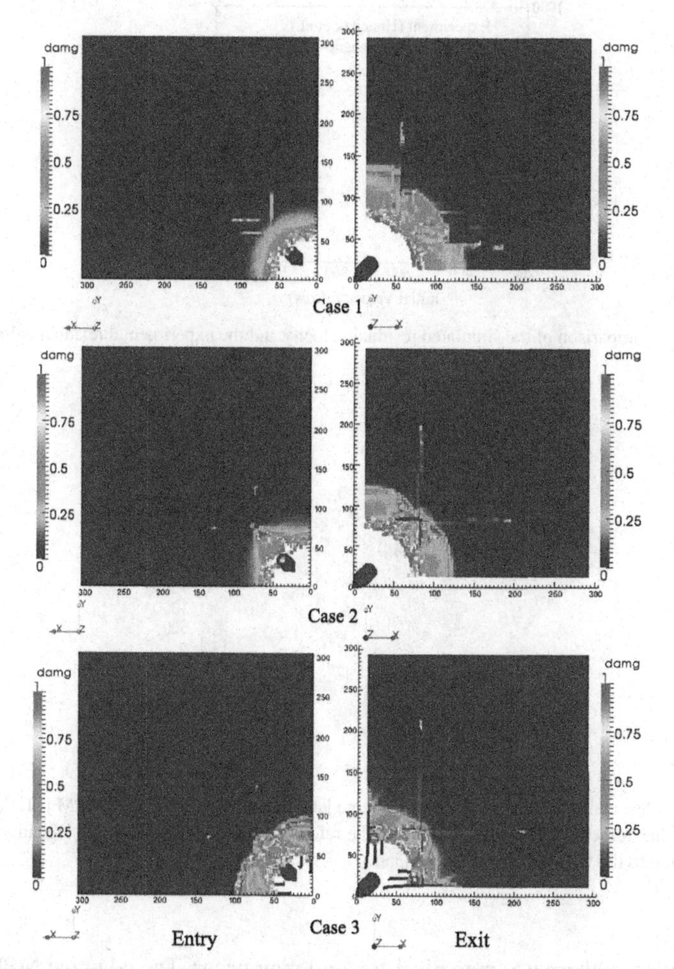

FIGURE 8.8 Comparison of the damage on the target surface for different simulation cases [39].

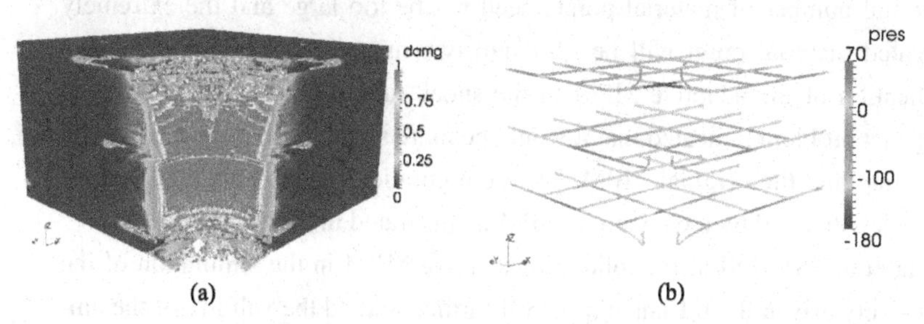

FIGURE 8.9 The configuration and damage of the target plate (left) and the deformation and fracture of the rebar network (right) at $t = 0.5$ ms when the projectile directly hits the rebar [39].

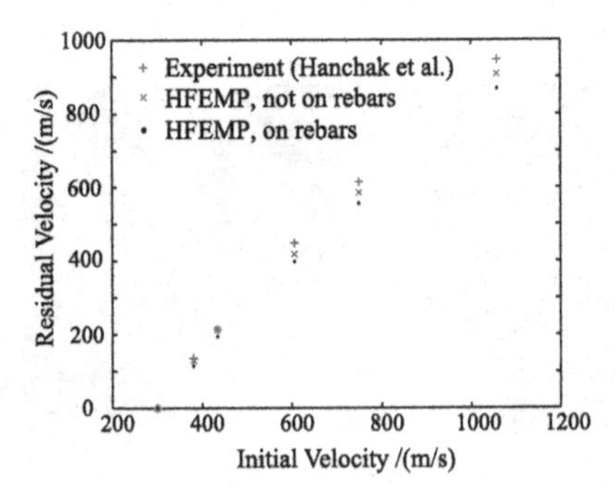

FIGURE 8.10 Comparison of the simulated residual velocity and the experimental residual velocity [39].

FIGURE 8.11 The configuration of 3D shaped charge jet obtained by the adaptive MPM [81,188] (color denotes the velocity). (For interpretation of the references to color in this figure legend, the reader is referred to the web version of this chapter.)

sion simulation. Otherwise, numerical fracture may occur. The adaptive MPM proposed by Ma et al. [81,188] can effectively avoid numerical fracture because the initial number of material points need not be too large and the extremely elongated material point will be split into two material points automatically. Application of the adaptive MPM in the shock tube problem and the shaped charge jet problem indicated that it would be more suitable in solving explosion problems than the original MPM. The configuration of the simulated shaped charge jet obtained by the adaptive MPM is illustrated in Fig. 8.11.

Ma et al. [81,188] also applied the adaptive MPM in the simulation of the explosively driven flyer. Lian et al. [189] further studied the validity of the empirical formula for the open-face sandwich and the flat sandwich configurations of the explosively driven flyer. The open-face sandwich configuration is shown

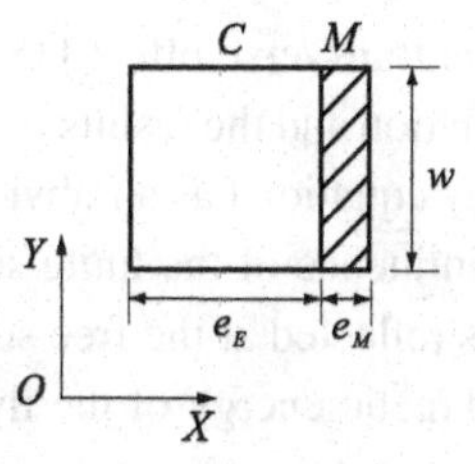

FIGURE 8.12 The open-face sandwich configuration of the explosively driven flyer [189].

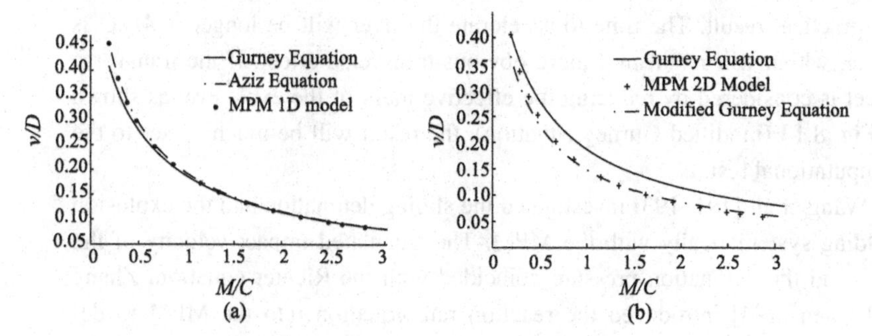

FIGURE 8.13 Comparison of the final velocity of the flyer obtained by the numerical results and empirical formulae [189]: (a) 1D model; (b) 2D model.

in Fig. 8.12. The Gurney equation gives the final velocity of the flyer by [50]

$$V = \sqrt{2E}\left[\frac{3}{1+5(M/C)+4(M/C)^2}\right]^{1/2}$$

where $\sqrt{2E}$ is the Gurney characteristic velocity, M and C are the masses of the flyer and explosive, respectively. It was reported that the Gurney equation is not accurate enough when M/C is small. Aziz et al. [190] improved the estimate with the following formulation

$$\frac{V}{D} = 1 - \frac{27}{16}\frac{M}{C}\left[\left(1+\frac{32}{27}\frac{C}{M}\right)^{1/2} - 1\right]$$

where D is the detonation velocity. The JWL equation of state and the Johnson–Cook strength model were used for the explosive and the flyer, respectively, and the details of the material parameters can be found in Lian et al. [189]. Both the one- and two-dimensional models were calculated to validate the above equations. The final velocities of the flyer obtained by the one-dimensional model and the two empirical formulae are plotted in Fig. 8.13. All three results agree well when M/C is large, while the result by the Gurney equation is obviously lower than the other two solutions when M/C is small. The two-dimensional model was employed to study the transverse effect [189]. Fig. 8.13 also compares the result by the Gurney equation and the results from the two-dimensional model. It is shown that the Gurney equation has an obvious deviation for all the M/C values, which reflects the influence of the finite size in the transverse direction. The compression wave is reflected at the free surface, and the reflected expansion wave will reduce the kinetic energy of the flyer. In other words, the chemical energy of the explosive cannot be effectively converted into the kinetic energy of the flyer so that the Gurney equation will yield a higher solution than the practical result. The time to accelerate the flyer will be longer if M/C is larger, which will result in a more obvious transverse effect. If the transverse effect is considered by reducing the effective mass of the explosive, as shown in Fig. 8.13 (modified Gurney equation), the result will be much closer to the computational result.

Wang et al. [191–193] investigated the sliding detonation and the explosion welding systematically with the MPM. The calculated impact velocity of the flyer and the detonation pressure coincided with the Richter equation. Zhang and Chen [194] introduced the reaction rate equation into the MPM to describe the ignition process, and they analyzed the shock-to-detonation process of the heterogeneous solid explosive. The influences of the fragment material and the thickness of the shielding plate on the critical detonating speed were studied. It was found that the critical speed for copper and steel were close, but that for tungsten was much lower. They also calculated the same problem with the Lagrangian FEM and ALE method. Mesh distortion was observed for both mesh-based methods, and using the erosion scheme in the FEM led to smaller fragments. Yang et al. [195] introduced the Gurson model and the random failure scheme into the MPM simulation, and they investigated the expansion and fragmentation process of metal shells under explosion loading. Both the material model based on the evolution of microscale damage and the material model with macroscale failure were employed. The Gurson model and the Tepla-f failure condition were used to describe the evolution of micro-cavities, while the J_2 plasticity theory and the random failure scheme with Weibull distribution were used to calculate the macroscale failure. Yang et al. [195] simulated the fragmentation process of typical metal shells, including the cylindrical shell and spherical shell, under explosion loading. Results indicated that the shell ruptured in a mixed mode, and both necking and shear instability were observed, as shown in Fig. 8.14. For the spherical shell, the accumulative mass of the fragments approximately obey the power law. An efficient scheme for fragment statistics was presented by Yang et al. [195] based on the background mesh. The results of shell fragmentation agreed well with the theoretical prediction and experiments, and it was also shown that the MPM simulation with the Gur-

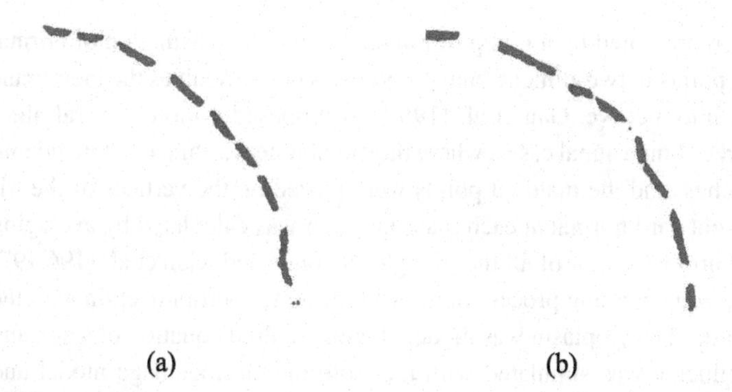

FIGURE 8.14 The configuration of the cylindrical metal shell subjected to inner detonation [195]: (a) the results obtained by the Gurson model; and (b) the results obtained by the random failure model.

son model can effectively predict the transition from ductile fracture to brittle fracture.

8.4 FLUID–STRUCTURE/SOLID INTERACTION

The challenges in simulating fluid–structure/solid interaction (abbreviated as fluid–structure or fluid–solid interaction in the context) lies in the nonlinear variation of the interface shape and the huge difference between the properties of different material phases. It is usually demanded to solve the entire problem simultaneously if the interactions between different material phases are strong. More complex phenomena may appear in some strong fluid–structure interaction problems. For example, the splitting, curling, and merging of free liquid surface bring strong nonlinearity into the liquid sloshing problem. Accurately describing the interactions, the pressure on the structure exerted by the fluid, for instance, is often desired in the engineering practice.

How to apply the MPM in fluid–structure interaction problems has been receiving special attention since the infant stage of the MPM. This is attributed to the fact that the MPM inherits the advantages of both Lagrangian and Eulerian methods. The MPM can be used in both the structure/solid region and the fluid region directly. If the fluid–solid interface satisfies the non-sliding condition, the inherent single-valued velocity field can automatically describe the interaction between different regions. York et al. [92] first adopted the two-dimensional MPM to simulate the interaction between the fluid and the membrane structure such as the airbag expansion. The calculation of outward normal vector of the membrane structure is important in the above simulation. If the structure is discretized with finite elements, the normal can be easily determined from the element geometry. Special scheme is required for the MPM simulation as the boundary is represented by a group of points, though the calculation of normal vector from points in two-dimensional problems is not difficult as the membrane degenerates into a curve. Gan et al. [196,197] proposed a more general algorithm for three-dimensional cases, where the membrane was first discretized into triangle patches, and the material points were placed on the vertices of the triangles. The outward normal at each material point was calculated by averaging the outward normal vector of all the triangles it connected. Gan et al. [196,197] analyzed the zona piercing process in intracytoplasmic sperm injection with the above method. The cytoplasm was modeled with the fluid equation of state, and the zona pellucida was simulated with the isotropic elastodamage model and the decohesion model. Hu et al. [94,95] simulated the aeroelastic problem with a unified MPM framework. They also discretized the structure region and fluid region both with material points, and different constitutive models were employed for different regions. The adaptive mesh refinement scheme was adopted in the fluid region to achieve better accuracy.

One essential difference between the MPM and FEM is that the Gaussian quadrature scheme is replaced by numerical integration with particles in the MPM, which will lead to accuracy loss in small deformation problems. Some investigators coupled the MPM with high-accuracy CFD solvers to satisfy the accuracy requirement for the fluid field. The key issue to effectively and seamlessly couple the MPM with other solvers is the interface treatment. Owing to the existence of the background mesh, the MPM can be more easily coupled with other Eulerian methods. Guilkey et al. [96] developed a fluid–structure coupling scheme based on the multi-material implicit continuous fluid Eulerian (ICE) solver where the variables of all the materials were assembled onto the background mesh and were solved from the fluid dynamic equations for multiple materials. A fractional stepping scheme was adopted in the ICE framework. One time step was split into the Lagrangian and Eulerian steps, whose flowchart was similar to the MPM flowchart, which made it easier to incorporate the MPM simulation. In Guilkey et al.'s work, the MPM was used to simulate the solid material, and the state variables of solid particles were mapped onto the background mesh nodes. Then the nodal equations were solved in an iterative way to achieve the equilibrium status. Finally, the stress, velocity, and the other variables of the material points were updated through the MPM flowchart, while the variables in the fluid region were updated by their corresponding schemes. Problems with strong fluid–solid interactions, such as the explosion of the energetic devices, were successfully simulated.

The immersed boundary (IB) method is an effective way to deal with fluid-structure interactions. Gilmanov and Acharya [198,199] combined the IB method and MPM to construct a numerical method for three-dimensional fluid–structure interaction problems. The equations in the fluid region was first

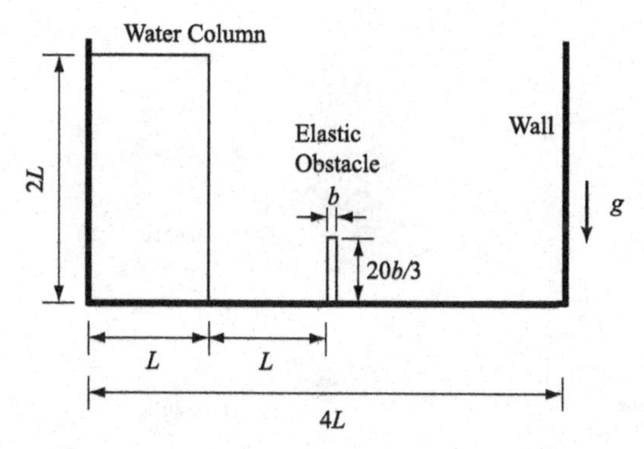

FIGURE 8.15 Schematic view of the collapse of the water column and the impact of water on the elastic obstacle [37].

solved in each time step so that the traction applied by the fluid on the solid surfaces could be obtained. Then the MPM was used to solve the deformation of the solid region under the aforementioned traction, and the variables of the material points were updated. After that the boundary velocity and the boundary pressure of the fluid region could be obtained from the solution of the solid region. The fluid region and solid region were solved alternatively with the above scheme. The surfaces of the solid region were discretized into triangular patches, and the mesh nodes adjacent to the patches were recognized as the fluid boundary nodes. The prescribed boundary conditions on the fluid boundary nodes were constructed by the information from material points in solid region and from the inner part of the fluid region. Such a method was also extended to the analysis of the fluid–solid–thermal coupling problems [199].

The deformation of the solid region is not large in many kinds of fluid–solid interaction problem, and the FEM can well simulate these solid regions. The coupled finite element material point (CFEMP) method is quite suitable for such fluid–solid interaction problems. Lian et al. [37] utilized the CFEMP method to successfully simulate the collapse of water column and the interaction between the water and the elastic obstacle. As shown in Fig. 8.15, the water column was initially at rest, and then it collapsed under the gravity. Thus, the water impacted on the elastic bar. The water was simulated with the MPM and the Mie–Grüneisen EOS, and the elastic bar was simulated with the FEM and the elastic constitutive model. The plane strain condition was applied to confine the movement in 2D.

The configurations of the system at different times are plotted in Fig. 8.16. The results calculated by the CFEMP method were consistent with those of the particle finite element method [200]. The time history of the deflection of the

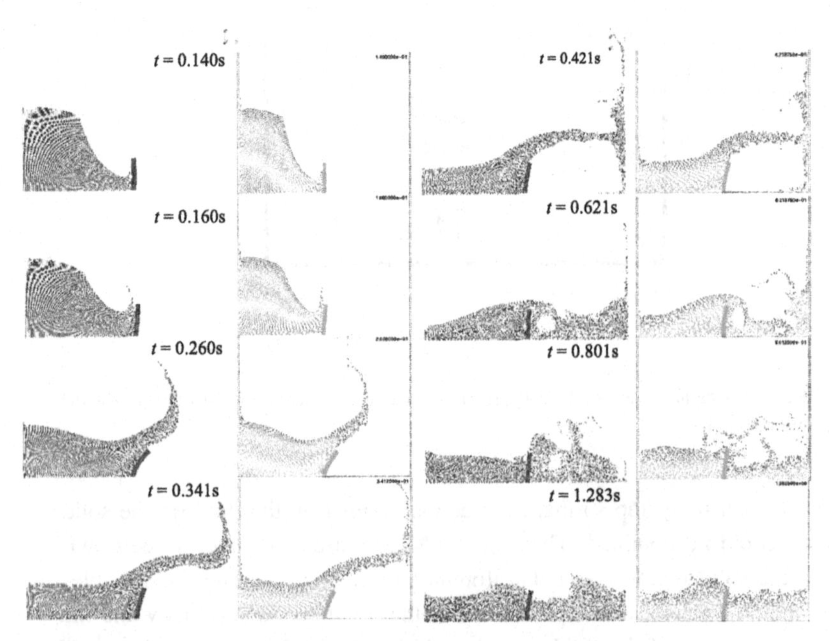

FIGURE 8.16 Comparison of the configurations at different times for the water collapse problem [37]. The sub-figures in the first and the third columns were obtained by the CFEMP method, and the sub-figures in the second and the fourth columns were obtained by the particle finite element method.

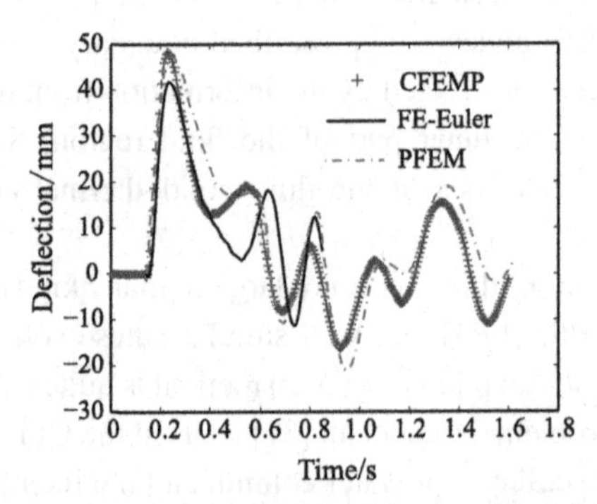

FIGURE 8.17 Comparison of the time history of the deflection of the left upper corner of the elastic obstacle [37]. FE-Euler stands for the solution obtained by the finite element-Eulerian coupled method.

upper left corner of the obstacle is plotted in Fig. 8.17. Results of the other two numerical methods, that is, the particle finite element method [200] and the finite element-Eulerian couple method based on the level set algorithm [201], are also plotted. All three results agree very well, but the CFEMP solution is closer to the solution of the particle finite element method.

Another category of fluid-structure coupling problems is that the fluid and solid occupy the same region, and the coupling between different materials is strong. One problem being widely focused on is the porous media problem composed by the solid skeleton and the pore fluid. Zhang and Wang [202] proposed a bi-phase material point method to analyze the dynamic problems of the saturated porous media. Two groups of material points were introduced in the bi-phase MPM, which was utilized to respectively describe the solid skeleton and the porous fluid. The solid-phase material point and the fluid-phase material point resided at the same place. In other words, each position actually had two material points. Then the material points were updated with the momentum equations of their own phase. The interaction between different phases was reflected by assigning the damping force proportional to the velocity difference between the two phases. Only one set of background mesh was used in the bi-phase MPM.

8.5 MULTISCALE SIMULATION

Many engineering problems are related to multiple spatial and temporal scales. Representative examples are the dynamic fracture evolution, mechanics of bones, and turbulence problem. Analyzing these problems across different scales, however, is still very challenging. In Chapter 7, a multiscale MPM has been presented, in which both hierarchical and concurrent approaches are used to simulate a certain type of problems. In this section, other recent advances in multiscale simulation are discussed with a focus on the MPM-related development.

The multiscale methods can be divided into two groups. The first is called the concurrent multiscale method whose essential idea is to employ the numerical methods applicable to the atomic scale and the continuum-based numerical methods (for example, the FEM, MPM, and FDM) for different regions, and all the regions are integrated concurrently. The MD method, molecular statics method, or the Monte Carlo method is adopted for the kernel region such as the regions around the crack tip or the shear band. The continuum-based methods are usually adopted for the far-field regions where the continuum assumption is valid. The concurrent multiscale methods can greatly increase the efficiency owing to the reduction of degrees of freedom in the far-field region, and the accuracy of the results is still satisfactory. The second group is called the sequential (or hierarchical or upscaling) multiscale method. Numerical simulations at a smaller scale will provide necessary material parameters for the simulations at a larger scale, while the results from a larger scale are the constraints for the simulations at a smaller scale. A typical example of the sequential multiscale method is first to calculate the potential parameters of classical MD with quantum mechanics, and then to run the simulation with classical MD to obtain the heat conductivity, and finally, solve the heat conduction problems with the FEM.

As a continuum-based meshfree particle method, the MPM is quite suitable to be combined with the MD method to construct concurrent multiscale methods. Guo and Yang [203] and Lu et al. [204] investigated the MD-MPM concurrent coupling multiscale method. The MPM region and the MD region provided the boundary condition for each other, and a transition region (or handshaking region) was used between the MD and the MPM regions. Their coupling schemes required material points to be located at the lattice sites in the transition region. A coarser background mesh was also adopted in the far-field region to save the computational cost. High-velocity impact problems of copper-to-copper and silicon-to-silicon at nanoscale was investigated with the concurrent method by Guo and Yang [203]. Lu et al. [204] used the multiscale method to simulate the necking process in the tension of nanowire.

The limitation of MD mainly comes from its huge computational cost. The MPM provides the ideas to improve the MD method for a larger applicable temporal scale. The MD results include both the low-frequency motions and the high-frequency motions. The low-frequency motions reflect the overall deformation, while the high-frequency part denotes the atomic vibrations around the atom sites. The high-frequency motions are not important in some problems, but their existence hinders the use of a larger time step size. The factor controlling the critical time step size in the MPM is the size of the background mesh instead of the distances between material points, which is because the equations of motions are constructed at the background mesh nodes and the variables of the material points are updated by the variables of the background mesh nodes. It has been widely validated that the time step of the MPM will not be decreased by the decrease of the distances between material points. Liu et al. [205] proposed the smoothed molecular dynamics (SMD) method by introducing a background mesh into the flowchart of molecular dynamics, as shown in Fig. 8.18. The atomic equations of motion are assembled onto the background mesh nodes, and the atomic variables are updated by the interpolation of the nodal increments after solving nodal momentum equations. The assembling and interpolation process converts the factor controlling the critical time step size to the size of the background mesh in the SMD method so that the available time step size of the SMD method can be much larger than that of the MD method [205]. The time step size of the SMD method can even be one order larger than the critical time step size of the MD method. Although the assembling and interpolation process in the SMD flowchart increases the computational cost in one time step, the overall computational cost will be much smaller than MD cost because the total number of time steps is greatly decreased. Examples indicated that the essence of the SMD method is to smooth out the high-frequency atomic motions that the

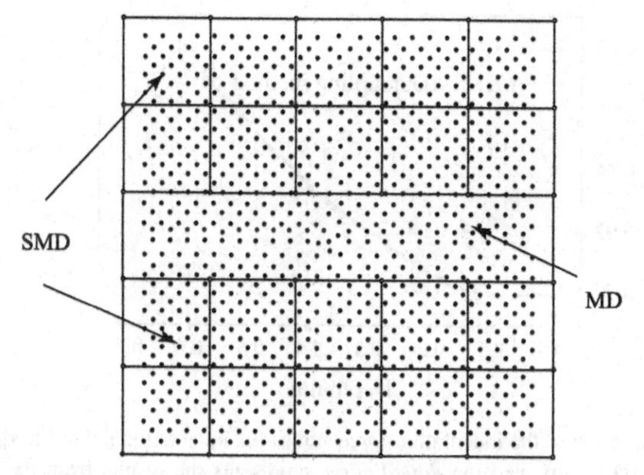

FIGURE 8.18 The background mesh of the SMD method and the coupling between the MD method and SMD method [205].

mesh cannot describe, which is why the name "smoothed molecular dynamics" was used.

The shortcoming of the SMD method is that the local atomic disorders would be smoothed in the assembling and interpolation process. If the evolution of local atomic disorders is important, the simulation accuracy would be affected. One solution scheme is to couple the MD method and SMD method, as shown in Fig. 8.18. The region containing atomic disorders is calculated with the MD method, and the far-field region containing smooth deformations is calculated with the SMD method. Multiple time steps are employed in the MD-SMD coupling method. Because the regions containing atomic disorders usually occupy a small part of the entire region, the overall computational time could still be well decreased as compared with the pure MD computation. Except for the assembling and interpolation process, the SMD flowchart is the same as the MD flowchart. The coupling between MD and SMD is therefore natural and straightforward. The atomic interactions between the atoms in the MD region and the atoms in the SMD region actually reflect the interaction between two regions, and the background mesh size near the interface does not have to be reduced to the lattice constant. He et al. [206] further improved MD-SMD coupling method with the adaptive scheme. The entire region is covered with background elements, and the elements inside SMD region will be automatically converted into MD region if local atomic disorders happen. The element in MD region will be converted back to SMD region if the atomic disorders propagate into other elements. Fig. 8.19 demonstrates the results obtained by the MD method and the MD-SMD coupling method for the nano-indentation problem. The yield point, which is denoted by a sudden decrease in the force-indentation depth curve, can be well captured by the coupling method. The MD computation is executed as

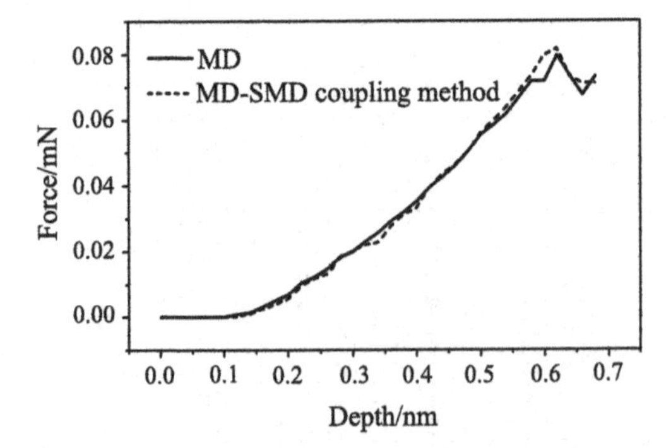

FIGURE 8.19 The variation of the indentation force versus the indentation depth [205]. The *solid curve* represents the MD results, and the *dotted curve* represents the results from the MD-SMD coupling method.

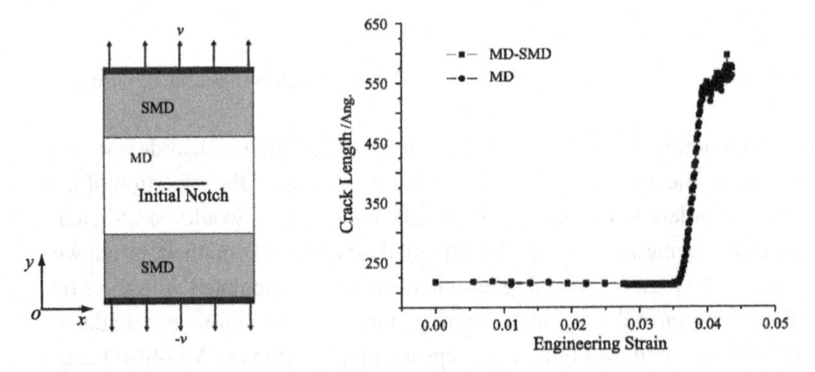

FIGURE 8.20 Tension of the pre-cracked plate (left) and the variation of the crack length versus the engineering strain (right) [207].

the dislocations occur in the region right beneath the indenter, and the SMD computation is used in the other regions.

Wang et al. [207,208] developed the parallel algorithm and the adaptive mesh refining scheme for the SMD method. Fig. 8.20 illustrates the pre-cracked plate subjected to uniform tension, which was simulated with the parallel MD-SMD coupling method. The MD method was used for the central cracked region, and the SMD method was used in the other regions. The crack length–strain curves obtained by the MD and coupling methods match each other well. The variation of the background mesh in the simulation of nano-indentation problem with the adaptive SMD method is shown in Fig. 8.21. The mesh for the region under the indenter was refined continuously when the indentation depth was increasing. The mesh size for the far-field region had the same value. Similar to other concurrent multiscale methods, the high-frequency parts of the atomic motions, if they are beyond the description capability of the background mesh, will be

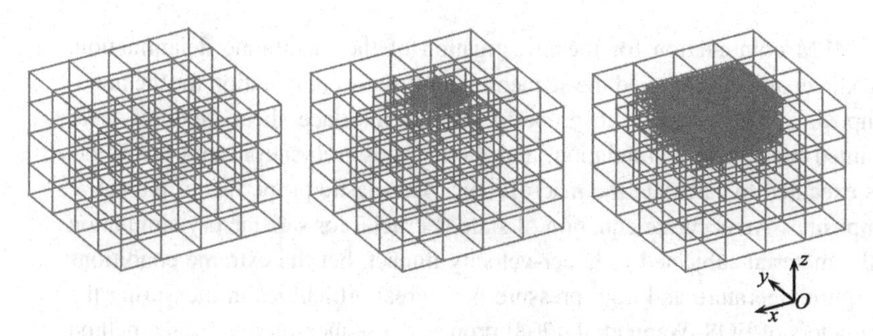

FIGURE 8.21 History of mesh refinement when the indentation depth increases in the nano-indentation example [208].

reflected back into MD region when they arrive at the MD-SMD interface. He et al. [206] proposed a seamless transition scheme to suppress the reflection of high-frequency motions based on scale decomposition. A damping force is applied to the high-frequency part of atomic motions before they approach the interface. Numerical examples indicated that the transition scheme can well absorb the high-frequency motions and avoid their influences on the solutions of MD region. The formulation of the transition scheme is concise because the scale decomposition is intrinsic in the assembling and interpolation process of SMD flowchart.

Criteria are crucial to the accuracy and efficiency in both the adaptive mesh-refinement scheme and the adaptive MD-SMD coupling scheme. Such a criterion can be constructed naturally in the SMD flowchart. The forces to update atomic displacements and velocities in the SMD method are different from the original MD atomic forces. Wang et al. [208] developed a criterion based on the average of the differences in the two types of forces in one element. If the average value is large, it is indicated that the mesh needs refinement. If the value is very small, the mesh can be coarsened to reduce the computational burden. He et al. [206] developed two robust criteria. One criterion is the average of the centro-symmetry parameter (CSP) value of all the atoms in one element. The other one is based on the average of the difference between the original atomic displacement and the SMD atomic displacement, which is defined as the displacement after assembling to the nodes and then being interpolated back. They found that the CSP and displacement criteria can both identify the regions with atomic disorders well, and that the displacement criterion is much more efficient.

The MPM has also been applied in sequential multiscale computations. Ayton et al. [209] utilized the non-equilibrium molecular dynamics (NEMD) to calculate the transfer coefficient for the lipid bilayers, which was then used in the MPM computation. Shen and Chen [210] combined the MD computation and MPM computation for the investigation of the membrane delamination. Borodin et al. [211] applied the sequential multiscale computation for the nanocomposites. The viscoelastic property and the interface shear modulus were obtained first by MD simulation, and then a stress relaxation with the MPM was executed to calculate the macroscopic viscoelastic properties of the entire composites. An accurate equation of state is crucial for satisfactory simulation of the material subjected to hyper-velocity impact, but the extreme conditions of high temperature and high pressure pose great difficulties in measuring the parameters of EOS. Wang et al. [208] proposed a sequential multiscale method for the hyper-velocity impact problems. MD simulation was used to calculate the corresponding values of a part of thermodynamic variables with the other variables being prescribed. A large number of MD simulations were carried out to obtain a group of state points. These state points can be used to construct an EOS in tabular form or can be used to calculate the parameters of an existing EOS. Wang et al. obtained the parameters of the GRAY EOS and simulated the hyper-velocity impact processes of several typical metals. The simulated Hugoniot curve and debris cloud configurations coincided with experimental results. Zhang et al. [212] combined the MPM and hybrid Monte Carlo (MC) method to study the evolution of the microstructure of the polycrystalline material. The Monte Carlo method was used to analyze the motion of the grain boundaries, and the MPM was used to calculate the stress field. The MPM-MC method was then extended to analyze the plastic behaviors of polycrystals.

The MPM has also demonstrated the potential in constructing micro- and meso-structures of novel materials. For instance, the foam material has attractive material properties such as high specific strength, high specific modulus, and nice capability to absorb energy. The stress–strain curve of the foam material contains three typical stages, the elastic, plateau, and densification stages. These are closely related to the local buckling of the cell walls, the yielding of the cell wall material, and the collapse of the cells. Reconstructing the cellular structures inside the foam material could improve the prediction of macroscopic material properties. But the randomly distributed cells with a variety of sizes inside the foam material make it almost impossible to discretize the cell wall structures with traditional mesh-based methods. In addition, the cell walls of the intrinsic structure often undergo large deformations under compression. The above issues make meshfree particle methods more appropriate than the FEM to model the intrinsic structure of the foam material. A large number of contacts between different cell walls and self-contacts of the cell walls may appear during the deformation, which requires efficient contact algorithms along with meshfree particle methods. Since the original MPM assures the non-interpenetration condition automatically and efficient contact algorithms can be constructed based on the background mesh, the MPM is more competitive in modeling the foam

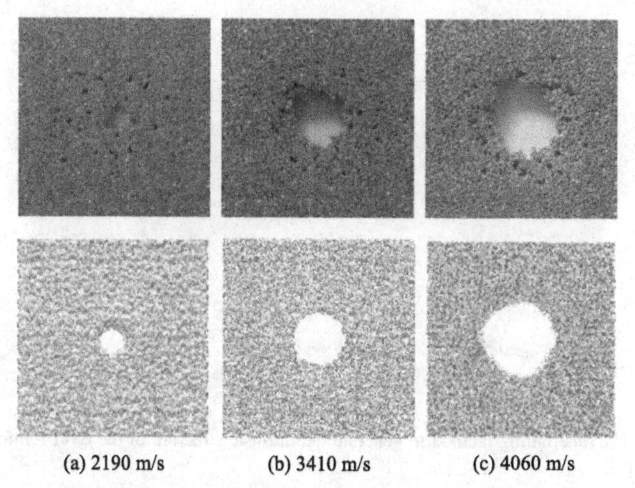

FIGURE 8.22 Comparison of the damage profiles of the aluminum foam layer under hyper-velocity impact [216]: (upper row) numerical results; (lower row) experimental results. The corresponding impact velocities are marked under their corresponding sub-figures.

material than other spatial discretization methods. Bardenhagen and his collaborators [213,214] constructed an intrinsic structure of the foam material with both the reconstruction from X-ray images and random generation, and they simulated the compression of a cubic representative volume element (RVE). Although only the hyper-elastic constitutive model was adopted in their simulation, the typical stress–strain curve with the plateau phase and densification phase was obtained. The results indicated that the porous ratio has important influences on the macroscopic properties of the foam material. Daphalapurkar et al. [215] reproduced the intrinsic structure of the close-celled PMI foam material from μ-CT images, and simulated the response of the RVE with the MPM. Although the calculated stress–strain curve had some differences from the experimental results, their numerical results clearly showed a stress–strain curve with three stages. The elastic buckling of the cell walls was observed even in the elastic phase of the macroscopic stress–strain curve. The intrinsic porous structure of the aluminum foam results in an excellent capability to scatter and fragment the space debris when the foam is used to shield the hyper-velocity impact from the space debris. The complex interaction between the space debris and the intrinsic structure also requires the formulation of a realistic intrinsic structure model. Gong et al. [216] reconstructed the intrinsic structure model from μ-CT images, and simulated the shielding performance of foam-based shielding structures under hyper-velocity impact. The filled Whipple structure and sandwich Whipple structure were investigated. Fig. 8.22 compares the damage profiles of the foam layer of the filled Whipple structure. The numerical and

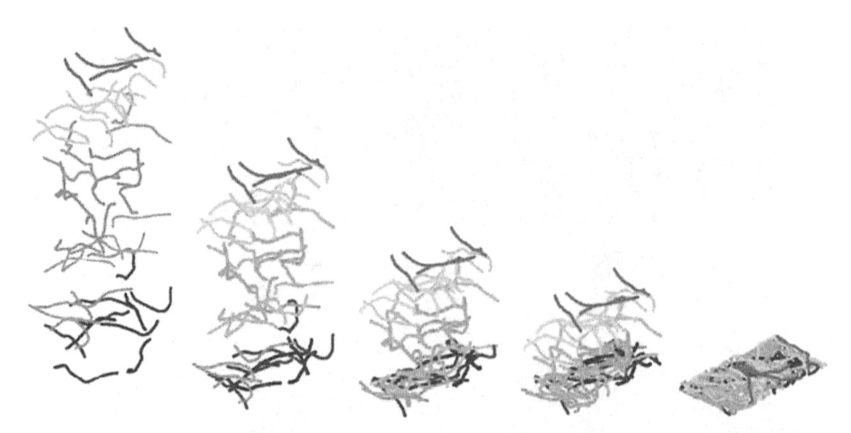

FIGURE 8.23 The tube-falling method to generate the intrinsic structure of the CNT-reinforced composites [218].

experimental results were consistent in both shape and size. Liu and Gong et al. [217] further investigated the influences of the relative density and strain rate effect of the cell wall material on the plateau stress level and the strain rate effect of the overall material. If all the pixels are reserved during the conversion from images to corresponding material points, the computational cost may be too high. Hence, a reduction in RVE size and merging of multiple points into one point were adopted in the above works to balance the computational cost and the numerical accuracy. It should be noted that the outcome of conversion and merging of material points depends on the particle feature of the MPM.

The carbon nanotube (CNT) reinforced composites have outstanding mechanical and electric properties, which is also attributed to their complex intrinsic structure. But only the model composed of quasi-straight tubes can be constructed with traditional mesh-based methods. If more realistic curved tubes are used, the penetration between different tubes will not be removed. Otherwise, the desired high volume ratio of CNTs will not be achieved. Wang et al. [218] proposed a tube-falling method based on the MPM. As shown in Fig. 8.23, a series of curved tube described by material points were created and piled up first. Then the tubes fell under the gravity and formed the reinforcement network. After the reinforcement network was filled with matrix points, the final micro-structure model for CNT-reinforced composites is generated. The tube-falling method fully utilized the advantage of the MPM that the contacts between different CNTs can be efficiently treated. The micro-structure model generated by the tube-falling method does not have interpenetration between tubes, and the volume ratio of CNTs is consistent with the actual value. Wang et al. [208, 218] investigated the mechanical and electric properties of CNT-reinforced composites based on the above model, and the influences of the volume ratio and connection properties between different CNTs on the macroscopic properties were described.

The aforementioned methods to investigate the macroscopic properties based on the simulation of intrinsic structure model or RVE with the MPM can be widely used for other materials. Shen et al. [219] studied the local plastic deformation of the polycrystals. Xue et al. [220] simulated larger grains directly from the intrinsic model, and simulated smaller grains by homogenization. They also applied such a method in the study of viscoelastic properties of an explosive.

8.6 BIOMECHANICS PROBLEMS

The complex nature of an organism makes it extremely challenging to understand the functionality of the organism from its internal structure. Even with very fine scanned images, modeling the internal structure of the organism with traditional mesh-based methods is still extremely difficult, and many assumptions and simplifications have to be introduced. The generated mesh requires careful smoothing to avoid bad mesh qualities. Similar to constructing micro- and meso-structures in the multiscale analysis, the conversion of CT-scanned images to the material points is convenient and efficient. The MPM is very suitable for large-scale simulation so that the MPM models can preserve many details and reflect the real situation. The deformations of the tissues are usually large, for which the MPM simulation is competitive. Investigating the biomechanics problems with the MPM should be appealing.

Based on the images provided by confocal microscopy, Guilkey et al. [221] constructed a material point model for the microvascular fragments. A large scale simulation with 13 million material points and 0.45 million background nodes was performed by a modified implicit MPM and 200 processors. Both the microvascular and the collagen gel were modeled with a neo-Hookean hyperelastic model. Simulation results indicated that the stress distribution was highly inhomogeneous, and the stress of the vascular was higher than that of the gel. The influences of the background mesh size and shear modulus of the vascular were also investigated. Ionescu et al. [222] discussed the simulation scheme for the failure of soft tissue with the MPM. They simplified the tissue to a composite model with unidirectional reinforced fibers. The matrix was modeled with a neo-Hookean model, and the stress of reinforced fiber was described with a piecewise function of the fiber elongation. Different failure criteria were assigned to the matrix and the fiber, where the matrix failure depended on the shear strain, and the fiber breakage depended on the tensile strain. They simulated the process in which a myocardial slab was penetrated by a bullet. The results indicated that the size of the wound exit was larger than the size of the wound

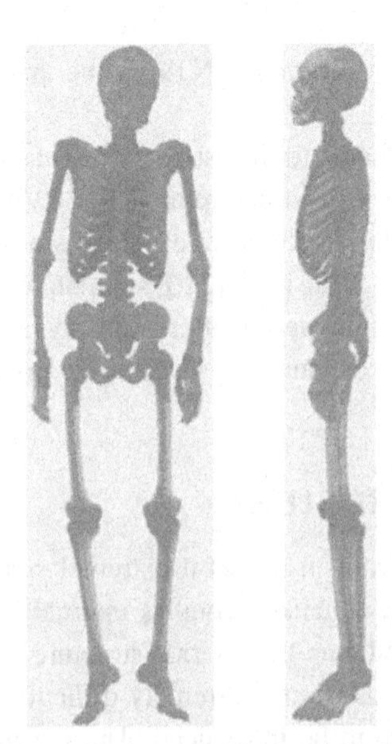

FIGURE 8.24 The MPM model of the human body generated by CT-scanned images: (left) front view; (right) side-view.

entrance, which coincided with the clinical report. When the projectile speed was only 50 m/s, the anisotropy resulted in an irregular wound shape. When the projectile speed was increased to 150 m/s, the wound shape was closer to a circle.

Zhou et al. [223–225] constructed a material point model for the entire human body based on CT images, and investigated the responses of the head and spine to impact loading. The material point model of the human body is shown in Fig. 8.24. They developed three kinds of head model to examine the proper boundary conditions to be applied. The three models were the simplified head model with free boundary condition (SHFr), the head model with muscles and free boundary condition (MHFr), and the head model with muscles and fixed shoulders (MHSFi). The responses of different models to the impact of a cylinder of the velocity 6.4 m/s demonstrated that whether the boundary was free or fixed did not affect the results if the impact time was less than 2 ms. However, the fixed shoulders would induce multiple impacts and affect the head damage significantly for the impact with a longer time. The muscles would distribute the stresses and reduce the damage. The injuries caused by the acceleration and deceleration were also simulated, and the impact velocity was from 6.4 to 19.2 m/s. Simulation results showed a direct injury at the impact region and an indirect injury near the eye socket. The injury mechanism of the acceleration and deceleration was found to be different. Zhou et al. [225] further constructed

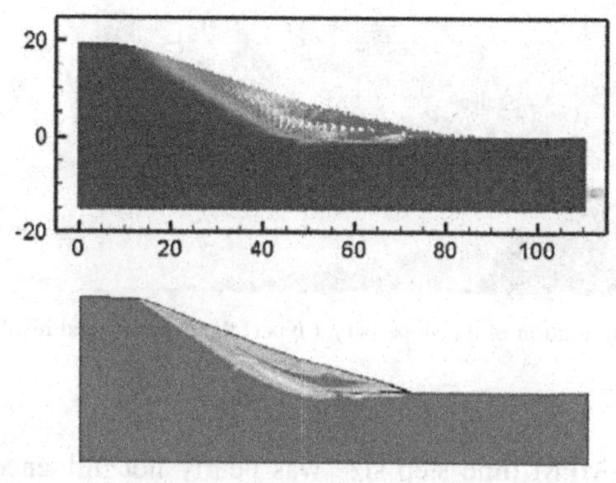

FIGURE 8.25 Failure of the sandy soil slope [120]: (upper) the MPM results; (lower) the FEM results.

a computational model for the impact of the human skeleton, which was composed of the head, cervical vertebra, thoracic vertebra, ribs, shoulder blades, lumbar vertebra, coccygeal vertebra, pelvis, ligaments, intervertebral discs, the seat back. They investigated the influence of the bone density on the responses of the human skeleton. It was found that the maximum acceleration at the head, breast, lumbar vertebra, cervical vertebra, and pelvis were all increased when the bone density was decreased. The magnitude of the acceleration of the cervical vertebra varied nonlinearly with the bone density, and the acceleration of the other parts varied almost linearly with the bone density.

8.7 OTHER PROBLEMS WITH EXTREME DEFORMATIONS

Large deformations of geological materials are often encountered in engineering problems. Since the MPM does not have mesh distortion, the applications of the MPM in geotechnical engineering have increased in recent years. Bardenhagen et al. [71] first employed the MPM to simulate grain materials. Cummins and Blackbill [104] developed an implicit material point method for grain materials. Coetzee and his collaborators [226,227] investigated the excavation process and the flow of grain materials with the MPM. They compared the results obtained by the MPM with a nonpolar continuum theory, the MPM with a polar continuum theory, the discrete element method (DEM), and the experiments. They found that the results calculated with the MPM were better, and the results calculated by the MPM with the polar theory were the best. Huang et al. [120] simulated the slope failure and the penetration of soil material with the MPM and the Drucker–Prager model. The final configuration of the failed sandy soil slope is shown in Fig. 8.25. The FEM and MPM configurations are very similar, and extremely large deformations can be observed at the surface of the slope. Large deformations led to a sharp decrease of the FEM time step size and mesh

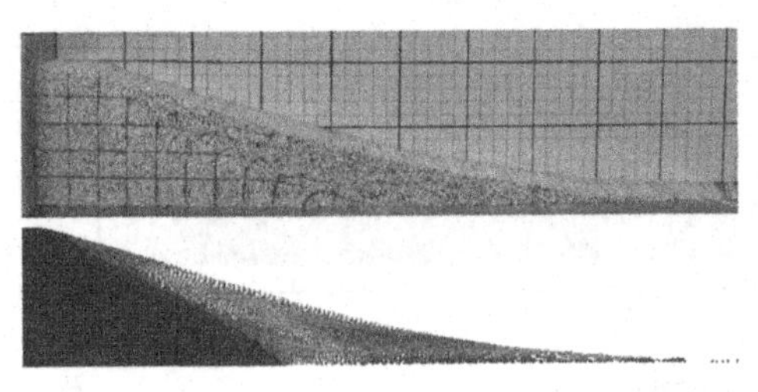

FIGURE 8.26 Final configuration of the slope [41]: (upper) the experimental results; (lower) the AFEMP results.

distortion, while the MPM time step size was nearly not influenced by large deformations. The overall computational time of the MPM was much smaller than that of the FEM. Anderson et al. [228] also simulated slope failure with the MPM, and they employed a simplified model to study the influence of slope failure on the structures built on the slope.

Small deformations exist in a large area in the slope failure problem such that it will be more efficient and accurate to employ the FEM for the small-deformation region. Lian et al. [41] analyzed the slope failure problem with the adaptive finite element material point (AFEMP) method. The entire slope was discretized with hexahedron elements, and the finite elements were converted into material points whenever their effective plastic strain exceeded the critical value. The Drucker–Prager model was used to model the sandy soil. The AFEMP and experimental results were consistent, as shown in Fig. 8.26. It should be noted that the experiment was equivalent to an indoor experiment with a piles of dry aluminum tubes carried out by Bui et al. [229]. The aluminum tubes were initially piled up to a cuboid shape, and they were kept stationary by a vertical baffle. The baffle was then removed, and the tubes collapsed to form a slope. The corresponding material properties of the soil could be derived from the interactions between the tubes. Grid lines were drawn on the background wall and on the initial outer surface of the tube pile. The movement of the grid lines could clearly indicate the slope surface and failure surface. The surface lines of the experimental results, the standard MPM results, and the AFEMP results are all compared in Fig. 8.27. Both the MPM and AFEMP produced identical to the experimental results, but the AFEMP fully utilized the advantages of the FEM in small deformation phase. The time step size and the CPU time of the standard MPM and AFEMP are listed in Table 8.3. The AFEMP is more efficient than the original MPM when the deformation is small. Although the total number of time steps is larger in the AFEMP computation, the total computational cost of the AFEMP is much less than that of the original MPM.

Wieckowski [230] applied the MPM in metal forming and metal cutting processes. Ambati et al. [231] simulated the metal cutting process with the

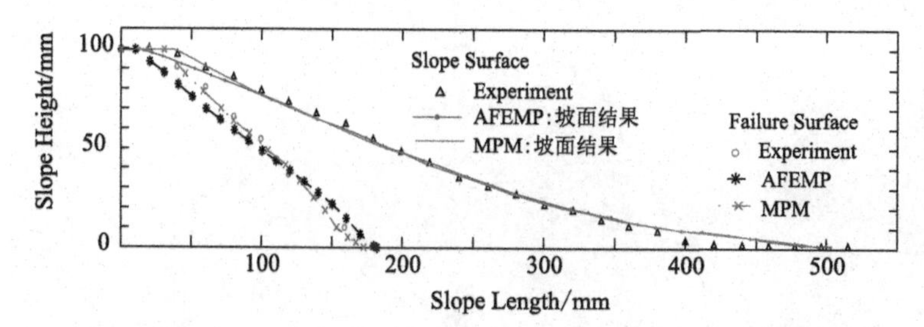

FIGURE 8.27 Comparison of the slope surfaces and the failure surfaces from the experimental results, the original MPM results and the AFEMP results [41].

TABLE 8.3 Comparison of the Time Step Size and the Efficiency Between the Original MPM and AFEMP [41]

Method	$\Delta t_{max}/\mu s$	$\Delta t_{min}/\mu s$	Total number of time steps	CPU time/s
AFEMP	48.41	22.85	47,402	1123
MPM	48.41	29.94	34,765	1791

GIMP method, and they thoroughly investigated the effects of friction and cutting depth. The results indicated that the MPM can well simulate the extremely large deformation during cutting. The localization phenomenon, i.e., the adiabatic shear band, can be captured by numerical simulation. Burghardt et al. [232] combined the MPM with a nonlocal theory to avoid the mesh dependency associated with local models. They proposed a scheme to calculate the nonlocal integration in the MPM. The results of compression and slope failure problems indicated that the nonlocal theory ensured the convergence of shear band size when the background mesh size was decreased. The mesh dependency was illustrated for local models for which the characteristic length of the shear band was decreased with the reduction in mesh size. Li et al. [233] studied the contact law by simulating the large deformation process near the contact region with the MPM.

Bibliography

[1] M.R. Benioff, E.D. Lazowska, Computational science: Ensuring America's competitiveness, Tech. Rep., President's Information Technology Advisory Committee, 2005.

[2] Simulation-based engineering science: Revolutionizing engineering science through simulation, in: Report of the National Science Foundation, USA, May 2006.

[3] J.A. Zukas, Introduction to Hydrocodes, Elsevier Science, Oxford, 2004.

[4] D.J. Benson, Computational methods in Lagrangian and Eulerian hydrocodes, Computer Methods in Applied Mechanics and Engineering 99 (1992) 235–394.

[5] D. Sulsky, Z. Chen, H.L. Schreyer, A particle method for history-dependent materials, Computer Methods in Applied Mechanics and Engineering 118 (1–2) (1994) 179–196.

[6] X. Zhang, Y.P. Lian, Y. Liu, Z. Xu, The Material Point Method (in Chinese), Tsinghua University Press, Beijing, 2013.

[7] http://www.paraview.org.

[8] M.L. Wilkins, Computer Simulation of Dynamic Phenomena, Springer, Heidelberg, 1999.

[9] C.A. Wingate, R.F. Stellingwerf, R.F. Davidson, M.W. Burkett, Models of high velocity impact phenomena, International Journal of Impact Engineering 14 (1993) 819–830.

[10] S.W. Attaway, M.J. Brown, K.H. Mello, M.W. Heinstein, J.W. Swegle, J.A. Ratner, R.I. Zadoks, PRONTO3D User's Instructions: A Transient Dynamic Code for Nonlinear Structural Analysis, 1998.

[11] S.W. Attaway, M.W. Heinstein, J.W. Swegle, Coupling of smooth particle hydrodynamics with the finite element method, Nuclear Engineering and Design 150 (1994) 199–205.

[12] J.O. Hallquist, LS-DYNA Theoretical Manual, 1998.

[13] J.O. Hallquist, LS-DYNA Keyword User's Manual, Livermore Software Technology Corporation, 2003.

[14] J.O. Hallquist, ANSYS/LS-DYNA Theoretical Manual, Livermore Software Technology Corporation, 2006.

[15] L.J. Hageman, J.M. Walsh, HELP, a multi-material eulerian program for compressible fluid and elastic–plastic flows in two space dimensions and time, vol. I, Contract Rep. AD0726459, Systems Science and Software, La Jolla, CA, 1971.

[16] J.M. McGlaun, S.L. Thompson, M.G. Elrick, CTH: A three-dimensional shock wave physics code, International Journal of Impact Engineering 10 (1990) 351–360.

[17] http://www.sandia.gov/cth/index.html.

[18] G.C. Bessette, R.L. Bell, R.A. Cole, C.T. Vaughan, L. Yarrington, S.W. Attaway, ZAPOTEC: A coupled Eulerian–Lagrangian computer code, methodology and user manual, version 1.0, Tech. Rep. SAND2003-3097, Sandia National Laboratories, Albuquerquc, NM, 2003.

[19] G.C. Bessette, Modeling coupled blast/structure interaction with ZAPOTEC: Benchmark calculations for the conventional weapon effects backfill (conweb) tests, Tech. Rep. SAND20044096, Sandia National Laboratories, Albuquerque, NM, 2004.

[20] J. Donea, A. Huerta, J.-P. Ponthot, A. Rodríguez-Ferran, Arbitrary Lagrangian–Eulerian methods, in: Encyclopedia of Computational Mechanics, John Wiley & Sons, 2004.

[21] F.H. Harlow, A machine calculation method for hydrodynamic problems, Tech. Rep. LAMS-1956, Los Alamos Scientific Laboratory, 1955.

[22] F.H. Harlow, The particle-in-cell computing method for fluid dynamics, Methods in Computational Physics 3 (1964) 319–343.

[23] W.F. Noh, CEL: A time-dependent, two-space-dependent, coupled Euler–Lagrange code, Methods in Computational Physics 3 (1964) 117–179.

[24] R.M. Frank, R.B. Lazarus, Mixed Eulerian–Lagrangian method, in: B. Alder, S. Fernbach, M. Rotenberg (Eds.), Methods in Computational Physics, Vol. 3: Fundamental Methods in Hydrodynamics, Academic Press, New York, 1964, pp. 47–67.

[25] J. Donea, S. Giuliani, J. Halleux, An arbitrary Lagrangian–Eulerian finite element method for transient dynamic fluid-structure interactions, Computer Methods in Applied Mechanics and Engineering 33 (1–3) (1982) 689–723.

[26] T.B. Belytschko, J.M. Kennedy, Computer models for subassembly simulation, Nuclear Engineering and Design 49 (1978) 17–38.

[27] T.J.R. Hughes, W.K. Liu, T.K. Zimmermann, Lagrangian–Eulerian finite element formulation for incompressible viscous flows, Computer Methods in Applied Mechanics and Engineering 29 (3) (1981) 329–349.

[28] F.H. Harlow, Hydrodynamic problems involving large fluid distortions, Journal of the ACM 4 (1) (1957) 137–142.

[29] F.H. Harlow, Fluid dynamics in group T-3 Los Alamos National Laboratory (LA-UR-03-3852), Journal of Computational Physics 195 (2004) 414–433.

[30] F.H. Harlow, J.E. Welch, Numerical calculation of time-dependent viscous incompressible flow of fluid with a free surface, Physics of Fluids 8 (1965) 2182–2189.

[31] S. McKee, M.F. Tomé, V.G. Ferreira, et al., The MAC method, Computers & Fluids 37 (8) (2008) 907–930.

[32] N.L. Johnson, The legacy and future of CFD at Los Alamos, Tech. Rep. LA-UR- 96-1426, 1996.

[33] A. Nishiguchi, T. Yabe, Second order fluid particle scheme, Journal of Computational Physics 52 (1983) 390–413.

[34] J.U. Brackbill, H.M. Ruppel, FLIP: A method for adaptively zoned, particle-in-cell calculations of fluid flows in two dimensions, Journal of Computational Physics 65 (1986) 314–343.

[35] J.U. Brackbill, D.B. Kothe, H.M. Ruppel, FLIP: A low-dissipation, particle-in-cell method for fluid flow, Computer Physics Communications 48 (1988) 25–38.

[36] X. Zhang, K.Y. Sze, S. Ma, An explicit material point finite element method for hyper velocity impact, International Journal for Numerical Methods in Engineering 66 (2006) 689–706.

[37] Y.P. Lian, X. Zhang, Y. Liu, Coupling of finite element method with material point method by local multi-mesh contact method, Computer Methods in Applied Mechanics and Engineering 200 (2011) 3482–3494.

[38] Y.P. Lian, Y. Liu, X. Zhang, Coupling of membrane element with material point method for fluid–membrane interaction problems, International Journal of Mechanics and Materials in Design 10 (2) (2014) 199–211.

[39] Y.P. Lian, X. Zhang, X. Zhou, Z.T. Ma, A FEMP method and its application in modeling dynamic response of reinforced concrete subjected to impact loading, Computer Methods in Applied Mechanics and Engineering 200 (17–20) (2011) 1659–1670.

[40] Y.P. Lian, X. Zhang, Y. Liu, Coupling between finite element and material point method for problems with extreme deformation, Theoretical and Applied Mechanics Letters 2 (2012) 021003.

[41] Y.P. Lian, X. Zhang, Y. Liu, An adaptive finite element material point method and its application in extreme deformation problems, Computer Methods in Applied Mechanics and Engineering 241–244 (1) (2012) 275–285.
[42] S. Li, W.K. Liu, Meshfree and particle methods and their applications, Applied Mechanics Reviews 55 (1) (2002) 1–34.
[43] L.B. Lucy, A numerical approach to the testing of the fission hypothesis, The Astrophysical Journal 8 (12) (1977) 1013–1024.
[44] R.A. Gingold, J.J. Monaghan, Smoothed particle hydrodynamics: Theory and application to nonspherical stars, Monthly Notices of the Royal Astronomical Society 18 (1977) 375–389.
[45] J.J. Monaghan, Smoothed particle hydrodynamics, Annual Review of Astronomy and Astrophysics 30 (1992) 543–574.
[46] G.R. Liu, M.B. Liu, Smoothed Particle Hydrodynamics: A Meshfree Particle Method, World Scientific, Singapore, 2003.
[47] S. Ma, X. Zhang, X.M. Qiu, Comparison study of MPM and SPH in modeling hypervelocity impact problems, International Journal of Impact Engineering 36 (2009) 272–282.
[48] S. Ma, Material point meshfree methods for impact and explosion problems, Ph.D. thesis, Tsinghua University, Beijing, 2009.
[49] J. Zukas, T. Nicholas, H. Swift, L. Greszczuk, D. Curran, Impact Dynamics, Krieger Publishing Company, Malabar, FL, 1992.
[50] M.A. Meyers, Dynamic Behavior of Materials, John Wiley & Sons, New York, 1994.
[51] J. von Neumann, R.D. Richtmyer, A method for the numerical calculation of hydrodynamical shocks, Journal of Applied Physics 21 (1950) 232.
[52] R. Landshoff, A numerical method for treating fluid flow in the presence of shocks, Tech. Rep. LA-1930, Los Alamos Scientific Laboratory, 1955.
[53] W. Fickett, W. Davis, Detonation – Theory and Experiment, Dover Publications, New York, 1979.
[54] D.L. Chapman, VI. On the rate of explosion in gases, Philosophical Magazine Series 5 47 (284) (1899) 90–104.
[55] J.C.E. Jouguet, Sur la propagation des réactions chimiques dans les gaz (On the propagation of chemical reactions in gases), Journal de Mathématiques Pures et Appliquées 6 (1) (1905) 347–425.
[56] J.C.E. Jouguet, Sur la propagation des réactions chimiques dans les gaz (On the propagation of chemical reactions in gases), Journal de Mathématiques Pures et Appliquées 6 (2) (1906) 5–85.
[57] D. Burgess, D. Sulsky, J.U. Brackbill, Mass matrix formulation of the FLIP particle-in-cell method, Journal of Computational Physics 103 (1) (1992) 1–15.
[58] G. Noh, K.J. Bathe, An explicit time integration scheme for the analysis of wave propagations, Computers & Structures 129 (2013) 178–193.
[59] K. Bathe, Finite Element Procedures, Prentice-Hall, New Jersey, 1996.
[60] T. Belytschko, W.K. Liu, B. Moran, K.I. Elkhodary, Nonlinear Finite Elements for Continua and Structures, 2nd edition, John Wiley & Sons, Chichester, 2014.
[61] B.M. Irons, Applications of a theorem on eigenvalues to finite element problems (cr/132/70), Tech. Rep., University of Wales, Department of Civil Engineering, Swansea, 1970.
[62] D. Flanagan, T. Belytschko, Simultaneous relaxation in structural dynamics, Journal of the Engineering Mechanics Division 107 (1981) 1039–1055.
[63] R. Courant, K. Friedrichs, H. Lewy, On the partial difference equations of mathematical physics, IBM Journal (1967) 215–234.
[64] C.E. Anderson, An overview of the theory of hydrocodes, International Journal of Impact Engineering 5 (1987) 33–59.
[65] S.G. Bardenhagen, Energy conservation error in the material point method for solid mechanics, Journal of Computational Physics 180 (2002) 383–403.

[66] J.A. Nairn, Material point method calculations with explicit cracks, Computer Modeling in Engineering & Sciences 4 (2003) 649–663.

[67] D. Sulsky, S.J. Zhou, H.L. Schreyer, Application of a particle-in-cell method to solid mechanics, Computer Physics Communications 87 (1–2) (1995) 236–252.

[68] Z. Chen, R. Brannon, An evaluation of the material point method, Tech. Rep. SAND 2002-0482, Sandia National Laboratories, Albuquerque, NM, 2002.

[69] A.R. York II, D. Sulsky, H.L. Schreyer, The material point method for simulation of thin membranes, International Journal for Numerical Methods in Engineering 44 (1999) 1429–1456.

[70] W. Hu, Z. Chen, A multi-mesh MPM for simulating the meshing process of spur gears, Computers & Structures 81 (2003) 1991–2002.

[71] S.G. Bardenhagen, J.U. Brackbill, D. Sulsky, The material-point method for granular materials, Computer Methods in Applied Mechanics and Engineering 187 (3–4) (2000) 529–541.

[72] S.G. Bardenhagen, J.E. Guilkey, K.M. Roessig, J.U. Brackbill, W.M. Witzel, J.C. Foster, An improved contact algorithm for the material point method and application to stress propagation in granular material, Computer Modeling in Engineering & Sciences 2 (4) (2001) 509–522.

[73] P. Huang, X. Zhang, S. Ma, X. Huang, Contact algorithms for the material point method in impact and penetration simulation, International Journal for Numerical Methods in Engineering 85 (4) (2011) 498–517.

[74] Z. Ma, X. Zhang, P. Huang, An object-oriented MPM framework for simulation of large deformation and contact of numerous grains, Computer Modeling in Engineering & Sciences 55 (1) (2010) 61–87.

[75] S.G. Bardenhagen, E.M. Kober, The generalized interpolation material point method, Computer Modeling in Engineering & Sciences 5 (6) (2004) 477–495.

[76] P.C. Wallstedt, J.E. Guilkey, An evaluation of explicit time integration schemes for use with the generalized interpolation material point method, Journal of Computational Physics 227 (2008) 9628–9642.

[77] A. Sadeghirad, R.M. Brannon, J. Burghardt, A convected particle domain interpolation technique to extend applicability of the material point method for problems involving massive de-formations, International Journal for Numerical Methods in Engineering 86 (2011) 1435–1456.

[78] A. Sadeghirad, R.M. Brannon, J.E. Guilkey, Second-order convected particle domain interpolation (CPDI2) with enrichment for weak discontinuities at material interfaces, International Journal for Numerical Methods in Engineering 95 (11) (2013) 928–952.

[79] D.Z. Zhang, X. Ma, P.T. Giguere, Material point method enhanced by modified gradient of shape function, Journal of Computational Physics 230 (2011) 6379–6398.

[80] M. Steffen, R.M. Kirby, M. Berzins, Analysis and reduction of quadrature errors in the material point method (MPM), International Journal for Numerical Methods in Engineering 76 (2008) 922–948.

[81] S. Ma, X. Zhang, Y.P. Lian, X. Zhou, Simulation of high explosive explosion using adaptive material point method, Computer Modeling in Engineering & Sciences 39 (2) (2009) 101–123.

[82] H.L. Tan, J.A. Nairn, Hierarchical, adaptive, material point method for dynamic energy release rate calculations, Computer Methods in Applied Mechanics and Engineering 191 (19–20) (2002) 2095–2109.

[83] J. Ma, H. Lu, B. Wang, S. Roy, R. Hornung, A. Wissink, R. Komanduri, Multiscale simulations using generalized interpolation material point (GIMP) method and SAMRAI parallel processing, Computer Modeling in Engineering & Sciences 8 (2) (2005) 135–152.

[84] J. Ma, H.B. Lu, R. Komanduri, Structured mesh refinement in generalized interpolation material point (GIMP) method for simulation of dynamic problems, Computer Modeling in Engineering & Sciences 12 (3) (2006) 213–227.

[85] P.F. Yang, Material point method study of localized failure, Ph.D. thesis, Tsinghua University, Beijing, China, 2013.

[86] Y.P. Lian, P.F. Yang, X. Zhang, F. Zhang, Y. Liu, P. Huang, A mesh-grading material point method and its parallelization for problems with localized extreme deformation, Computer Methods in Applied Mechanics and Engineering 289 (1) (2015) 291–315.

[87] J.M. McDill, J.A. Goldak, A.S. Oddy, M.J. Bibby, Isoparametric quadrilaterals and hexahedrons for mesh-grading algorithms, Communications in Applied Numerical Methods 3 (2) (1987) 155–163.

[88] J. Lysmer, R.L. Kuhlemeyer, Finite dynamic model for infinite media, Journal of the Engineering Mechanics Division 95 (1969) 859–877.

[89] Dassault Systèmes, Abaqus Analysis User's Manual.

[90] L.M. Shen, Z. Chen, A silent boundary scheme with the material point method for dynamic analyses, Computer Modeling in Engineering & Sciences 7 (3) (2005) 305–320.

[91] L.T. Tran, J. Kim, M. Berzins, Solving time-dependent PDEs using the material point method, a case study from gas dynamics, International Journal for Numerical Methods in Fluids 62 (7) (2010) 709–732.

[92] A.R. York II, D. Sulsky, H.L. Schreyer, Fluid-membrane interaction based on the material point method, International Journal for Numerical Methods in Engineering 48 (2000) 901–924.

[93] W.Q. Hu, Z. Chen, Model-based simulation of the synergistic effects of blast and fragmentation on a concrete wall using the MPM, International Journal of Impact Engineering 32 (12) (2006) 2066–2096.

[94] P. Hu, L. Xue, K. Qu, K. Ni, Unified solver for modeling and simulation of nonlinear aeroelasticity and fluid-structure interactions, in: Proceedings of AIAA Atmospheric Flight Mechanics Conference, 2009, pp. 2009–6148.

[95] P. Hu, L. Xue, S. Mao, R. Kamakoti, H. Zhao, N. Dittakavi, Material point method applied to fluid-structure interaction (FSI)/aeroelasticity problems, in: Proceedings of 8th AIAA Aerospace Sciences Meeting Including the New Horizons Forum and Aerospace Exposition, 2010.

[96] J.E. Guillkey, T.B. Harman, B. Banerjee, An Eulerian–Lagrangian approach for simulating explosions of energetic devices, Computers & Structures 85 (2007) 660–674.

[97] J. Li, Y. Hamamoto, Y. Liu, X. Zhang, Sloshing impact simulation with material point method and its experimental validations, Computers & Fluids 103 (2014) 86–99.

[98] C. Mast, P. Mackenzie-Helnwein, P. Arduino, G. Miller, W. Shin, Mitigating kinematic locking in the material point method, Journal of Computational Physics 231 (16) (2012) 5351–5373.

[99] J.J. Monaghan, Simulating free surface flows with SPH, Journal of Computational Physics 110 (2) (1994) 399–406.

[100] F. Zhang, X. Zhang, Y.P. Lian, Y. Liu, Incompressible material point method for free surface flow, Journal of Computational Physics (2016), in press.

[101] R.P. Fedkiw, T. Aslam, B. Merriman, S. Osher, A non-oscillatory Eulerian approach to interfaces in multimaterial flows (the ghost fluid method), Journal of Computational Physics 152 (2) (1999) 457–492.

[102] F. Gibou, R.P. Fedkiw, L.T. Cheng, M. Kang, A second-order-accurate symmetric discretization of the Poisson equation on irregular domains, Journal of Computational Physics 176 (1) (2002) 205–227.

[103] D. Enright, D. Nguyen, F. Gibou, R. Fedkiw, Using the particle level set method and a second order accurate pressure boundary condition for free surface flows, in: ASME/JSME 2003 4th Joint Fluids Summer Engineering Conference, American Society of Mechanical Engineers, 2003, pp. 337–342.

[104] S.J. Cummins, J.U. Brackbill, An implicit particle-in-cell method for granular materials, Journal of Computational Physics 180 (2002) 506–548.

[105] J.E. Guilkey, J.A. Weiss, Implicit time integration for the material point method: Quantitative and algorithm comparisons with the finite element method, International Journal for Numerical Methods in Engineering 57 (2003) 1323–1338.

[106] D. Sulsky, A. Kaul, Implicit dynamic in the material-point-method, Computer Methods in Applied Mechanics and Engineering 193 (12–14) (2004) 1137–1170.

[107] T. Belytschko, W.K. Liu, B. Moran, Nonlinear Finite Elements for Continua and Structures, John Wiley & Sons, Chichester, 2000.

[108] O. Axelsson, Iterative Solution Methods, Cambridge University Press, 1996.

[109] D. Knoll, D. Keyes, Jacobian-free Newton–Krylov methods: A survey of approaches and applications, Journal of Computational Physics 193 (2) (2004) 357–397.

[110] D.A. Knoll, W.J. Rider, A multigrid preconditioned Newton–Krylov method, SIAM Journal on Scientific Computing 21 (1999) 691.

[111] A. Nair, S. Roy, Implicit time integration in the generalized interpolation material point method for finite deformation hyperelasticity, Mechanics of Advanced Materials and Structures 19 (6) (2012) 465–473.

[112] Y.P. Lian, X. Zhang, F. Zhang, X.X. Cui, Tied interface grid material point method for problems with localized extreme deformation, International Journal of Impact Engineering 70 (2014) 50–61.

[113] Y.P. Lian, P.F. Yang, X. Zhang, F. Zhang, Y. Liu, P. Huang, A mesh-grading material point method and its parallelization for problems with localized extreme deformation, Computer Methods in Applied Mechanics and Engineering 289 (1) (2015) 291–315.

[114] http://qt.nokia.com/.

[115] http://www.vtk.org/.

[116] http://www.cmake.org/.

[117] http://www.tecplot.com/.

[118] G.R. Johnson, T.J. Holmquist, Evaluation of cylinder-impact test data for constitutive model constants, Journal of Applied Physics 64 (8) (1988) 3901–3910.

[119] A.J. Piekutowski, M.J. Forrestal, K.L. Poormon, T.L. Warren, Peroration of aluminum plates with ogive-nose steel rods at normal and oblique impacts, International Journal of Impact Engineering 18 (1996) 877–887.

[120] P. Huang, Material point method for metal and soil impact dynamics problems, Ph.D. thesis, Tsinghua University, 2010.

[121] D.P. Flanagan, T. Belytschko, A uniform strain hexahedron and quadrilateral and orthogonal hourglass control, International Journal for Numerical Methods in Engineering 17 (1981) 679–706.

[122] X. Zhang, T.S. Wang, Y. Liu, Computational Dynamics (in Chinese), 2nd edition, Tsinghua University Press, Beijing, 2007.

[123] S.J. Hanchak, M.J. Forrestal, E.R. Young, J.Q. Ehrgott, Perforation of concrete slabs with 48 Mpa (7ksi) and 140 Mpa (20ksi) unconfined compressive strengths, International Journal of Impact Engineering 12 (1) (1992) 1–7.

[124] Z.P. Chen, X.M. Qiu, X. Zhang, Y.P. Lian, Improved coupling of finite element method with material point method based on a particle-to-surface contact algorithm, Computer Methods in Applied Mechanics and Engineering 293 (2015) 1–19.

[125] T. Belytschko, J.I. Lin, A three-dimensional impact-penetration algorithm with erosion, Computers & Structures 25 (1) (1987) 95–104.

[126] J. Hallquist, G. Goudreau, D. Benson, Sliding interfaces with contact-impact in large-scale Lagrangian computations, Computer Methods in Applied Mechanics and Engineering 51 (1–3) (1985) 107–137.

[127] D.J. Benson, J.O. Hallquist, A single surface contact algorithm for the post-buckling analysis of shell structures, Computer Methods in Applied Mechanics and Engineering 78 (2) (1990) 141–163.

[128] R. Zhao, O. Faltinsen, J. Aarsnes, Water entry of arbitrary two-dimensional sections with and without flow separation, in: 21st Symposium on Naval Hydrodynamics, 1997.

[129] Z. Chen, H.L. Schreyer, Formulation and computational aspects of plasticity and damage models with application to quasi-brittle materials, Tech. Rep. SAND95-0329, Sandia National Laboratories, Albuquerque, NM, 1995.

[130] L. Malvern, Introduction to the Mechanics of a Continuous Medium, Prentice-Hall, Englewood Cliffs, NJ, 1969.

[131] O.C. Zienkiewicz, S. Valliappan, I.P. King, Elasto-plastic solutions of engineering problems 'initial stress', finite element approach, International Journal for Numerical Methods in Engineering 1 (1) (1969) 75–100.

[132] G.C. Nayak, O.C. Zienkiewicz, Elasto-plastic stress analysis. A generalization for various contitutive relations including strain softening, International Journal for Numerical Methods in Engineering 5 (1) (1972) 113–135.

[133] D.J.R. Owen, E.F. Hinton, Finite Element in Plasticity: Theory and Practice, Pineridge Press, Swansea, 1980.

[134] M.L. Wilkins, Calculations of elastic plastic flow, Methods in Computational Physics 3 (1964) 211–263.

[135] R.D. Krieg, D.B. Krieg, Accuracies of numerical solution methods for the elastic–perfectly plastic model, Journal of Pressure Vessel Technology 99 (4) (1977) 510–515.

[136] J.C. Simo, R.L. Taylor, Consistent tangent operators for rate-independent elastoplasticity, Computer Methods in Applied Mechanics and Engineering 48 (1) (1985) 101–118.

[137] J.C. Simo, T.J.R. Hughes, Computational Inelasticity, Springer, New York, 1998.

[138] B. Moran, M. Ortiz, C.F. Shih, Formulation of implicit finite element methods for multiplicative finite deformation plasticity, International Journal for Numerical Methods in Engineering 29 (3) (1990) 483–514.

[139] G.R. Johnson, W.H. Cook, A constitutive model and data for metals subjected to large strains, high strain rates, and high temperatures, in: Proc. of the 7th Intern. Symp. on Ballistics, 1983, pp. 541–547.

[140] G.R. Johnson, W.H. Cook, Fracture characteristics of three metals subjected to various strains, strain rates, temperatures and pressures, Engineering Fracture Mechanics 21 (1) (1985) 31–48.

[141] W.T. Koiter, Stress-strain relations, uniqueness and variational theorems for elastic–plastic materials with a singular yield surface, Quarterly of Applied Mathematics 11 (1953) 350–354.

[142] Itasca Consulting Group, Inc., Minneapolis, MN 55415, USA, FLAC-Fast Lagrangian Analysis of Continua User's Manual, Version 5.0, 2005.

[143] J.A. Zukas, W.P. Walters, Explosive Effects and Applications, Springer, Berlin, 1998.

[144] E.D. Giroux, HEMP user's manual, Tech. Rep. UCRL-51079, University of California, Lawrence Livermore National Laboratory, 1973.

[145] G.K. Batchelor, An Introduction to Fluid Dynamics, Cambridge University Press, Cambridge, 1967.

[146] J.P. Morris, P.J. Fox, Y. Zhu, Modeling low Reynolds number incompressible flows using SPH, Journal of Computational Physics 136 (1) (1997) 214–226.

[147] B.M. Dobratz, LLNL explosive handbook, properties of chemical explosives and explosive simulants, Tech. Rep. UCRL-52997, University of California, Lawrence Livermore National Laboratory, 1981.

[148] B.L. Boyce, S.L.B. Kramer, H.E. Fang, T.E. Cordova, et al., The sandia fracture challenge: Blind round robin predictions of ductile tearing, International Journal of Fracture 186 (1–2) (2014) 5–68.

[149] Z. Chen, Continuous and discontinuous failure modes, Journal of Engineering Mechanics 122 (1) (1996) 80–82.

[150] E. Bazant, Z.P. Chen, Scaling of structural failure, Applied Mechanics Reviews 50 (10) (1997) 593–627.

[151] P. Yang, Y. Gan, X. Zhang, Z. Chen, W. Qi, P. Liu, Improved decohesion modeling with the material point method for simulating crack evolution, International Journal of Fracture 186 (1–2) (2014) 177–184.

[152] H. Lu, N.P. Daphalapurkar, B. Wang, S. Roy, R. Komanduri, Multiscale simulation from atomistic to continuum – coupling molecular dynamics (MD) with the material point method (MPM), Philosophical Magazine 86 (2006) 2971–2994.

[153] Y. Liu, H. Wang, X. Zhang, A multiscale framework for high-velocity impact process with combined material point method and molecular dynamics, International Journal of Mechanics and Materials in Design 9 (2013) 127–139.

[154] Z. Chen, Y. Han, S. Jiang, Y. Gan, T. Sewell, A multiscale material point method for impact simulation, Theoretical and Applied Mechanics Letters 2 (2012) 051003.

[155] Z. Chen, S. Jiang, Y. Gan, S. Oloriegbe, T. Sewell, D. Thompson, Size effects on the impact response of copper nanobeams, Journal of Applied Physics 111 (2012) 113512.

[156] S. Jiang, Z. Chen, Y. Gan, S. Oloriegbe, T. Sewell, D. Thompson, Size effects on the wave propagation and deformation pattern in copper nanobars under symmetric longitudinal impact loading, Journal of Physics D: Applied Physics 45 (2012) 475305.

[157] Z. Chen, S. Jiang, Y. Gan, H. Liu, T. Sewell, A particle-based multiscale simulation procedure within the material point method framework, Computational Particle Mechanics 1 (2014) 147–158.

[158] S. Jiang, Z. Chen, T.D. Sewell, Y. Gan, Multiscale simulation of the responses of discrete nanostructures to extreme loading conditions based on the material point method, Computer Methods in Applied Mechanics and Engineering 297 (2015) 219–238.

[159] Y.J. Guo, J.A. Nairn, Three-dimensional dynamic fracture analysis using the material point method, Computer Modeling in Engineering & Sciences 16 (3) (2006) 141–155.

[160] J.A. Nairn, Numerical implementation of imperfect interfaces, Computational Materials Science 40 (2007) 525–536.

[161] Y. Guo, J.A. Nairn, Calculation of j integration and stress intensity factors using the material point method, Computer Modeling in Engineering & Sciences 6 (2004) 295–308.

[162] S.G. Bardenhagen, J.A. Nairn, H.B. Lu, Simulation of dynamic fracture with the material point method using a mixed J-integral and cohesive law approach, International Journal of Fracture 170 (2011) 49–66.

[163] N.P. Daphalapurkar, H. Lu, D. Coker, R. Komanduri, Simulation of dynamic crack growth using the generalized interpolation material point (GIMP) method, International Journal of Fracture 143 (1) (2007) 79–102.

[164] Z. Chen, L.M. Shen, Y.W. Mai, Y.G. Shen, A bifurcation-based decohesion model for simulating the transition from localization to decohesion with the MPM, Zeitschrift für Angewandte Mathematik und Physik 56 (2005) 908–930.

[165] H.L. Schreyer, D. Sulsky, S.J. Zhou, Modeling delamination as a strong discontinuity with the material point method, Computer Methods in Applied Mechanics and Engineering 191 (2002) 2483–2507.

[166] Z. Chen, W. Hu, L.M. Shen, X. Xin, R. Brannon, An evaluation of the MPM for simulating dynamic failure with damage diffusion, Engineering Fracture Mechanics 69 (2002) 1873–1890.
[167] Z. Chen, R. Feng, X. Xin, L. Shen, A computational model for impact failure with shear-induced dilatancy, International Journal for Numerical Methods in Engineering 56 (2003) 1979–1997.
[168] D. Sulsky, H.L. Schreyer, MPM simulation of dynamic material failure with a decohesion constitutive model, European Journal of Mechanics. A, Solids 23 (2004) 423–445.
[169] Y.G. Zheng, Y.X. Gu, Z. Chen, Numerical simulation of thin film failure with MPM, Chinese Journal of Theoretical and Applied Mechanics 38 (3) (2006) 347–355.
[170] Z. Chen, Y. Gan, J.K. Chen, A coupled thermo-mechanical model for simulating the material failure evolution due to localized heating, Computer Modeling in Engineering & Sciences 26 (2) (2008) 123–137.
[171] L.M. Shen, A rate-dependent damage/decohesion model for simulating glass fragmentation under impact using the material point method, Computer Modeling in Engineering & Sciences 49 (1) (2009) 23–45.
[172] D. Sulsky, K. Peterson, Toward a new elastic–decohesive model of Arctic sea ice, Physica D 240 (2011) 1674–1683.
[173] V. Gputa, S. Rajagopal, N. Gupta, A comparative study of meshfree methods for fracture, International Journal of Damage Mechanics 20 (2011) 729–751.
[174] B. Wang, V. Karuppiah, H. Lu, S. Roy, R. Komanduri, Two-dimensional mixed mode crack simulation using the material point, Mechanics of Advanced Materials and Structures 12 (2005) 471–484.
[175] F.A. Gilabert, V.C. Cantavella, E. Sanchez, G. Mallol, Modelling fracture process in ceramic materials using the material point method, Europhysics Letters 96 (2011) 24002.
[176] F.A. Gilabert, M. Dal Bó, V. Cantavella, E. Sánchez, Fracture patterns of quartz particles in glass feldspar matrix, Materials Letters 72 (2012) 148–152.
[177] S. Ma, Material point method for 3d hypervelocity impact simulation, Master's thesis, Tsinghua University, 2005.
[178] D. Sulsky, H.L. Schreyer, Axisymmetric form of the material point method with applications to upsetting and Taylor impact problems, Computer Methods in Applied Mechanics and Engineering 139 (1–4) (1996) 409–429.
[179] Y.X. Wang, Z. Chen, H.W. Zhang, M. Sun, Response of multi-layered structure due to impact load using material point method, Engineering Mechanics 24 (12) (2007) 186–192.
[180] D.Z. Zhang, Q. Zou, W.B. VanderHeyden, X. Ma, Material point method applied to multiphase flows, Journal of Computational Physics 227 (2008) 3159–3173.
[181] F. Li, J.Z. Pan, C. Sinka, Modelling brittle impact failure of disc particles using material point method, International Journal of Impact Engineering 38 (2011) 653–660.
[182] S. Ma, X. Zhang, X.M. Qiu, Three-dimensional material point method for hypervelocity impact, Explosion and Shock Waves 26 (3) (2006) 273–278.
[183] P. Huang, X. Zhang, S. Ma, Shared memory OpenMP parallelization of explicit MPM and its application to hypervelocity impact, Computer Modeling in Engineering & Sciences 38 (2008) 119–147.
[184] X.F. Pan, A.G. Xu, G.C. Zhang, J.S. Zhu, Generalized interpolation material point approach to high melting explosive with cavities under shock, Journal of Physics D: Applied Physics 41 (1) (2008) 015401.
[185] X. Aiguo, Z. Guangcai, Y. Yangjun, Z. Jianshi, Numerical study on porous materials under shock, Chinese Journal of Theoretical and Applied Mechanics 42 (6) (2010) 1138–1148.
[186] A.G. Xu, G.C. Zhang, H. Li, et al., Dynamical similarity in shock wave response of porous material: From the view of pressure, Computers & Mathematics with Applications 61 (2011) 3618–3627.

[187] W.D. Chen, F. Zhang, W.M. Yang, Simulation of spall fracture based on material point method, in: 11th International Conference on Fracture and Damage Mechanics, 2012.

[188] S. Ma, X. Zhang, Adaptive material point method for shaped charge jet formation, Chinese Journal of Solid Mechanics 30 (5) (2009) 504–508.

[189] Y.P. Lian, X. Zhang, X. Zhou, S. Ma, Y.L. Zhao, Numerical simulation of explosively driven metal by material point method, International Journal of Impact Engineering 38 (2011) 237–245.

[190] A.K. Aziz, H. Hurwitz, H.M. Sternberg, Energy transfer to a rigid piston under detonation loading, Physics of Fluids 4 (1961) 380–384.

[191] Y.X. Wang, Z. Chen, M. Sun, Numerical simulation of slippage detonation by material point method – MPM, Mechanics in Engineering 29 (3) (2006) 20–25.

[192] Y.X. Wang, Numerical analysis of explosion/impact response involved multiple materials using the material point method, Ph.D. thesis, Dalian University of Technology, 2006.

[193] Y.X. Wang, H.G. Beom, M. Sun, S. Lin, Numerical simulation of explosive welding using the material point method, International Journal of Impact Engineering 38 (2011) 51–60.

[194] Z. Zhang, W. Chen, W. Yang, The material point method for shock-to-detonation transition of heterogeneous solid explosive, Explosion and Shock Waves 31 (1) (2011) 25–30.

[195] P.F. Yang, Y. Liu, X. Zhang, X. Zhou, Y.L. Zhao, Simulation of fragmentation with material point method based on Gurson model and random failure, Computer Modeling in Engineering & Sciences 85 (3) (2012) 207–236.

[196] Y. Gan, Z. Chen, A study of the zona piercing process in piezodriven intracytoplasmic sperm injection, Journal of Applied Physics 104 (2008) 044702.

[197] Y. Gan, Z. Chen, S. Montgomery-Smith, Improved material point method for simulating the zona failure response in piezo-assisted intracytoplasmic sperm injection, Computer Modeling in Engineering & Sciences 73 (1) (2011) 45–76.

[198] A. Gilmanov, S. Acharya, A hybrid immersed boundary and material point method for simulating 3D fluid-structure interaction problems, International Journal for Numerical Methods in Engineering 56 (2008) 2151–2177.

[199] A. Gilmanov, S. Acharya, A computational strategy for simulating heat transfer and flow past deformable objects, International Journal of Heat and Mass Transfer 51 (2008) 4415–4426.

[200] S.R. Idelsohn, J. Marti, A. Limache, E. Onate, Unified Lagrangian formulation for elastic solids and incompressible fluids: Application to fluid-structure interaction problems via the PFEM, Computer Methods in Applied Mechanics and Engineering 197 (19–20) (2008) 1762–1776.

[201] E. Walhorn, A. Kölke, B. Hübner, D. Dinkler, Fluid-structure coupling within a monolithic model involving free surface flows, Computers & Structures 83 (2005) 2100–2111.

[202] H.W. Zhang, K.P. Wang, Material point method for dynamic analysis of saturated porous media (iii): Two-phase material point method, Chinese Journal of Geotechnical Engineering 32 (4) (2010) 507–513.

[203] Z. Guo, W. Yang, MPM/MD handshaking method for multiscale simulation and its application to high energy cluster impacts, International Journal of Mechanical Sciences 48 (2006) 145–159.

[204] H. Lu, N.P. Daphalapurkar, B. Wang, Multiscale simulation form atomistic to continuum – coupling molecular dynamics (MD) with the material point method (MPM), Philosophical Magazine 86 (20) (2006) 2971–2994.

[205] Y. Liu, X. Zhang, K.Y. Sze, M. Wang, Smoothed molecular dynamics for large step time integration, Computer Modeling in Engineering & Sciences 20 (3) (2007) 177–191.

[206] N. He, Y. Liu, X. Zhang, Molecular dynamics-smoothed molecular dynamics (md-smd) adaptive coupling method with seamless transition, International Journal for Numerical Methods in Engineering (2016), http://dx.doi.org/10.1002/nme.5224, in press.

[207] H.K. Wang, Parallelization of the smoothed molecular dynamics coupled with MD, Master's thesis, Tsinghua University, 2007.

[208] H.K. Wang, Material point method for metal and carbon nanotube composite materials, Ph.D. thesis, Tsinghua University, 2011.

[209] G. Ayton, S.G. Bardenhagen, P. McMurtry, D. Sulsky, G.A. Voth, Interfacing continuum and molecular dynamics: An application to lipid bilayers, Journal of Chemical Physics 114 (15) (2001) 6913–6924.

[210] L. Shen, Z. Chen, A multi-scale simulation of Tungsten film delamination from Silicon substrate, International Journal of Solids and Structures 42 (2005) 5036–5056.

[211] O. Borodin, D. Bedrov, G.D. Smith, J. Nairn, S. Bardenhagen, Multiscale modeling of viscoelastic properties of polymer nanocomposites, Journal of Polymer Science. Part B, Polymer Physics 43 (2005) 1005–1013.

[212] L.Z. Zhang, T. Bartel, M.T. Lusk, Parallelized hybrid Monte Carlo simulation of stress-induced texture evolution, Computational Materials Science 48 (2010) 419–425.

[213] S.G. Bardenhagen, A.D. Brydon, J.E. Guilkey, Insight into the physics of foam densification via numerical simulation, Journal of the Mechanics and Physics of Solids 53 (3) (2005) 597–617.

[214] A.D. Brydon, S.G. Bardenhagen, E.A. Miller, G.T. Seidler, Simulation of the densification of real open-celled foam microstructures, Journal of the Mechanics and Physics of Solids 53 (2005) 2638–2660.

[215] N.P. Daphalapurkar, J.C. Hanan, N.B. Phelps, H. Bale, H. Lu, Tomography and simulation of microstructure evolution of a closed-cell polymer foam in compression, Mechanics of Advanced Materials and Structures 15 (2008) 594–611.

[216] W.W. Gong, Y. Liu, X. Zhang, H.L. Ma, Numerical investigation on dynamical response of aluminum foam subject to hypervelocity impact with material point method, Computer Modeling in Engineering & Sciences 83 (5) (2012) 527–545.

[217] Y. Liu, W. Gong, X. Zhang, Numerical investigation of influences of porous density and strain-rate effect on dynamical responses of aluminum foam, Computational Materials Science 91 (2014) 223–230.

[218] H.K. Wang, Y. Liu, X. Zhang, The carbon nanotube composite simulation by material point method, Computational Materials Science 57 (2012) 23–29.

[219] Y.-L. Shen, W. Li, D.L. Sulsky, H.L. Schreyer, Localization of plastic deformation along grain boundaries in a hardening material, International Journal of Mechanical Sciences 42 (2000) 2167–2189.

[220] L.P. Xue, O. Borodin, G.D. Smith, J. Nairn, Micromechanics simulations of the viscoelastic properties of highly filled composites by the material point method (MPM), Modelling and Simulation in Materials Science and Engineering 14 (4) (2006) 703–720.

[221] J.E. Guilkey, J.B. Hoying, J.A. Weiss, Computational modeling of multicellular constructs with the material point method, Journal of Biomechanics 39 (2006) 2074–2086.

[222] I. Ionescu, J.E. Guilkey, M. Berzins, Simulation of soft tissue failure using the material point method, Journal of Biomechanical Engineering 128 (2006) 917–924.

[223] S.Z. Zhou, X. Zhang, H.L. Ma, Numerical simulation of human head impact using the material point, International Journal of Computational Methods 10 (4) (2013) 1350014.

[224] S.Z. Zhou, X. Zhang, H.L. Ma, Effects from different head models on dynamic responses of the head impact using material point method, Journal of Medical Biomechanics 28 (1) (2013) 74–79.

[225] S.Z. Zhou, X. Zhang, H.L. Ma, Numerical simulation on response of human spine with different bone mineral density to landing impact, Journal of Medical Biomechanics 28 (3) (2013) 1–5.

[226] C.J. Coetzee, P.A. Vermeer, A.H. Basson, The modelling of anchors using the material point method, International Journal for Numerical and Analytical Methods in Geomechanics 29 (2005) 879–895.

[227] C.J. Coetzee, A.H. Basson, P.A. Vermeer, Discrete and continuum modelling of excavator bucket filling, Journal of Terramechanics 44 (2) (2007) 177–186.

[228] S. Andersen, L. Andersen, Modelling of landslides with the material-point method, Computational Geosciences 14 (2010) 137–147.

[229] H.H. Bui, R. Fukagawa, K. Sako, S. Ohno, Lagrangian meshfree particles method (SPH) for large deformation and failure flows of geomaterials using elastic–plastic soil constitutive model, International Journal for Numerical and Analytical Methods in Geomechanics 32 (2008) 1537–1570.

[230] Z. Wieckowski, The material point method in large strain engineering problems, Computer Methods in Applied Mechanics and Engineering 193 (39–41) (2004) 4417–4438.

[231] R. Ambati, H. Yuan, X.F. Pan, X. Zhang, Application of material point method for cutting process simulations, Computational Materials Science 57 (2011) 102–110.

[232] J. Burghardt, R. Brannon, J. Guilkey, A nonlocal plasticity formulation for the material point method, Computer Methods in Applied Mechanics and Engineering 225–228 (2012) 55–64.

[233] F. Li, J.Z. Pan, C. Sinka, Contact laws between solid particles, Journal of the Mechanics and Physics of Solids 57 (2009) 1194–1208.

Index

Printed in the United States
By Bookmasters